Biophotonics and Coherent Systems in Biology

Alexander Gurwitsch (first row, 3rd from the right) and his associates near the Taurida University laboratory, where mitogenetic radiation was discovered (1924).

Participants of the 3rd International Gurwitsch Conference near Gurwitsch's former laboratory (2004).

L.V. Beloussov
V.L. Voeikov
V.S. Martynyuk

Biophotonics and Coherent Systems in Biology

With 118 Illustrations

Springer

L.V. Beloussov
Biology Faculty of
M.V. Lomonosov Moscow State University
Moscow, Russia 119899
morphogenesis@yandex.ru

V.L. Voeikov
Biology Faculty of
M.V. Lomonosov Moscow State University
Moscow, Russia 119899
vvl@soil.msu.ru

V.S. Martynyuk
NAS Ukraine
Crimean Scientific Center, Simferopol
2 Vernadsky Ave.
Crimea, Ukraine 95007
csc@science-center.net

Library of Congress Control Number: 2005932555

ISBN-10: 0-387-28378-1 e-ISBN-10: 0-387-28417-6
ISBN-13: 978-0387-28378-4 e-ISBN-13: 978-0387-28417-0

Printed on acid-free paper.

9 8 7 6 5 4 3 2 1

springer.com

EDITORIAL PREFACE

The 3rd Alexander Gurwitsch Conference on Biophotonics and Coherent Systems in Biology was held from September 27th to October 2nd, 2004. Contrary to the first two conferences from the same cycle which took place at Moscow State University in 1994 and 1999,[1,2] the latter one was hosted by V.I. Vernadsky Taurida National University (Simferopol, Crimea, Ukraine). In no case was this occasional. Modern Taurida University (re-established under this name a few years ago) regards itself as an inheritor of the same name institution that was opened in 1918 and existed as a University until 1924 (when it was renamed as a Pedagogical Institute). In many respects, the first Taurida University was a remarkable organization. Under severe conditions of a starting post-revolutionary civil war in Russia, when normal research and educational activity in the main centers (such as St. Petersburg and Moscow) became almost impossible, Taurida University succeeded in collecting a brilliant company of professors and students who did not want to emigrate from Russia but were willing to continue their activities in their native country. Among them was a famous geochemist and philosopher, Vladimir Vernadsky, who was for some time the Rector of Taurida University and who gave his name to the modern Taurida University. Soon he became a close friend of Alexander Gurwitsch, who was elected as a Professor of Histology of this University already in 1918. It took meanwhile almost a year for him together with his family to reach a relatively peaceful Crimean land by going from starving Petrograd (former and later St. Petersburg) through an enormous territory of Russia and Ukraine already separated into several fighting estates. Although the situation in Crimea was also quite far from idyllic, and a civil war with all of its shortages and cruelties soon reached this area, Taurida University could provide much more academic freedom and cooperation between its outstanding members than any other institution in those days in Russia. For Gurwitsch, who originated from Ukraine, wonderful nature of the Crimean peninsula was also a source of inspiration. The first few years of his work in Taurida University turned out to be extremely fruitful. Then he made his famous "onion experiment", which opened a door to a miraculous world of biophotonics and electro-magnetic biology, and gave a first sketch of his "embryonic field" theory. It was also amazing how rapidly developed the biophotonic studies in Taurida University and how soon they became known to the worldwide scientific community. The main reason was that not only the professors, but also the students of this University were outstanding. Some of them continued to work in this field for their whole life. To be mentioned among them is a later well-known cytologist, Semen Zalkind, and a biophysicist, Gleb Frank, who became a member of the Soviet Academy of Sciences and the founder of the main center for biophysical research in the Soviet Union and now in Russia, the Institute of Biophysics in Puschino.

Being eager to revive these glorious traditions, the authorities of the modern Taurida University started the Conference by a ceremony of opening a common Gurwitsch-Frank memorial desk at the main University entrance. The ceremony was preceded by a special University session with Dr. V. Lavrov lecturing about the history of Taurida University in Gurwitsch's times and Prof. V. Voeikov's lecture about Gurwitsch's main works. By a miraculous occasion (nobody arranged it intentionally), this ceremony took place exactly on Alexander Gurwitsch's 130th birthday!

On the same day, an unforgettable enterprise was the excursion to the house where Gurwitsch's lab (and his family flat) was located and where he made his onion experiments. The beautiful villa safely survived the Second World War and was only slightly redesigned. We, the conference participants, made a group photo at almost the same place where the Gurwitsch group was photographed exactly 80 years before (see two photographs on the frontispiece).

The Conference collected several dozen participants from Russia, Ukraine, several European countries, USA, and Israel. Unfortunately, due to traveling problems, several potential participants, including IIB members, could not personally attend the conference. However, they presented the contributions that we included into the volume. Taken together, they give a representative picture of the modern state of biophotonics and the related branches of biology and biophysics.

By the Editors view, the main novel feature of the 3rd Gurwitsch Conference, as compared with the previous ones, is the extension of biophotonics from its traditional optical wavelength range toward that including smaller electromagnetic frequencies and stationary fields. In other words, biophotonics becomes a part of a common science that may be called the electromagnetic biology. Such an extension is far from being formal: a main conceptual basis of this new trend of science is to a great extent borrowed from the modern biophotonic studies. This relates most of all to the concept of coherence. It is this concept that permits to explain the biological effects not only of the UV and optical wavelengths range, but also those of much smaller frequencies. The idea of coherent regimes of molecular interactions as well as the related views and experimental findings seem to be of an utmost importance and heuristic power not only for electromagnetic biology *per se*, but also for the cell and organismic physiology.

Although several papers from this volume are treating different matters, some of which are only indirectly linked with biophotonics in *sensu stricto*, we decided not to subdivide the entire volume into different sections. By arranging the papers, we put in the beginning those completely or partly devoted to the biophoton emission. These were followed by the papers treating electromagnetic fields, and at the end of the volume we put the contributions not related to electromagnetic events but associated with the concept of a coherence in its broader sense, including even sociological and philosophical aspects.

One of the aims of the 3rd Alexander Gurwitsch Conference was to emphasize the links between the pioneer Crimean experiments and modern biophotonics. Accordingly, we found it suitable to end the Conference volume by a brief tribute to the person who did more than everybody else for promoting a continuity of this research line – Professor Anna Alexandrovna Gurvich (1909-1993).

Together with all the Conference participants, we express our deep thanks to the Rector of Taurida University, Professor N.V. Bagrov, to the Deputy Rector, Professor V.N. Berzhansky, and to all the members of the Organizing Committee of the 3rd Alexander Gurwitsch Conference for their wonderful acceptance at the land of Crimea, making the conference a remarkable scientific and cultural event. Our special gratitude comes to Dr. N.D. Vilenskaya who took on herself the burden to format the whole volume. We thank also Mr. A. Johnson and Mrs. K. Zimmer from Springer for their help in issuing this book.

REFERENCES

1. L.V. Beloussov and F.-A. Popp, eds., Biophotonics. Non-equilibrium and Coherent Systems in Biology, Biophysics and Biotechnology (Bioinform Services Co., Russia, 1995).
2. L.V. Beloussov, F.-A. Popp, V. Voeikov and R. van Wijk, eds. Biophotonics and Coherent Systems. Proceedings of the 2nd Alexander Gurwitsch Conference and Additional Contributions (Moscow University Press, Moscow, 2000).

CONTENTS

1. **FROM MITOGENETIC RAYS TO BIOPHOTONS** 1
V.L. Voeikov and L.V. Beloussov

2. **PHOTON SUCKING AS AN ESSENTIAL PRINCIPLE OF BIOLOGICAL REGULATION** 17
F.-A. Popp and W. Klimek

3. **QUANTUM SQUEEZED STATE DESCRIPTION OF SPECTRAL DECOMPOSITIONS OF A BIOPHOTON SIGNAL AND THE POSSIBILITY OF REMOTE INTERVENTION** 33
R.P. Bajpai

4. **BIOLOGICAL STRUCTURE AS A CONVERTER OF COHERENT RADIATION** 47
A. Budagovsky, O. Budagovskaya, and I. Budagovsky

5. **THE OSCILLATION BEHAVIOR OF THE DELAYED LUMINESCENCE OF PLANT LEAVES** 65
Yu Yan, F.-A. Popp, S. Sigrist, D. Schlesinger, A. Dolf, Zhongchen Yan, S. Cohen, A. Chotia, and D. Busch

6. **HUMAN GENOME REALIZATION AT THE VIEWPOINT OF PHYSICS OF THE ALIVE** 75
S.P. Sit`ko

7. **FUNDAMENTAL ROLE OF WATER IN BIOENERGETICS** 89
V.L. Voeikov

8. **THE HYDROPHOBIC-HYDROPHILIC BALANCE IN WATER SOLUTIONS OF PROTEINS AS THE POSSIBLE TARGET FOR EXTREMELY LOW FREQUENCY MAGNETIC FIELDS** 105
V.S. Martynyuk and Yulia V. Tseyslyer

9. **FEATURES OF REACTIONS OF BIOLOGICAL AND PHYSICO-CHEMICAL SYSTEMS TO EXTERNAL FACTORS** 123
I.A. Stepanyuk

10. NEWS AND VIEWS IN UVA-LASER–INDUCED ULTRAWEAK DELAYED LUMINESCENCE OF CULTURED MAMMALIAN CELLS **129**
H.J. Niggli, S. Tudisco, G. Privitera, L.A. Applegate, A. Scordino, and F. Musumeci

11. ULTRAWEAK PHOTON EMISSION AS A TOOL FOR ANALYSING COLLECTIVE PROCESSES IN CELLS AND DEVELOPING EMBRYOS **139**
L.V. Beloussov

12. DISTANT INTERACTION DURING GERMINATION OF *BACILLUS SUBTILIS* SPORES **159**
Y.A. Nikolaev, G.I. El'-Registan, and S.B. Desu

13. INFLUENCE OF RADIOFREQUENCY EMF ON THE YEAST *SACCHAROMYCES CEREVISIAE* AS MODEL EUKARYOTIC SYSTEM **167**
E.N. Gromozova, and S.I Voychuk

14. SPATIAL CHARACTERIZATION OF HUMAN ULTRA-WEAK PHOTON EMISSION **177**
R. Van Wijk, M. Kobayashi, and E.P.A. Van Wijk

15. INFLUENCE OF ELECTROMAGNETIC FIELDS OF EXTREMELY DIFFERENT FREQUENCY DIAPASON ON INFRADIAN RHYTHMS OF PHYSIOLOGICAL PROCESSES **191**
N.A. Temuryants, V.S. Martynyuk, and E.N. Chuyan

16. ABSORPTION AND EMISSION OF PHOTONS BY COLLAGEN SAMPLES **203**
T.G. Troshina, N.N. Loochinskaia, E. Van Wijk, R. Van Wijk, and L.V. Beloussov

17. INTERACTION OF PLANT SHOOTS AND ROOTS: DYNAMICS AND STABILITY **213**
N.V. Budagovskaya

18. CORRELATION OF THE "NEAR-ZONE EFFECT" AMPLITUDE DYNAMICS WITH SOLAR-GEOPHYSICAL INDICES **225**
T.A. Zenchenko, A.A. Konradov, and K.I. Zenchenko

19. FREE WILL AND VIOLATION OF PHYSICAL LAWS **235**
M. Lipkind

20. VERNADSKY'S NOOSPHERE AND SLAVOPHILE SOBORNOST' 279
M. Bischof

21. A LIFE THAT LINKED MITOGENETIC RAYS AND BIOPHOTONS 299
L.V. Beloussov, M. Lipkind, F.-A. Popp, and V.L. Voeikov

INDEX 303

1
FROM MITOGENETIC RAYS TO BIOPHOTONS

Vladimir L. Voeikov and Lev V. Beloussov*

> Persons with the most complete scientific worldview for their time always belong to "scientific heretics", rather than to representatives of the scientific mainstream. Contemporaries are unable to distinguish them from those deluded... Their opinions do not attract our attention or arouse our dissatisfaction and rejection.
>
> Academician Vladimir Vernadsky
>
> "On the Scientific Worldview", 1904

1. NON-ACCIDENTAL DISCOVERY OF MITOGENETIC RADIATION

In the full swing of a civil war, when terror, hunger, and ruin were reigning in Russia, a new university admitted first students at the far South of the country - in Crimea. Taurida University remained at that time the only self-governed, free university in the whole of Russia. Outstanding Russian scientists, those who did not want to leave their country, gathered there. Such prominent figures as physicists Alexander Ioffe, Igor Tamm, geologist Vladimir Obruchev, and biologists Michail Zavadovsky, Alexander Lubischev were among them. In 1920, academician Vladimir Vernadsky was elected Rector of Taurida University by the Professor Counsel.

The Chair of Histology of the Medical and Natural Sciences faculties was headed by Professor Alexander Gawrilowich Gurwitsch. He was born in 1874 in the Ukrainian town of Poltava, and after graduation from the Medical faculty of Munich University worked for nearly a decade in German and Swiss leading biology centers. He was deeply interested in one of the most enigmatic problems in biology - in morphogenesis - and desired to understand the mechanism of emerging of complex tissues and organs, of organisms with unique architecture from rather primitively structured embryo cells.

In 1906, A.G. Gurwitsch moved to St. Petersburg were he was elected professor of histology at Bestoughev Women College. In 1912, he published an

---

* Faculty of Biology, Lomonosov Moscow State University, Moscow, Russia, 119234.

original theory of embryonic development in which for the first time in biology the term "field" belonging to the realm of physics was used (1). According to Gurwitsch's Morphogenetic Field theory, behaviour of both individual cells and organ rudiments is controlled by a field of forces common to all the elements of an embryo. This field regulates behaviour of individual cells in a developing embryo, routes their movements, controls their divisions and differentiation, and evolves itself with embryo growth. Gurwitsch was cautious enough and did not specify the physical nature and initial sources of this field.

Morphogenetic Field theory explained many facts of the process of embryogenesis, allowed to predict further stages of morphogenesis based on the analysis of the actual disposition of cells in the embryo, and attracted great interest. But the number of the opponents of the Field theory much exceeded that of its supporters mainly because the theory did not keep within the postulate that had already become the cornerstone of physiology and biochemistry of the XXth century: all biological phenomena including the process of development of a living organism result from mere summations of usual chemical processes. Gurwitsch considered that experimental evidence can persuade his opponents, but after the Bolshevic revolution it was impossible to think of experimental work in Petersburg, and he moved to Crimea.

In the process of embryonic development, the number of cells increases as they divide. Gurwitsch questioned which factor or factors cause cell reproduction. He studied statistical distribution of mitoses, that is of cells being in the process of division into two, and came to the conclusion that mitosis occurs when two independent events coincide. The first of them is the resumption of cell maturation. A cell should synthesize all the components necessary both for the process of mitosis and for daughter cells. Gurwitsch defined this condition dependent on the given cell activity - the "possibility factor". But even a mature cell does not enter mitosis unless it is triggered by some external impulse, "realization factor", which can originate from the organism to which the cell belongs.

Reflecting upon the nature of the external signal, Gurwitsch noticed that in symplasts (in tissue regions containing many nuclei non-separated by cell membranes), all the nuclei either divide synchronously or are involved in common waves of mitotic divisions. Meanwhile in tissues consisting of uninucleated cells, these always divide asynchronously. This pointed out that a cell membrane (or cell-environment interface) was an organ perceiving external signals for cell divisions. For defining the molecular structures of cell surfaces respon-responsible for such a perception, Gurwitsch used a term "receptor" long before it was established in the modern molecular biology.

For getting more precise information about these receptors properties, Gurwitsch studied the dependence between the length of onion root cells and frequency of cell divisions. If assuming that: (1) the cell elongation is associated with random insertion of new non-receptive surface particles between the receptors and (2) the perception of mitotic signals is proportional to density of receptors (as should be expected if the signal is a kind of a soluble chemical "hormone"), such dependence should be inversely proportional and fit the first order hyperbola. However, he found that it was and much steeper than the first order hyperbolic one (Fig. 1). This dependence indicated that "receptors" do not act as independent units. Rather, they create a kind of a holistic (cooperative) system, which may have the properties of a resonating contour. From this Gurwitsch concluded that the perceived signal should be a physical factor triggering cell division on the principle of resonance. For example, this factor could be photons because of their wave nature. In his later papers, Gurwitsch numerously emphasized that such a conclusion was no more than tentative, but it gave him the idea of his famous "onion experiment" which, as we can see, in no way was occasional.

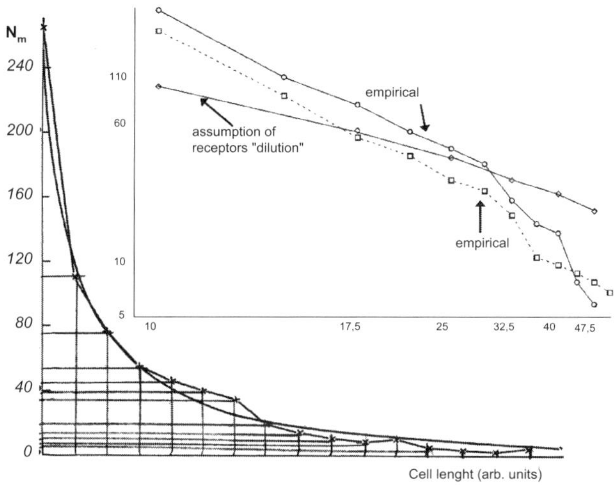

Figure 1. Dependence of the number of mitoses in an onion root (N_m – ordinates) on cell lengths (arbitrary units, abscissa). Left bottom plot – original Gurwitsch's data in linear coordinates and their best fit. Upper plot – two sets of experimental data and theoretical curve on the assumption of receptors "dilution" with a cell elongation in log-log coordinates.

2. BIOPHYSICAL PECULIARITY OF MITOGENETIC RADIATION

In 1923, Gurwitsch performed the crucial experiment for the evaluation of his hypothesis that photons were triggering cell divisions (Fig. 2). A tip of an onion root - the inducer - was directed at the wall of another onion root - the detector. After they had been kept for some time in this configuration, the number of mitoses at the detector side facing the tip of the inducer root significantly exceeded that at the opposite one. If a glass plate was introduced between the inducer and the detector, there was no stimulation of mitotic activity. A quartz plate shielding the tip of the inducer root did not interfere with its action. If the tip of the inducer was aimed at the metal mirror in such a way that its reflection fell onto the wall of the detector, stimulatory action again was observed.

These results could be explained neither by chemical nor by mechanical action of one root on another. The most plausible explanation of the effect was the following: a living organism can emit photons that stimulate cell divisions. These photons belong to the ultraviolet region of the spectrum since quartz but

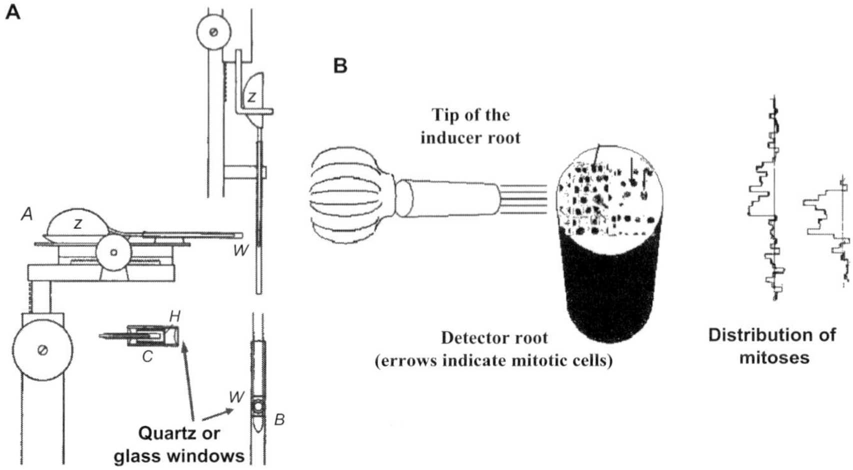

Figure 2. Schematic of "onion" experiment of Gurwitsch. **A**. Installation of an inducer root (horizontal) and a detector root (vertical) on moving tables of microscopes. *Z* – onion bulbs, *C* – tip of the inducer root fixed in an air-tight chamber, *H* and *W* – quarts or glass windows. **B**. Sketch of experimental results evaluation. A detector root was sliced below and above the "irradiated" zone and an excess of mitotic cells on the left (irradiated) or right (non-irradiated) sides of the root on each slice was calculated. Two indented lines at the left illustrate the results of two representative experiments. Significant excess of mitotic cells on the left side over the average distribution was observed in the irradiated region.

not glass is transparent for them. That is why photonic emission from a tip of the root stimulating mitoses in another one was named "mitogenetic radiation" (MGR) (3).

It is well-known that UV-light is hazardous for living cells. However, when the light beam of an UV-lamp was attenuated several thousand times, the number of mitoses increased. Thus, the conclusion that UV-photons induce the performance by a living cell of its major function - reproduction - had been proved. It also turned out that the effect of light on living systems strongly depends on its intensity and duration of action: excessive illumination resulted in suppression rather than stimulation of cell divisions.

In 1924, Professor Gurwitsch was elected the head of Histology Department at the Medical Faculty of Moscow State University, and the investigation of the new phenomenon was continued there. It was shown that MGR is produced by various animal and plant tissues, by microorganisms. As regards onion roots, it turned out that when a root was cut off the base of a bulb, it immediately ceased to emit MGR. On the other hand, the basal membrane of the onion bulb and even minced tissue was an effective source of MGR. From this two important conclusions could be gained. First, the original source of MGR was an onion bulb tissue where intensive respiration took place. Second, as it was highly unlikely that photons originating in the onion basement membrane could travel along the root without being absorbed, one should conclude that waves of electronic excitation can propagate in a living tissue on macroscopic distances analogous to a chain burning in a Bickford's fuse. The reality of dissipation- and radiationless propagation of electronic excitation was later confirmed by Gurwitsch and colleagues in *in vitro* experiments and in physical-chemical aqueous model systems (see below).

Besides onion root cells, many other capable to divide cells could serve as MGR detectors, but an ordinary yeast culture turned out to be the most convenient test-system. It was irradiated with an MGR source at a lag-period, and the surplus in the number of mitoses in it over the control culture was calculated at the early stage of exponential growth. If cells were irradiated already at the phase of exponential growth, or at the stationary phase, no effect was observed. On the other hand, an yeast culture itself was the most efficacious source of MGR at the exponential stage of its growth, when the intensity of cell divisions was the highest. From this it followed, that MGR emerged due to high metabolic activity and that MGR in its turn induced metabolic processes. From the technical point of view, investigation of MGR needed understanding of physiology of both biological detectors and emitters of this radiation, and a lot of studies which allegedly refuted MGR existence were methodologically ignorant. Though rather elaborate, this method was very sensitive and served as an excellent tool for MGR studies in the following years (4).

An important observation was made by Gurwitsch's student Gleb Frank (he later became a Member of the USSR Academy of Sciences, founded and headed the Institute of Biophysics of the USSR Academy of Sciences). Frank made spectral analysis of MGR from different sources using plates with east cultures as detectors. UV nature of MGR was again proved: cell divisions could be induced by any photons in the region between 190 and 280 nm in the darkness and up to 326 nm if the detector cell population was illuminated by even a dim visible light. MGR spectra from various sources looked as sets of distinct bands. The latter varied in width from 25 to 0.5 nm. Each individual source of MGR produced a spectrum with the unique set of bands (Fig. 3).

This allowed obtaining "finger-print" spectra for several enzymatic reactions, to identify substances capable to scatter or to re-emit MGR of specific wavelengths. Physiological changes of a particular biological source were followed with spectral changes of its emission. MGR spectral analysis became one of the most informative methods in Gurwitsch's laboratory (5).

Though the notion of most properties of MGR was obtained based on its biological effects, in the 1930s UV-radiation emitted by some biological objects and chemical reactions was successfully registered in several laboratories using physical detectors - modified Geiger-Mueller counters. Their photocathodes were made of materials having maximal light sensitivity in the range of 190-280 nm and practically insensitive to visible light - copper, magnesium, aluminium, or their compounds. When such a material absorbs UV-photons it emits photoelectrons that trigger a gas discharge in a counter (6, 7, 8). It was shown with these counters that intensity of UV-photon emission from developing frog eggs or a nerve-muscle preparation excited with an electrical current is very small: 10-10000 photons (equivalent of 10^{-10} - 10^{-8} erg) per 1 sec from 1 cm^2 of the emitting object.

When such weak light beam is divided by a spectrograph prism into multiple bands, the intensity of the narrowest of them should barely exceed single photons per 1 second. If so, then single photons are able to induce an "epidemic" of mitoses in a yeast colony or in another biological MGR detector.

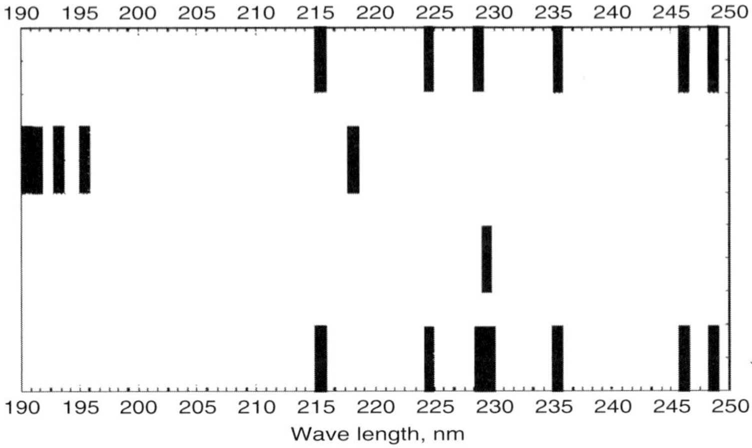

Figure 3. Spectra of MGR emitted by different reaction systems *in vitro*: 1st raw – hydrolysis of a nucleic acid or lecitin by phospatase, 2d raw – sugar fermentation, 3d raw – reaction of glycin autoxidation induced with glycin solution irradiation with MGR, 4th raw – same as above after addition of sodium phosphate.

Such a conclusion is to be made since only an "epidemic" of mitosis can be registered as a significant effect over the normal number of them in the control non-irradiated sample. How is it possible that a single photon can produce such a strong effect? Gurwitsch suggested that amplification of the initial signal is caused by secondary emission: a cell that happened to catch a photon becomes a secondary emitter of MGR. It may not enter the mitosis itself, but serves to "multiply" photons by a branching chain reaction mechanism. Later this suggestion was proved experimentally (9).

3. SOURCES OF ENERGY FOR EMERGENCE OF MITOGENETIC RADIATION

All living systems exist at mild temperatures and their bioenergetic facilities are generally considered to be satisfied by portions of energy stored in ATP molecules not exceeding 0.5 eV (7-10 kcal/mol). Energy of quanta of MGR reaches >8 eV (150 kcal/mol). This discrepancy for many decades was one of the obstacles for acknowledgement of the reality of MGR by biochemical and biophysical community. However, the preliminary answer to the question of the ultimate source of such high density energy that is released as mitogenetic radiation was received by Gurwitsch and his colleagues as long ago as in the 1930s (10).

It was discovered that MGR accompanied enzymatic and non-enzymatic hydrolytic and glycolytic reactions *in vitro*, as well as usual chemical reactions such as base-acid neutralization, redox reactions, and even dissolution of salt crystals in water or sol-gel transitions in aqueous solutions if and only if aqueous solutions in which the reactions were running were contacting air. For hydrolytic reactions, illumination of the reaction system with blue-green or shorter wavelength light was also necessary for MGR emergence. Gurwitsch calculated the energy balance of such reactions and came to the conclusion that if energy for oxygen activation is provided, MGR arousal is not forbidden by the laws of thermodynamics. Let us take as an example hydrolytic reaction catalyzed by urease:

$$(NH_2)_2C{=}O + H_2O \rightarrow 2NH_3 + CO_2$$

This reaction is nearly thermoneutral, though energy of activation (or creation by the enzyme of an appropriate condition for reagents interaction) is needed for conversion of urea and water into ammonia and carbonate. In any case to convert reagents into products, the former should be dismantled into radicals:

$2H_2N\bullet$, $\bullet\bullet C{=}O$, $2H\bullet$ and $\bullet O\bullet$ (dots here designate unpaired electrons).

To break down urea to radicals and water to atoms, a total of 120+220=340 kcal/mol is needed. When these radicals recombine to 2 NH_3 and CO_2 approximately the same quantity of energy is released (87+87 = 174 kcal/mol and 167 kcal/mol, a total of 341 kcal/mol), thus this *chemical* reaction cannot provide energy for MGR. Indeed it does not emit MGR in the absence of oxygen. However, if energy no less than 118 kcal/mol is provided for oxygen molecule decomposition to two atoms $\bullet O\bullet$, "side" reactions may occur in which radicals arise from decomposition of initial reagents and oxygen atoms:

$$\bullet O\bullet + H_2O \rightarrow H_2O_2 + 51 \text{ kcal/mol}$$
$$H_2N\bullet + \bullet NH_2 \rightarrow H_2N - NH_2 \text{ (hydrazin)} + 69 \text{ kcal/mol}$$

This energy fully compensates energy expenditure for urea decomposition, and two radicals, to be more precise, two bi-radicals, carbonyl and oxygen atom, are left. Upon recombination of these bi-radicals, when four electrons simultaneously cancel each other's spins, an energy quantum, equivalent to the so-called "vacuum ultraviolet", is released:

$$\bullet\bullet C{=}O + \bullet O\bullet \rightarrow CO_2 + 167 \text{ kcal/mol } (\cong h\nu \text{ 170 нм})$$

If such high energy packages are generated in a milieu where appropriate fluorescents may be excited by them, all the array of photons with $\lambda \geq 170$ nm may be emitted. Indeed, as it can be seen from spectrograms presented in Figure 3, the short wavelength MGR photons with $\lambda = 192$ may be registered.

A necessary condition for the realization of such a scenario is splitting of oxygen molecule into two atoms with an energy quantum equivalent of a photon of $\lambda \leq 235$ nm (118 kcal/mole). Indeed illumination of the reaction systems in which hydrolytic reactions proceed with visible light was needed for the emergence of MGR, and Gurwitsch suggested that only those visible photons were effective, energy of two of which could be pooled to the energy of one UV-photon with $\lambda \leq 235$ nm. If this suggestion was right, illumination of solution with violet or blue-green light (up to $\lambda \leq 470$ nm) will be effective, while its illumination with more long-wave light will not support MGR emission. Experiments fully confirmed this hypothesis. Besides, application of ultrasensitive analytical methods based on the studies of MGR fluorescent spectra of solutions emitting MGR revealed there traces of ••C=O and H_2N• free radicals [see the description of this method in (11)]. Thus, the hypothesis that urea and water may decompose to free radicals has been confirmed.

Generally, the same mechanism is responsible for MGR emission from other reactions of enzymatic hydrolysis (e.g., proteolysis, nucleolysis, etc.). This phenomenon has been unrecognized and is neglected until now because "side reactions" of oxygenation serving the ultimate source of energy for MGR represent such a minor part of all chemical transformations taking place in the reaction systems that their input cannot be observed using customary calorimetric approaches.

Until recently, skeptics could call in question the possibility of oxygen splitting due to two-photon excitation on irradiation of the solution with visible light. In the 1990s, such a physical phenomenon was definitely demonstrated. It was shown that under appropriate conditions even a evanescent photonic wave could cause two-photonic excitation of a fluorescent compound (12).

Another important peculiarity of proteolytic reactions leading to the emergence of MGR is water oxidation to H_2O_2 with active oxygen. Recently, it turned out that such an "unthinkable" reaction takes place in aqueous solution provided by the availability of active oxygen and specifically organized water (13, see also Voeikov, this volume). An important consequence of water oxidation accompanying hydrolytic reactions in the presence of active oxygen is that these normally thermoneutral reactions become the source of not only MGR but also of high density energy that may be stored in the form of H_2O_2 or other metastable peroxides and be used for the performance of other forms of functional work.

Glycolysis is an exception among other *in vitro* enzymatic and non-enzymatic reactions serving the source of MGR. Glycolytic reaction emitted MGR without highlighting with visible light, though the presence of oxygen was still needed. However, glycolysis unlike hydrolysis is an exothermic reaction, and oxygen may be activated in its course without external sources of energy, in particular, when glucose or other hexose split into two trioses, which tend gradually to convert into methylglyoxal (14). Methylglyoxal is one of the most active carbonyls activating oxygen especially in the presence of amine compounds (the latter should be present in the zymase preparation used by Gurwitsch to catalyze glycolysis) (15).

Hydrolytic and glycolytic reactions are the major catabolic reactions in any living organism. So if oxygen is available and the means of its activation efficiently operate, this sources of high density energy should operate in an organism permanently. Energy of electronic excitation may be used "as such" - for cell division triggering when this is needed, as energy of activation for low probability biochemical reactions; it may be degraded to lower frequency levels by fluorescent compounds, and it may also pool in cells and tissues in some peculiar and up to now not completely understood form.

4. SPECTRAL PROPERTIES OF MGR

As mentioned above, spectral analysis of MGR was accomplished by G.M. Frank soon after MGR discovery. As it can be seen in Figure 3, spectra of different enzymatic reactions *in vitro* consist of several narrow bands, and the width of some of them is as narrow as 1 nm, a feature very unusual for spectra of fluorescence recorded from aqueous solutions at room temperatures, where band widths usually reach several dozens nm. Another characteristic feature of MGR spectra is their specificity: for example, spectrum of nuclease reaction is the same, whether a nucleic acid or a phospholipid lecitin is taken as a substrate, and it has nothing in common with the spectrum of sugar fermentation. Initially Gurwitsch and co-workers interpreted these spectra as characteristic for a specific chemical or biochemical reaction, but later they understood that spectra were characteristic of some particular low molecular substances present in the reaction system, to be more precise, characteristic of specific chemical residues of these low molecular substances.

In particular, MGR spectra of phosphatase reaction is characteristic of a phosphate released from the substrate independent of a substrate nature. Indeed, addition of sodium phosphate into the reaction system where autoxidation of glycin proceeds and which is characterized with a very simple MGR spectrum results in the appearance of all the bands that are typical for a nuclease reaction. When glucose was added to a reaction system where urea was hydrolyzed with urease, the bands characteristic of glucose fermentation in addition to the bands characteristic of urease reaction appeared, while addition of urea to the reaction system where fermentation was going on resulted in enrichment of the spectrum with the bands representative for the urease reaction.

Hence, mitogenetic spectrum of reaction systems in which there go by reactions accompanied with the release of very high density quanta (e.g., λ=170 nm as in recombination of ••C=O + ••O) reflects excitation and fluorescence of simple low weight substances such as glucose and of chemical residues such as $-NH_2$, =C=O, -OH, and two forms of a peptide bond, R—CO—NH—R' and R—C(OH)=N—R'. This discovery allowed performing deep analysis of mechanisms of biochemical reactions beyond the reach of other methods, even highly sophisticated modern ones.

The very possibility to perform MGR spectral analysis of enzymatic and chemical reactions in the case when both concentrations of fluorescent compounds in solutions as well as the intensity of exciting radiation were extremely low argues that low intensity energy may propagate in aqueous solutions without dissipation for large distances. This property of water and aqueous system started to be acknowledged only recently (see Voeikov, this volume).

However, MGR spectral analysis recognizing particular substances and simple processes could be efficiently applied only to *in vitro* reaction systems or to some biological systems where catabolic processes were dominating over anabolic, such as freshly taken out blood, fresh excised tumor, and some others. In a lot of studies of complex biological tissues, MGR spectra did not reflect fluorescence of simple substances, and signified emission of some other entities. This applies to the so-called "degradation radiation".

4. DEGRADATION RADIATION AND NON-EQUILIBRIUM MOLECULAR CONSTELLATIONS

Pooling of MGR in living tissues was ascertained by Gurwitsch when he discovered a form of radiation which he somewhat awry defined as "degradation radiation". Unlike MGR releasing in the course of oxidative processes, degradation radiation arises from tissues (e.g., liver or muscle) in response to physiological stimulus or stress. For example, liver lay bare on a body of an alive mouse does not emit MGR spontaneously though all the biochemical processes giving rise to appropriate electronic excitation proceed in it. However, nearly immediately after an animal is injected with cocaine or glucose, a wave of MGR emerges. Same happens if liver is irritated with a weak electrical impulse, or mechanically disturbed, or just sprinkled with ice-cold water. Irrespective of the nature of the first irritation, a subsequent one cannot provoke a second wave of degradation radiation for a certain period of time.

Spectra of degradation MGR are very different from that of homogenous systems. For example, bands characteristic of specific fluorophores may be identified in spectra of spontaneous MGR from yeast cultures and they are relatively constant. Spectra of degradation MGR of yeast differ not only from that obtained in a resting state, but they differ significantly for different strains of yeast, they change depending on the nature of a factor inducing degradation MGR, on the physiological state of the living system, and on the state of its development.

The latter is well illustrated by A.A. Gurwitsh (16). She studied the properties of MGR of the naked baby rabbit muscle *in vivo* in different times of postnatal development. Spontaneous MGR was measured at a body temperature and degradation after the muscle was doused with cold physiological solution. Evolution of intensity of both kinds of MGR and of the spectrum of MGR are presented in Figure 4.

It can be seen that in the course of postnatal development intensity of spontaneous MGR decreases, while that of degradation MGR increases. During this period the processes related to muscle tissue functioning become progressively consolidated so that by the 15th day, intensity of degradation MGR reaches its maximum. Exactly by this time the posture and coordination of movements of an animal stabilizes, indicating that the processes related to muscle tissue functioning reached maximal interrelation. Spectrum of MGR also changes in a specific way. At the early stage of development, it looks like spectra of spontaneous MGR from systems containing many different flourophores. Later the number of spectral lines decreases, they widen, until only one wide line is left which has no correlation with known fluorophores.

All the properties of degradation MGR indicate that it differs in its origin from that of spontaneous one from homogenous systems. As already mentioned, the latter originates due to remission of energy released in reactions of radical recombination by fluorophores present in reaction systems. Properties of degradation MGR suggest that it arises due to disintegration of some preexisting objects retaining easily mobilizing energy of electronic excitation. Gurwitsch named these presumed objects "non-equilibrium molecular constellations". He supposed that they represent groups of excited macromolecules kept together

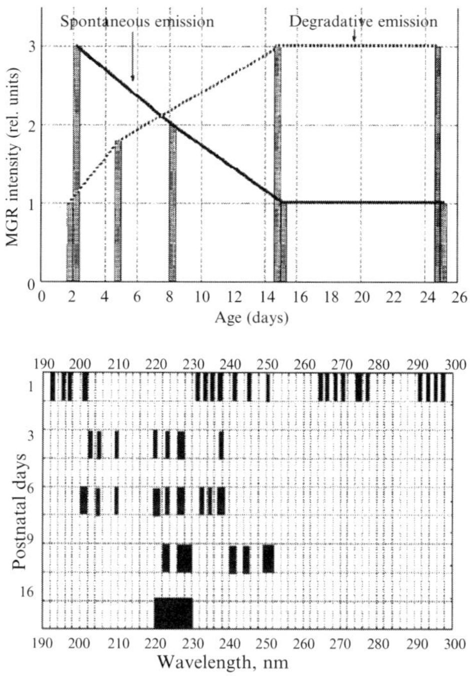

Figure 4. Top: evolution of intensity of spontaneous and degradative MGR of a baby rabbit muscle *in vivo* in different terms after birth (intensity was evaluated as a reverse of threshold period of irradiation by a muscle of yeast culture for obtaining a mitogenetic effect). Bottom: MGR spectra of a muscle (recorded at normal temperature).

due to constant energy circulation along their common energy levels. "Constellations" are fundamentally different from usual molecular associations and clusters. Components of the latter are kept together with different types of chemical bonds; energy is released when bonds lock, and sufficient energy inflow is needed for their dissociation. On the contrary, molecular constellations are sustained due to constant energy inflow, and any variations of energy supply let alone its blockade results in dispersion of constellations with release of energy retained by them. Lability of constellations precludes their revealing in fixed and even faulty biological material, and only the methods of studies of living cells and cellular systems similar to mitogenetic analysis may provide an insight of the existence of such dynamic structures.

If a constellation is disturbed by any means and loses its potential energy, the next distortion would not bring about a flash of degradation MGR until new constellations are formed. From the biological point of view, there is nothing extraordinary of such a behaviour of a constellation. Existence of refractory periods for excitable (more precise - irritable) tissues is well-known. There are many ways to trigger a nervous impulse (discharge), and until the critical value of membrane potential is restored, next irritation would not induce new impulse.

Gurwitsch applied to constellations the notion of Nobelist Albert Szent-Gyorgyi of migration of energy along the common electronic levels of protein molecules (17). But he broadened it to the possibility of migration of excitation energy along the constellations consisting of different molecules and also considered the possibility of energy quanta summation in different localities to the levels enough to emit UV-photons. Developing this concept, he stressed that because of univocacy of energetic and spatial parameters of constellations, fluctuations in energy migration should result in spatial realignments of constellations. That is why he stated that at the molecular level of living systems "… it is wrong to oppose the notion of a structure to the notion of a process. The only correct approach to living systems is an approach to them as to the structured processes, flowing in molecular complexes widely different in the degree of their lability" (18).

The possibility of high density energy transfer to macroscopic distances has been experimentally demonstrated using mitogenetic analysis by A.A. Gurwitsch. She has shown that if one fills a narrow capillary with a dilute protein solution and expose it to MGR from one end, no emission may be registered from the opposite end under usual conditions. However, if the capillary is placed in the longitudinal (parallel to its axis) electrical field (50 v/m), it becomes a MGR conductor. Same effect is observed if the protein solution flows in a capillary at a rate of 1 m/sec. A.A. Gurwitsch supposed that this phenomenon could be explained by a thread-like form of protein molecules, and their alignment along the capillary axis under the action of electrical field or a fluid flow, that increases the probability of energy transfer from one molecule to the next one. It is interesting that the rate of photon "diffusion" in this experimental system was around 30-32 m/sec, which is very close to the rate of a nerve impulse traveling along an axon.

It also follows from this experiment that pumping of constellations with energy is a necessary but not sufficient condition of their emergence. As the elements of constellations cannot mutually orient in them due to usual chemical bonding, there should exist an external vectorial factor that imposes a certain spatial arrangement to the elements of constellations. In an example with protein solution able to conduct MGR, the role of such factor was played by an electrical field or a fluid flow.

As constellations are postulated to be the most fundamental necessary condition for the existence of living matter ("structured processes"), the uninterrupted existence of such vectorial factor of dynamic nature is also to be postulated. That is why Gurwitsch's theory of biological and cellular fields - an imprescriptible property of all living systems - cannot be considered without referring to his experimental work in "mitogenetic biology". However, Gurwitsch's theory of biological field cannot be considered here, and the reader may inquire of other sources for more information on it (19).

Thus, degradation radiation is a signature of extremely non-equilibrium state of a living tissue implying that its energy potential difference with the environment is equivalent to many thousands of degrees.

5. OBLIVION OF MITOGENETIC BIOLOGY AND ITS STEADY RECOVERY

Gurwitsch's discovery attracted much attention. In the late 1920s, he was nominated for the Nobel prize (one of those who suggested his nomination was a world-famous physicist L.A. Mandelshtam). Many laboratories and researchers in the USSR, Germany, France, Italy, and Japan began to experiment with MGR. Mitogenetic biology reached its zenith by the middle of 1930s. From 1923 and up to 1939 hundreds of papers and a dozen comprehensive reviews on MGR appeared (for original sources see 20, 21, 22, 23). However, by the beginning of the 1940s the generally positive worldwide attitude to mitogenesis was ousted with indifference and even hostility. In 1943, Gurwitsch noted in the article devoted to the 20th anniversary of MGR discovery: "Many observers from different countries who did not dare even to try a single experiment and who are hardly acquainted with the current literature on the MGR problem, claim that the number of the works with negative results steadily grows while the number of confirming papers declines. The discrepancy of such statements with the real state of things is so striking, that if similar statements were made on ordinary, rather than scientific question, those who make them had to answer according to the law. One should just turn to the recent review by Maxia (Italy) to realize that hundreds of papers with data confirming MGR effects are opposed by less than a couple of dozens of reports with negative results"(23). However WWII destroyed all the European research centres where MGR was studied and independent works on this problem came to a halt.

In 1945, Gurwitsch founded and headed the Moscow Institute of Experimental Biology of the Academy of Medical Sciences of the USSR. In a short period of time, he and his associates managed to obtain a lot of new data. But this activity was soon interrupted. In 1948, when Lysenko regained full power in Soviet biology, Gurwitsch was dismissed. After Alexander Gawriliwich had passed away in 1954, the problem of MGR was forgotten for decades, and all his discoveries were considered to be doubtful.

At the beginning of the 1960s, Professor B.N. Tarusov and his associates at the Department of Biophysics of Moscow State University resumed studies of ultra-weak light emission from living organisms. They used photoelectronic multipliers rather than bio-tests for registering this radiation. But physical detectors still had much lower sensitivity than bio-tests. What is even more important is that most photomultipliers used register photons in much wider spectral range extending to green and even red part of the spectrum, while biological test-systems respond by mitotic reaction only to UV-photons. As it turned out photon emission in the visible range is much more intensive than in UV-range, and visible photons carry different information than UV-photons. Possibly that was the reason that the existence of Gurwitch's MGR was neither confirmed nor rejected using photomultipliers (unlike Geiger-Mueller counters used in the 1930s), though in general the ability of living systems to emit ultra-weak light discovered by Gurwitsch had been confirmed.

Professor Tarusov and his colleagues suggested that ultra-weak biological emission is the immediate result of free radical reactions, particularly of lipid peroxidation reactions and of recombination of active forms of oxygen. According to their concept, photon emission is just a by-product of such reactions, and photons do not play any significant functional role (24). This point of view is still dominant in biophysics, biochemistry, and physiology.

In the 1970s, the German physicist Friz-Albert Popp turned to the practically forgotten works of A.G. Gurwitsch (25). Using highly sensitive photoelectronic equipment for measuring ultra-weak biological emission, he confirmed many results of the predecessors. Popp was the first to attract modern theories of quantum physics for the analysis of these results. He discovered that ultra-weak light emission of biological systems of both plant and animal origin is highly coherent in the whole range of its detection - from UV to the red part of electromagnetic wave spectrum. In other words, living systems behave as if they were lasers. However, their radiation intensity is many orders of magnitude weaker and the degree of its coherence may be many orders of magnitude higher than that of technical lasers. Besides this, biological light is intrinsically polychromatic unlike that of a laser beam.

Why do biological radiations have such properties? To answer this question Popp and his colleagues attracted quantum-physical theory of a well-known American physisist R. Dicke as well as modern theories of cavity quantum electrodynamics and of coherent electromagnetic field (26). According to Dicke's theory, two oscillators residing in the "coherent volume" of each other (this volume is calculated basing on Heisenberg's uncertainty principle, and its diameter may significantly exceed the wavelength of emitted radiation) are constantly coupled. If both oscillators are initially in an excited state, they transit into the ground one simultaneously, hence coherent radiation is emitted. Dicke's theory of radiation corresponds with that of Planck's, like Prigogine's non-reversible thermodynamics corresponds with classical thermodynamics. Cavity quantum electrodynamics is the further development of Dicke's theory. It states that if excited oscillators are in a cavity with reflecting walls, where strong coupling between oscillators and their radiation field is established, spontaneous emission by the system may be either suppressed or become highly enhanced and coherent. But how do these purely physical theories and models correspond with biological phenomena?

It is possible that coherent radiation from the whole organism reflects coherency - that is interrelationship, cooperativity of molecular constellations. They are distributed all over an organism but at the same time represent elements of the common to all of them coherent field. Organism's coherency means that an event occurring in one particular part is immediately an event for the whole organism. It has been demonstrated in Popp's laboratory that not only single organisms, but also their communities, like daphnia in a small aquarium, or seeds put in one vessel, or a yeast culture, or animal cell suspensions, behave as coherent emitters. Taking into consideration specific properties of biological electromagnetic radiation, Popp suggested for it a new term - "Biophotons".

Popp's experimental and theoretical works supported Gurwitsch's theory of "non-equillibrium molecular constellations" as of a molecular collective being in an excited state in a common for all its elements field of energy. More than that: new data and its interpretation considerably expanded this concept, since whole organisms and even their communities turned out to be similar "constellations".

Spectral analysis of biophotons made by Popp also strongly favoured and expanded Gurwitsch's theory. It turned out that there is a strong deviation of spectral distribution of biophotons from radiation of inanimate objects - the ideal example of the latter is "black body radiation". Biophoton emission intensity is in the first approximation constant within the whole range of its detection - from UV to red light. That means that occupational probability of all energy levels - from electronic to translational ones - is practically the same, as if a living system has a tremendously high inter- and intramolecular "temperature".

Thus it is proved now that besides mitogenetic rays discovered by Gurwitsch, living organisms emit coherent and very weak light in the higher wavelength range. Gurwitsch had proved that ultraviolet photons carry out important function - they trigger cell division. And what is the function of biophotons belonging to other parts of the spectrum? There is no definite answer to this question at this time, but more and more evidence points out that they may serve for biological information transfer. Ultra-weak photon emission in the range from UV to near IR of electromagnetic spectrum from living cells and chemical reactions in aqueous media (27) affect activity of enzymes (28), activity and morphology of cells and tissues (29), regulate locomotion and mutual orientation of cultured cells (30). Back-reflected photons emitted during respiratory burst in human blood affect the intensity of this immune reaction by a feed-back mechanism (31).

Thus there are many indications that the period of oblivion of works and name of the great Russian biologist Alexander Gavrilovich Gurwitsch is passing away.

REFERENCES

1. A.G. Gurwitsch, Die Vererbung als Verwirklichungsvorgang, *Biologische Zentralblatt* **32,** 458-486 (1912).
2. A.Gurwitsch, Die Natur des spezifischen Erregers der Zellteilung, *Arch. Entwicklungsmech.* **100,** 11-40, (1923).
3. A. Gurwitsch, Physikalisches uber mitogenetische Strahlen, *Arch. Entwicklungsmech.* **103,** 490-498 (1924).
4. A. Gurwitsch, *Mitogenetic Emission.* (Gos. Med. Izdat., Moscow, 1932).
5. A. Gurwitsch and L. Gurwitsch, *Mitogenetic Radiation.* (The All-Union Institute of Experimental Medicine Publishing House, Leningrad, 1934).
6. S.Ya. Zalkind, Contemporary state of the problem of mitogenetic biology, *Usp. Sovrem. Biologii,* **7,** 50-66; 216-233 (1937).
7. R. Audubert, Emission from chemical reactions, *Uspekhi chimii.* **7**: 1858-1883 (1938). (Translated into Russian from: *Angew. Chemie,* **51,** 153 (1938).
8. A.I. Rabinerson and M.V. Filippov, Emission of short UV-photons from chemical reactions, *Zhurnal Fizicheskoi Chimii,* **9,** 688-701 (1938).
9. A.G.Gurwitsch and L.D. Gurwitsch, Peculiarities of chain reactions and common energy levels in biological systems, *Acta Physicochimica U.R.S.S.* **16,** 288-295 (1942).
10. A. Gurwitsch and L. Gurwitsch, *Mitogenetic Radiation. Physical-Chemical Basisand Applications in Biology and Medicine,* (Medgiz, Moscow, 1945).
11. V. Voeikov, Mitogenetic radiation, biophotons, and non-linear oxidative processes in aqueous media, in: *Integrative Biopysics. Biophotonics,* edited by F.-A. Popp, and L. Beloussov (Kluwer Academic Publishers, Dordrecht, 2003), pp. 331-360.
12. I. Gryczynski, Z. Gryczynski, and J.R. Lakowicz, Two-photon excitation by the evanescent wave from total internal reflection, *Anal. Biochem.* **247,** 69-76 (1997).
13. X. Xu, R.P. Muller, and W.A. Goddard 3rd., The gas phase reaction of singlet dioxygen with water: a water-catalyzed mechanism. *Proc. Nat. Acad. Sci. USA.* **99,** 3376-3381, (2002).
14. S.A. Phillips, and P.J. Thornalley, The formation of methylglyoxal from triose phosphates. Investigation using a specific assay for methylglyoxal. *Eur. J. Biochem.*, **212:** 101-105 (1993).
15. H.S. Yim, S.-O. Kang, Y.C. Hah, P.B. Chock, and M.B. Yim, Free radicals generated during the glycation reaction of amino acids by methylglyoxal. A model study of protein-cross-linked free radicals, *J Biol Chem,* **270,** 28228-28233. (1995).
16. A.A. Gurvich, Mitogenetic radiation as an evidence of nonequilibrium properties of living matter, in: *Recent Advances in Biophoton Research and its Applications,* edited by F-A. Popp, K-h. Li, and Q. Gu ed. (World Scientific, Singapore, 1992), pp. 457-468.

17. A. Szent-Gyorgyi, Towards a new biochemistry, *Science,* **93,** 609 (1941).
18. A. Gurwitsch, *Theory of Biological Field* (Sovetskaya Nauka, Moscow, 1944).
19. M. Lipkind, Can the vitalistic enthelechia principle be a working instrument? in: *Recent Advances in Biophoton Research and its Applications,* edited by F-A. Popp, K-h. Li, and Q. Gu ed. (World Scientific, Singapore, 1992), pp. 469-490.
20. O. Rahn, *Invisible Radiations of Organisms.* (Verlag von Gebruder Borntraeger, Berlin, 1936).
21. A.A. Gurwitsch, *Problem of Mitogenetic Radiation as an Aspect of Molecular Biology.* (Medicina, Leningrad, 1968).
22. A.G. Gurwitsch, *Principles of Analytical Biology and of the Theory of Cellular Fields.* (Nauka, Moscow, 1991).
23. A.G. Gurwitsch and L.D. Gurwitsch, Twenty years of mitogenetic radiation: emergence, development, and perspectives, *Uspekhi Sovremennoi Biologii,* **16,** 305-334 (1943). *(English translation: 21st Century Science and Technology. Fall, 1999; 12, No 3: 41.)*
24. B.N. Tarusov, I.I. Ivanov, and Yu.M. Petrusevich, *Ultra-weak Emission of Biological Systems* (Moscow State University Press, Moscow, 1967).
25. F.-A. Popp, Coherent photon storage in biological systems, in *Electromagnetic Bio-Information,* edited by F.-A. Popp, U. Warnke, H. Konig, and W. Peschka (Urban & Schwarzenberg, Munchen, Wien, Baltimor, 1989), pp. 144-167.
26. F.-A. Popp, Some essential questions of biophoton research, and probable answers. in: *Recent Advances in Biophoton Research,* edited by F-A. Popp, K-h. Li, and Q. Gu ed. (World Scientific, Singapore, 1992), pp. 1-46.
27. J. Slawinski, Luminescence research and its relation to ultraweak cell radiation, *Experientia.* **44,** 559-571, (1988).
28. G. Cilento, Photobiochemistry without light, *Experientia.* **44,** 572-576, (1988).
29. V.P. Galantsev, S.G. Kovalenko, A.A. Moltchanov, and V.I. Prutskov, Lipid peroxidation, low-level chemiluminescence and regulation of secretion in the mammary gland, *Experientia.* **49**, 870-875, (1993).
30. G. Albrect-Buehler, Changes of cell behavior by near-infrared signals, *Cell. Motil. Cytoskeleton.* **32**, 299-304, (1995).
31. V.L. Voeikov, R.R. Asfaramov, E.V. Bouravleva, C.N. Novikov, and N.D. Vilenskaya, Biophoton research in blood reveals its holistic properties, *Indian J. Exp. Biol.* **43,** 473-482, (2003).

2
PHOTON SUCKING AS AN ESSENTIAL PRINCIPLE OF BIOLOGICAL REGULATION

Fritz-Albert Popp and Wolfgang Klimek*

1. INTRODUCTION

The term *photon sucking* we understand as the *active* absorption of light. Contrary to passive absorption, this means that light becomes partially reabsorbed as soon as it is emitted or reemitted by the tissue under study. A typical example is displayed in Fig. 1.

The first observations of "photon sucking" in living tissues can be traced back to the strange phenomenon of oscillations around the relaxation curve of delayed luminescence (Popp et al., 1981). After confirmation of these findings by Chwirot et al. (1987), Schamhart and van Wijk (1987) observed some kind of photon-induced photon absorption in normal cell cultures of sufficiently high cell density, whereas this effect disappeared completely in tumor cell cultures (Fig. 2). As shown by Scholz et al. (1988), these effects are strongly correlated with the degree of coherence of the reemitted photons (Fig. 3). An even deeper understanding of this phenomenon was provided by the dissertation of M. Galle (1993). Figure 4 shows evidence of maxima and minima of biophoton emissions that were documented in populations of daphnia (and other animals), dependent on their average distances. These interference structures could be assigned to long-range interactions of the living organisms, establishing the organization of swarming, or, in more general terms, the "Gestaltbildung," of cell populations and the basis of intercellular communication. Beloussov (1997) pointed to photon sucking effects in eggshells, which behave rather differently depending on whether they are bound to their eggs or isolated. He generalized the results in case of embryonic batches of neurula stage frog embryos and loach embryos (Beloussov and Louchinskaia, 1998; Beloussov, 2002), and he established the connection to organization and communication of biological systems.

Further indications of photon sucking can be gathered from the experimental results of Vogel et al. (1998), who showed under our guidance in our laboratory that some bacteria suck up light from their nutrition medium (Fig. 5).

* Fritz-Albert Popp and Wolfgang Klimek, International Institute of Biophysics, Landesstiftung Hombroich, Raketenstation, Kapellener Strasse o.N., D-41472 Neuss, Germany.

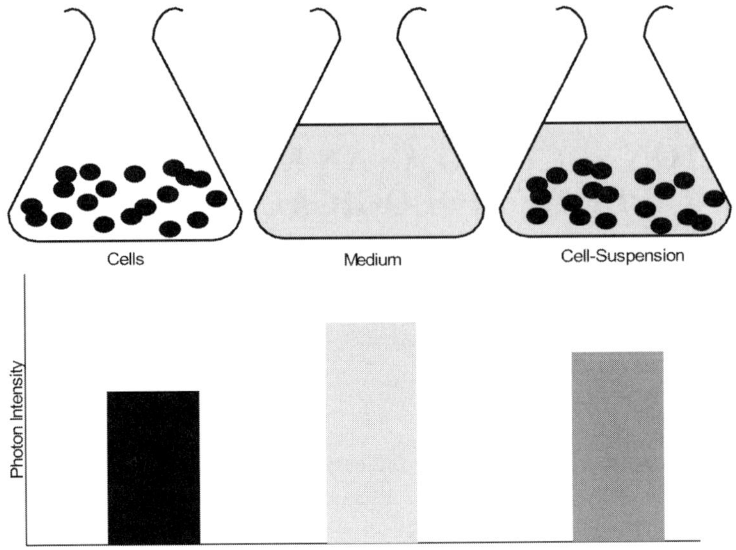

Figure 1. Cells without medium have, say, a photon emission intensity I_c. The medium shall display an intensity I_m. Both together, cells+medium, emit photons of an intensity $I < I_c+I_m$. The difference I_c+I_m-I is highly significant, indicating active absorption (sucking) of the cells within the medium.

Thus, the experimental evidence for photon sucking is already quite reliable. Therefore, the time is ripe for models of explanation, which are, however, at the present time, more or less the same, but different in the approach that is used.

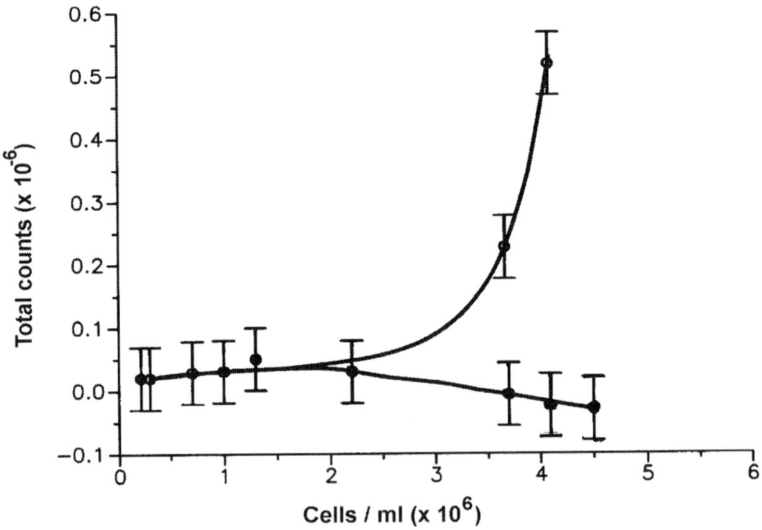

Figure 2. "Delayed luminescence" from tumor cells (upper curve) and normal cells (lower curve), as measured by Schamhart (1997). The normal cells suspended in medium display "induced absorption of photons" with increasing cell density. Tumor cells show in contrast non-linear increase of photon intensity with increasing cell density.

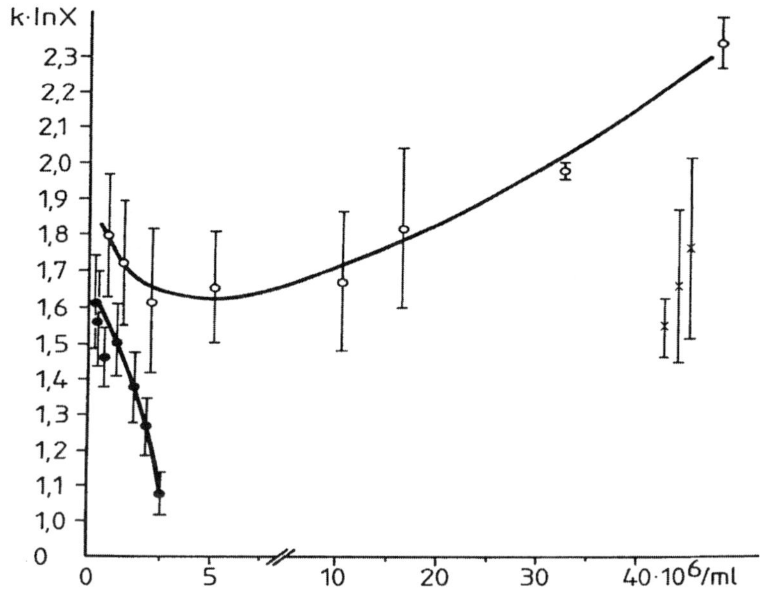

Figure 3. The decay parameter of the hyperbolic approximation that is adjusted to the relaxation dynamics of the afterglow of different cell suspensions after exposure to weak white light illumination is shown versus cell density. The lower curve displays the improvement of hyperbolic relaxation of normal amnion cells with increasing cell density. The upper curve shows the opposite dependence exhibited by malignant Wish cells. The three measurements at the right side of the figure correspond to the nutritive medium alone (Scholz et al., 1988).

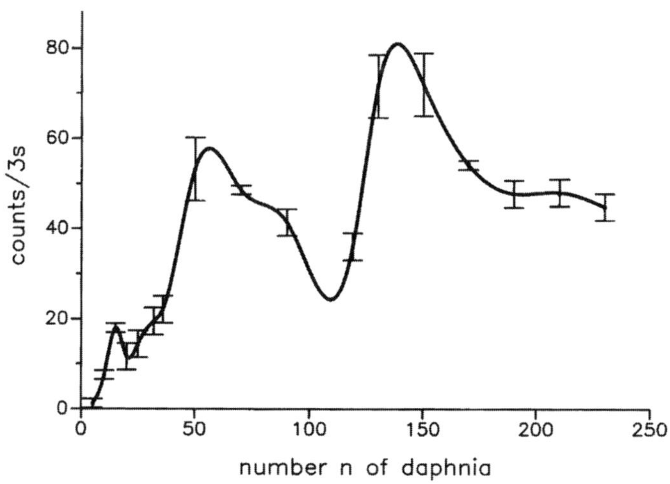

Figure 4. Mean values of the photon intensity of adolescent daphnia in 15 ml volume with the weighted standard deviation. Instead of the expected continuous increase of photon intensity with increasing number of daphnia, one measures interference-like changes (Galle, 1993).

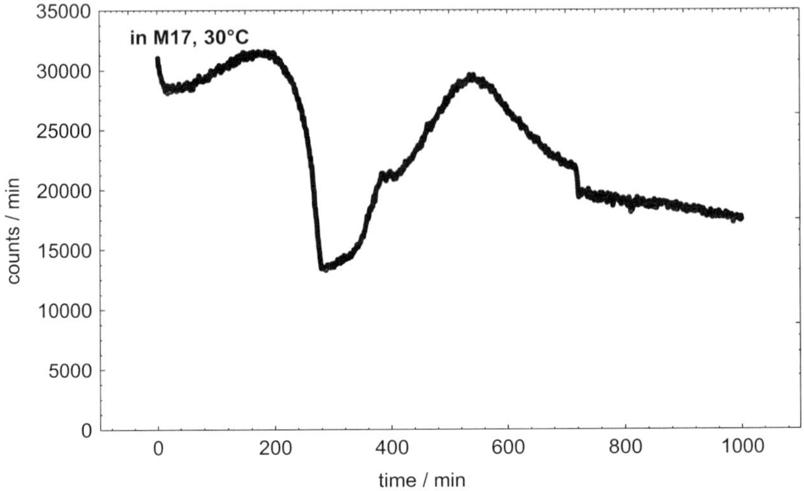

Figure 5. Growing bacteria in culture medium, that by oxidative reactions always emits light, absorb from a definite density on the light of the medium. For higher densities, this absorbance may decrease again (Vogel et al., 1998).

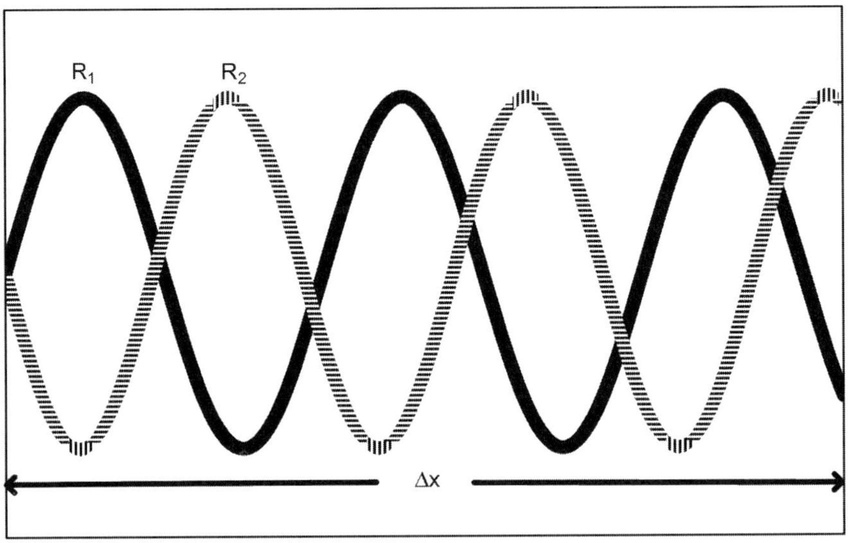

Figure 6. Zone Δx of destructive interference of two superimposing waves with different phase relations.

2. SIMPLE MODELS

Measured photons are the result of localized energy exchange of electromagnetic waves with the photon counting detector. By careful consideration using the uncertainty principle, it is never completely wrong to model them as waves. Let us start with the simple example of two wave trains of just the same wavelength λ superimposing in a way that they interfere destructively (Fig. 6). Obviously, this is a process where energy disappears over the zone just where destructive interference takes place. The question arises whether this local process of energy annihilation is possible, and over what region it is allowed, if it could take place at all. Actually, there is no reason to reject the existence of destructive interference, because it belongs to the generally accepted and much-observed physical principles of all interference effects. However, a simple calculation shows that destructive and constructive interference will always become balanced in such a way that the energy conservation law is not violated, as has been shown by Popp (1992). Actually, according to the fundamentals of Quantum Theory, it is never possible to superimpose two waves without any uncertainty in the wavelengths. One of the waves shall have the wavelength λ, the second the wavelength $\lambda+\Delta\lambda$. Then it is clear that the superposition over the distance of one wavelength will always lead to an increasing phase difference $\varphi(\Delta\lambda)$ between the waves. Destructive interference over n_u consecutive wavelengths will take place only as long as

$$n_u\Delta\lambda < \lambda/2 \tag{1}$$

As soon as with increasing number n_u Eq. (1) is not satisfied, the phase difference between the superimposing waves leads to constructive interference for the next n_u wavelengths.

It is well-known that a photon of wavelength λ carries the momentum

$$p = h/\lambda \tag{2a}$$

with the uncertainty

$$|\Delta p| = |\Delta\lambda|\, h/\lambda^2 \tag{2b}$$

Because n_u is the number of wavelengths that superimpose over a distance Δx of destructive interference, we can rewrite the position uncertainty in the form

$$\Delta x \equiv n_u\lambda \tag{3}$$

By the insertion of (3) and (2b) into (1), we find finally that

$$\Delta x\Delta p < h/2 \tag{4}$$

However, Eq. (4) obviously violates the uncertainty principle because always $\Delta x\Delta p \geq h$. We learn from this that

- destructive (and correspondingly also constructive) interference can always take place without any confinement, but only over limited regions of space and time,
- the energy conservation law and the uncertainty relation are mutually dependent.

Photon sucking takes place in the region of destructive interference where photons are really trapped. This effect would always be observed in a system, if on the outside essentially destructive interference took place while for reasons of energy conservation the inside had then to be consequently subject to constructive interference.

An example how this can be realized has been documented in a previous paper about phase conjugation effects in biology (Popp and Chang, 2000). Necessary for this phenomenon is a nonlinearly electrically polarized double layer (i.e., a biological membrane or exciplex structure) with a small distance d between the layers such that

$$d<\lambda \tag{5}$$

where λ is the wavelength under study.

We arrive at the most stable state of the system if and only if the double layer gets a nodal point of incoming waves, where on one side destructive and on the other side constructive interference has to take place. This discontinuity in the impinging wave trains is at the same time the source of a sucking force K_λ,

$$K_\lambda = - n_\lambda \lambda F h(c/\lambda)(Q\lambda/d), \tag{6}$$

where n_λ is the spectral photon density, F the surface area of the layer, h the Planck's constant, and Q the (dimensionless) resonator value. It is worthwhile to mention that $n_\lambda \lambda F h(c/\lambda)$ is just the radiation pressure that works in the opposite direction as the sucking force and is $(Q\lambda/d)$-times smaller.

This sucking force may provide, for instance, the phototropism of plants or the aggregation of cells and many other related biological phenomena.

3. COHERENCE AND PHOTON SUCKING

The simple examples above already tell us that the coherence of photons plays a fundamental role in the possible photon sucking. Take a coherence length $L \ll D$, where D are relevant biological dimensions, and then photon sucking certainly cannot play a relevant role. However, for $L \approx D$, this effect has always to be taken into account. When including infrared light, microwaves, and even radio waves, there are manifold possibilities where this effect may play an important biological role for phenomena that are not known or understood or are in question at present. In order to reveal the basic and general character of photon sucking, we have to extend our investigation to quantum theory and to find possible photons that become actively stored without violation of the energy conservation law, in line with the laws of quantum theory.

Several times we pointed to Dicke´s theory (Dicke, 1954) of sub- and superradiance that could be assigned to destructive and constructive interference, respectively. Dicke developed it as a quantum theoretical model, and it is worthwhile to note that (5) is just the same condition as has been provided for the application of his model. Because biological systems are "optically thick" media, (5) is always relevant.

However, the most accurate model is the description of "photon sucking" in terms of coherent states (Popp et al., 2002). Let us start with a well-known Hamiltonian **H** that keeps coherent states coherent. According to Glauber (1963) and Mehta (1966), this Hamiltonian generally takes the following form

$$\mathbf{H} = F(t)\mathbf{a}^{+}\mathbf{a} + G(t)\mathbf{a}^{+} + G^{*}(t)\mathbf{a} + B(t), \quad (7)$$

where **a+**, **a** are the creation- and annihilation operators, respectively, and F, B are real functions, while G is a complex function. For photons of frequency ω we have F(t)= ħω. Take $|\beta>$ as a coherent eigenstate of (7), then we write for

$$<\beta|\mathbf{H}|\beta> = E \quad (8a)$$

$$<\beta|\mathbf{a+a}|\beta> = n \quad (8b)$$

Let us introduce the resonator value

$$Q = \omega(E-B)/[\partial/\partial t(Fn)] \quad (9)$$

The Q-value is a measure of the storage capacity of the system under consideration. From (8b) we get

$$n= n_{ch} + n_c = (E-B)/F + (G\beta^* + G^*\beta)/F, \quad (10)$$

where the first term can be assigned to a chaotic part of a photon number n_{ch} and the second certainly to a coherent part n_c.

For the coherent part we provide

$$<\partial/\partial t(n_c)>\approx 0, \quad (11)$$

describing an oscillatory behavior.

Take F as a constant value and insert F of (10) into (9), then we obtain and consequently

$$F = \omega/(\partial/\partial t(n))(E-B)/Q = (E-B)/n + (G\beta^*+G^*\beta)/n,$$

$$Q = \omega\, n/(\partial/\partial t(n_{ch}))\ 1/(1-(G\beta^*-G^*\beta)/(E-B)) \quad (12)$$

In contrast to chaotic states, the Q-value of coherent states provides what we call "active photon storage," which may be called "photon sucking" or "photon trapping" . Actually, as soon as $(G\beta^*-G^*\beta)$ is oscillating around (E-B), the resonator becomes rather active in "sucking" photons for $Q<0$, highest storage capacity for $Q \rightarrow\infty$, and even transparency for $Q \rightarrow 1$.

After these plausibility considerations, it is certainly necessary to show evidence of this sucking process by taking account of the accurate solutions of

Eq. (7). This has been done in previous papers (Popp et al., 2002), but is worthwhile to repeat the results at least as far as they reflect directly the effects of photon sucking.

A particular solution of (7) can be expressed in terms of coherent states $|\alpha(t)\rangle$, which are eigenstates of the annihilation operator **a** according to

$$\mathbf{a}\,|\alpha(t)\rangle = \alpha\,|\alpha(t)\rangle, \tag{13}$$

where α is the field amplitude of this state under study.

We showed that under conditions of homeostatic regulation, Eq. (7) can be split into two parts, the first one concerning the photon number n, and the second one responsible for the field amplitude $\alpha(t)$.

Taking $F = \hbar\omega(t)$, the solution for n(t) follows the equation

$$\omega(\partial/\partial t(n)) + (\partial/\partial t(\omega))n + 1/\hbar[\partial/\partial t(B(t))] = 0 \tag{14a}$$

In addition to the trivial solution n = const., where ω is a constant, we obtain under ergodic conditions a non-trivial solution

$$\omega(t) = \omega(0)/(1+\upsilon t), \tag{14b}$$

where υ is a constant.

Under these conditions of homeostatic regulation, $\alpha(t)$ has to be derived from a function G that satisfies the equation

$$\partial/\partial t G(t) + i\omega(t)G(t) = 0 \tag{15a}$$

This result explains again the well-known fact that delayed luminescence relaxation functions follow a hyperbolic decay law (Popp and Li., 1993, Bajpai et al., 1998).

For G(t) we then get

$$G(t) = G(0)\exp(-i\ln(1+\upsilon t)) \tag{15b}$$

This solution provides the oscillation with linearly increasing period. It can be observed as taking place around the hyperbolic relaxation of the photon emission after external light illumination.

It could be shown that it is not possible to see these oscillations in monochromatic fields with the same field amplitudes $\alpha(t)$. A necessary condition for these oscillations is the coupling of at least two modes, in order to get destructive interference for mutual photon sucking between the modes, as has been shown by Popp (2002) (Figs. 7a, 7b).

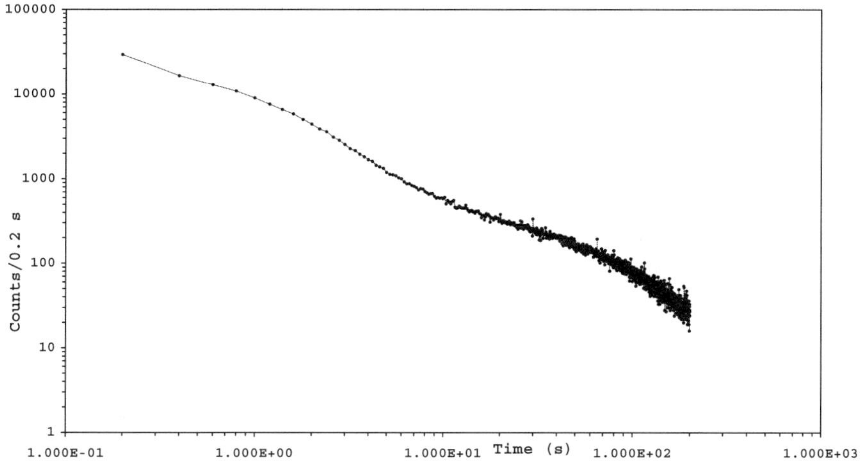

Figure 7a. Delayed luminescence of *Acetabularia acetabulum* after 10s white-light illumination by a tungsten lamp (150W). Courtesy of Rafael Moreno (IIB, Neuss). The relaxation displays oscillations around the hyperbolic decay behaviour (Popp et al., 2002).

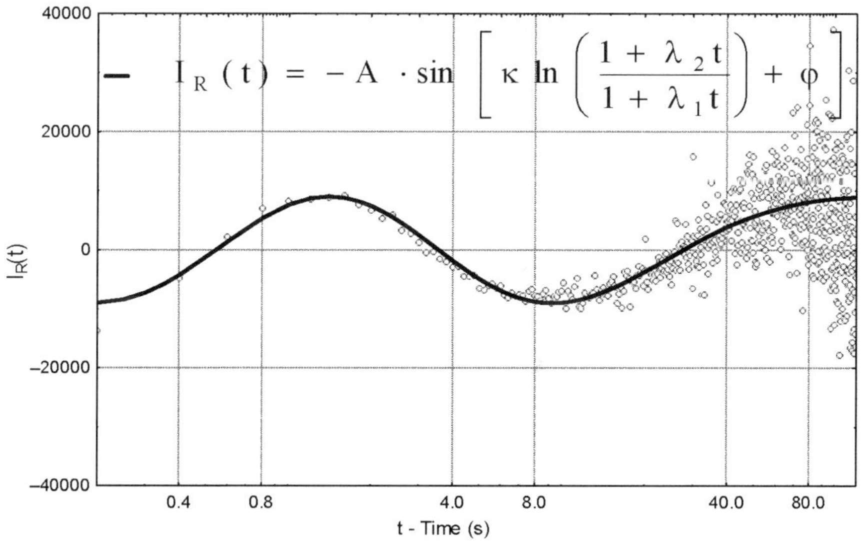

Figure 7b. The oscillations around the hyperbolic relaxation function can be calculated analytically (Popp et al., 2002).

The most general solution of the intensity I(t) of delayed luminescence under ergodic conditions and by providing homeostatic effects takes the form (Popp 2005):

$$I(t) = I(0)\{\prod_j(1+b_j t)^{P_j}\}\sin\{[\sum_j \gamma_j \ln(1+\kappa_j t)]+\Phi\} \tag{15c}$$

The investigations have demonstrated that by superimposition of coherent modes in biological systems, there is extremely high stability of frequency and phase, which is likely a basis of biological communication in terms of frequency and phase modulations of the signals.

4. MOLECULAR ANALYSIS

We are unsatisfied with the solution of the problem as long as we are not seeing the molecular interaction between field and matter. If "photon sucking" takes place, there must be excited states of molecular matter that are subject to photon absorption for sufficiently long periods. In order to trap photons, the excited state has to display a higher stability than the state before absorption. This condition can be fulfilled by an energy gap between the former and the latter electronic state. The rather strange condition has to be satisfied that, despite the absorbing of a photon, the latter state should have a lower energy than the former one. How is this possible?

We would like to show now that by a phase transition from chaotic to coherent states and vice versa, it is actually possible to construct such a mechanism. Let us start with a Hamiltonian $\mathbf{H_0}=\hbar\omega\mathbf{a^+a}$ that creates number states of photons of energy $\hbar\omega$. Influenced by coherent photons we switch on a coherent field $\mathbf{H_1} = G\mathbf{a}^+ + G^*\mathbf{a}$ and add it to the Hamiltonian $\mathbf{H_0}$, in order to get the complete Hamiltonian $\mathbf{H} = \mathbf{H_0} + \mathbf{H_1}$. Our goal is to compare the solutions before and after absorption by comparing that of the original Hamiltonian with the complete one. For simplicity, we take a basis set of two number states /0> and /1> of energy 0 and $\hbar\omega$. We provide then

$$\mathbf{a}/0> = 0/0>;\ \mathbf{a}/1> = /0>;\ <0/\mathbf{a}^+ = <1/ \tag{16a}$$

The elements of the secular matrix are then

$$<0/\mathbf{H_0}/0>=0;\ <1/\mathbf{H_0}/1>=\hbar\omega;\ <0/\mathbf{H_0}/1>=0;\ <0/\mathbf{H_1}/0>=0;\ <0/\mathbf{H_1}/1>=G^*;$$
$$<1/\mathbf{H_1}/0> = G; <1/\mathbf{H_1}/1>=0; \tag{16b}$$

The solution of the Schrödinger equation in terms of this basis set gets the form

$$/\Psi>=c/0> + ((1-c^2)^{1/2})/1> \tag{16c}$$

and by solving $H/\Psi>= E/\Psi>$ we arrive after straightforward calculation at the following results

$$E = ½\,[\hbar\omega +/- ((\hbar\omega)^2 + 4G^*G)^{1/2})] \tag{17a}$$

$$c^2 = (G^*G)/[G^*G + \hbar\omega] \tag{17b}$$

The solution, which can be understood completely also in classical terms, shows evidence that

- as soon as a coherent field is switched on, the photon gets stored in a new ground state with lower energy than the former ground state,

- the energy gap increases with increasing G*G and provides photon sucking,
- the conservation of energy is due to an excited state that has a higher energy than the former one,
- Time-dependent perturbation theory shows that the system starts to oscillate between the new ground state and the other states of the system.

The energy gap ΔE between the two states is according to (17a)

$$\Delta E = [(\hbar\omega)^2 + 4G^*G]^{1/2} \qquad (17c)$$

It is understandable from this point of view why the oscillation frequency Ω of the "delayed luminescence oscillations" gets smaller and smaller during the relaxation. Note that the coupling factor of the states will increase with the product of the amplitudes of the different modes of the stored photon field ($\alpha_0{}^*\alpha_1$ - $\alpha_0\alpha_1{}^*$). They may likely increase proportionate to the number n_0 of stored photons. Consequently, the oscillation parameter λ follows the same hyperbolic law as Eq. (15a), but with an imaginary exponent:

$$n(t)=n(0)/(1+\lambda t)^{i\gamma} = \exp(-i\,\gamma \ln(1+\lambda t)) \qquad (18)$$

This kind of photon sucking may work permanently in all biological systems. It becomes a relevant and even macroscopic effect because the oscillation frequency will always take low enough values after some time. Consequently, a considerable part of photons of all possible frequencies may get trapped and stored. The system can take the role of a most fundamental process, as a source of sensitivity, rhythmic tuning, as well as of long-range communication within the organism and with the environment. Actually, the fact that not only the visible range is included and that this process is connected to communication channels within the body shall be shown now in the example of infrared radiation.

5. PHOTON CONDUCTION (LIGHT PIPING) AND PHOTON SUCKING IN HUMAN TISSUE

Some years ago, Mandoli and Briggs (1982) evoked attention by impressive examples of "light piping in plant tissues." Smith (1982) showed that the results could be understood only in terms of coherent radiation that is conducted along light fibers in biological tissues. Not long ago, Schlebusch, Maric-Ohler, and Popp (Schlebusch et al., 2005) were successful in demonstrating that at least in the wavelength range from 3 to 5 micrometers, channels of light emission appeared on the body, and it turns out that they are mirrored by what the ancient Chinese introduced as "meridians." Figure (8a) displays an example of the "bladder meridian" after spinal moxibustion. Figure (8b) documents a part of the stomach meridian on the face of a patient with a facial paralysis without any external influence. The molecular basis of this effect can be described in terms of a differential equation that has been derived by Thomas (1961) from Non-Equilibrium Thermodynamics in the presence of a radiation field. He formulated the basic equation in rather simple terms

$$dI_v/d\tau_v = I_v - S_v \tag{19a}$$

where

$$dI_v = h\nu\,[-I_v B_{LU} n_L + I_v B_{UL} n_U + A_{UL} n_U]\,dl/(4\pi) \tag{19b}$$

I_v is the spectral intensity of radiation that travels through a medium with n_U molecules in the upper excited state and n_L molecules in the lower one. B_{LU} and B_{UL} are Einstein`s coefficients of induced absorption and induced emission, respectively, and A_{UL} is Einstein´s coefficient of spontaneous emission. dl is the infinitesimal distance of the light path. Thomas defined the optical depth τ_v and the source function S_v as follows:

$$d\tau_v = -h\nu\, B_{LU} n_L\,(1-X)dl/4\pi \tag{19c}$$

$$S_v = 2h\nu^3/c^2\,(1/X-1)^{-1}, \tag{19d}$$
$$\text{where } X = g_L n_U/g_U n_L \tag{19e}$$

g_L and g_U are the degeneracy factors of the lower and upper excited states, respectively.

Equation (19a) is the simple result of (19b) after insertion of (19c), (19d), and (19e). For simplicity, we confine to a homogeneous medium.
In the case of Figs. (8a) and (8b), which are confirmed by our experimental results of infrared photography, we can immediately see the solution I_v of the form

$$I_v = I_v(0)\, f(\delta) \tag{20a}$$

where δ is the vertical distance from the radiation channel (the "meridian"). $I_v(0) = S_v(0)$ along the meridian line. This provides that $I_v(0)$ does not change along $\delta = 0$, where

$$\partial I_v/\partial \tau_v = 0 \tag{20b}$$

$f(\delta)$ describes the change of Iv in the perpendicular direction. It is obvious that in case of the bladder meridian (Fig. 8b), where $\partial I/\partial\delta < 0$, also

$$\partial f/\partial \delta < 0 \tag{20c}$$

This describes a bright optical soliton, as it does not change its shape along the direction $\delta = 0$, where it displays its maximum intensity.

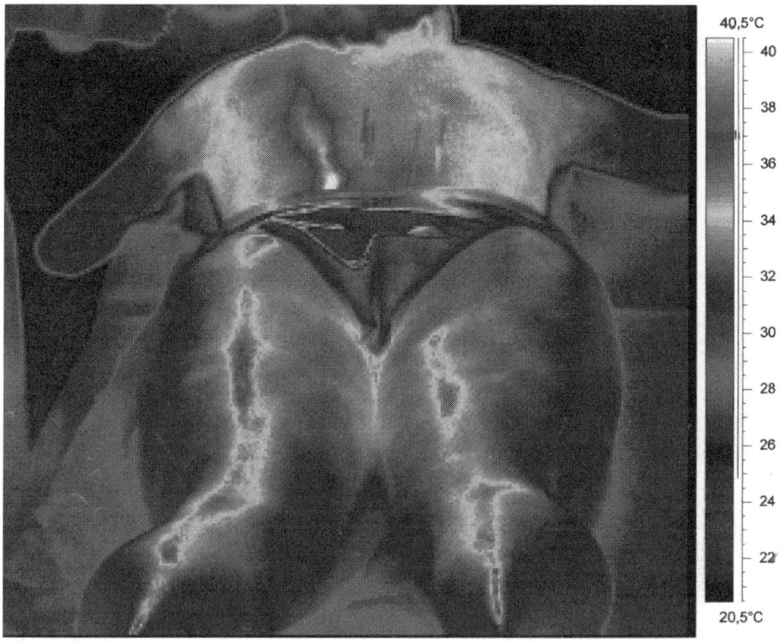

Figure 8a. Documentation of a part of the bladder meridians by moxibustion. They can be described in terms of bright solitons (see color plate).

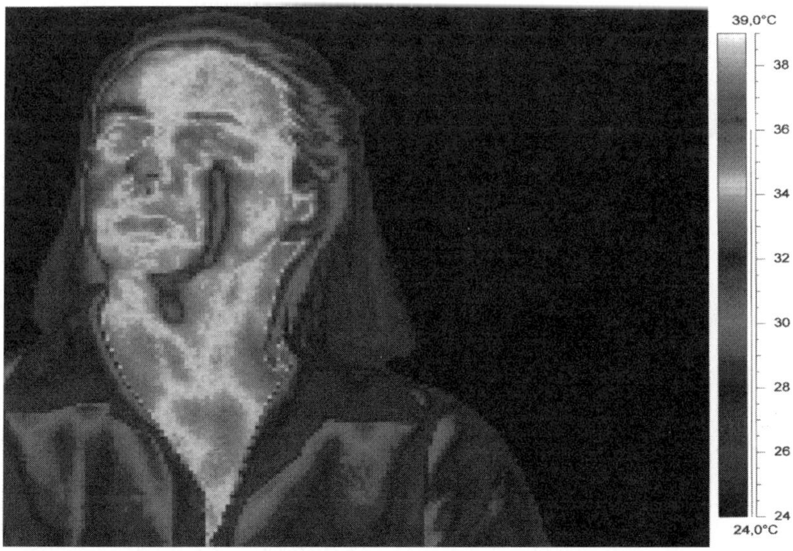

Figure 8b. Documentation of a part of the stomach meridian. There is no external excitation. This case corresponds to a black soliton where photon sucking from the surrounding tissue may take place. (see color plate).

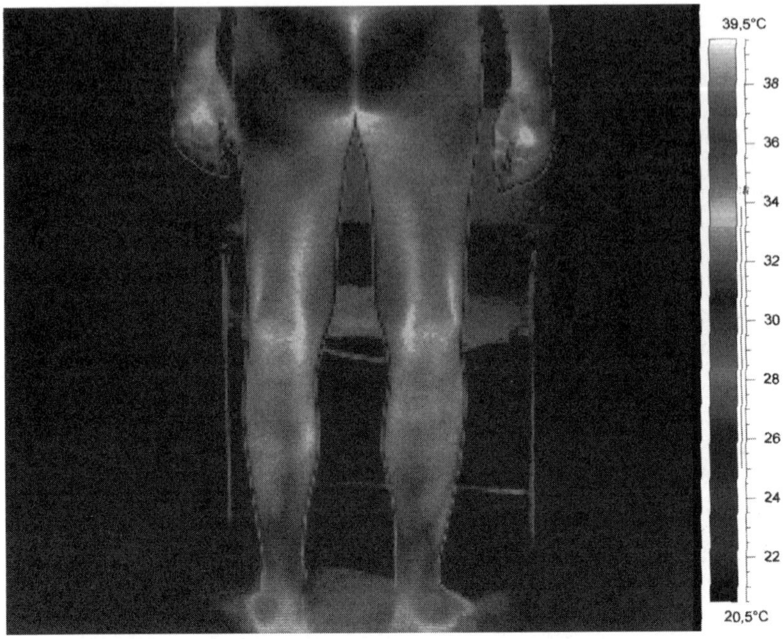

Figure 8c. Documentation of the superposition of bright and black solitons, giving rise to transparency. One can see three bones of the legs. (see color plate).

For the case of the stomach meridian, which can be seen in Fig. (8b) on the face (in a pathological case without external illumination), I_v increases with increasing vertical distance δ from the meridian:

$$\partial f/\partial\delta > 0 \qquad (20d)$$

Because its center is certainly "colder" than its environment, it describes a so-called black soliton which "sucks up" light from its environment.

After insertion of (20a) into (19a), we arrive generally at

$$I_v(\delta) = S_v(0)/[1-1/f(\delta)(\partial f/\partial\delta)(\partial\delta/\partial\tau_v)] \qquad (20e)$$

Take for S(0) (19d) and evaluate (19c), then we can write also

$$Iv(\delta) = 2\pi h v^3/c^2/(1/(X(0)-1)[1-4\pi/f(\delta)(\partial f/\partial\delta)/(B_{LU}n_U(1-X))] \qquad (20f)$$

In view of (20c), I_v can decrease with increasing δ only if $\partial\delta/\partial\tau_v>0$. From Eq. (19c), with $l = \delta$ we learn that this is possible only if the occupation X according to (19c) follows

$$g_L n_U/(g_U n_L) > 1 \qquad (20g)$$

This means that the meridians after moxibustion are understandable only if the electronic states of the corresponding molecules along the meridians and in the vicinity of the meridians are over-occupied. The radiation emitted from these areas should be at least partially coherent, where the matter may show optical anisotropy, as the requirement of X>1 is valid only for the direction perpendicular to the channels.

Just the same requirement of (20g) holds also for the black solitons of Fig. (8b). Actually, Eq. (20d) provides an increase of $I_v(\delta)$ if and only if $\partial\delta/\partial\tau_v>0$, and consequently X>1. But photons are sucked up instead of being emitted in these meridians. However, in both cases of Figs. (8a) and (8b), concerning bright and black solitons, the matter is electronically over-occupied and approaches a zone of $X\approx1$ at large distances from the meridians, where the matter becomes more and more transparent. Figure (8c) displays such a case where the transparency becomes evident.

Because $S_v(0)$ may take positive (Fig. 8a) as well as negative values (possibly in Fig. 8b), X may take values higher than 1 as well as values lower than 1. For X = 1, we arrive at equal occupation of the lower and upper electronic states. Actually, the transition from X<1 to X>1 is at the same time a phase transition from chaotic to coherent radiation, corresponding to the LASER threshold in technical devices. Note that for X=1, the dependency on l may disappear completely. It impressively displays the holistic character of this mechanism.

The most interesting region of X is ½<X<2, because this corresponds first to the occupation of exciplexes, passing the threshold and arriving at the most sensitive zones of photon sucking. Note here that in view of (20f), the second term is symmetric in X and 1/X. It is very likely that the molecular mechanisms of life are based on these fundamental functions, which have been discussed several times before by Popp (Popp et al., 1994). The real situation reflects the permanent superposition of emitting and sucking zones such that the mean value $<df/d\delta>$ tends to get zero. Local (non, random) fluctuations of $I_v>S_v$ and $I_v<S_v$ may work then for extremely sensitive but homeostatic regulation of the body.

We would like to thank the members of the IIB who are always helpful and guiding friends in putting these models of biophotons forward, and in particular Prof. Beloussov for the introduction to and progress of Biophotonics in the scientific community. We would also like to refer to a paper about the holistic character of biophoton emission that has been published recently by Yan (Yan et al., 2005). We like to thank also the Familie-Ernst-Wendt Stiftung (Colonia) for financial support.

6. REFERENCES

Bajpai, R.P., Kumar, S., and Sivadasan, V.A., 1998, Biophoton emission in the evolution of squeezed state of frequency stable damped oscillator, *Appl.Math.Comp.* **93**: 227.

Beloussov, L.V., Popp, F.A., and Kazakova, N.J., 1997, Ultraweak irradiations of hen eggs and embryos:non-additive interaction of two emitters and stable non-equilibricity, *Ontogenez (Russ. J. Devel. Biol.)* **28**: 377-388.

Beloussov, L.V., and Louchinskaja, N.N., 1998, Biophoton emission from developing eggs and embryos: non-linearity, wholistic properties and indications of energy transfer, in: *Biophotons*, J. J. Chang, J. Fisch and F.A. Popp, eds., Kluwer Academic Publishers, Dordrecht-Boston- London, pp. 121-141.

Beloussov, L.V., 2002, Exploring the dynamic background of the developmental processes and cell reactions with the use of ultraweak photon emission, *BioSystems* **68**/2-3: 199-212.

Chwirot, B., Dygdala, R.S., and Chwirot, S., 1987, Quasi-monochromatic-light-induced photon emission from microsporocytes of larch showing oscillator decay behaviour predicted by an electromagnetic model of differentiation, *Cytobios* **47**: 137-146.

Dicke, R.H., 1954, coherence in spontaneous radiation processes, *Phys. Rev.* **93**: 99-110.

Galle, M., 1993, Untersuchungen zum dichte- und zeitabhängigen Verhalten der ultraschwachen Photonenemission von pathogenetischen Weibchen des Wasserflohs Daphnia magna, *Inaugural-Dissertation*, Naturwissenschaften (Biologie), Universität Saarbrücken.

Glauber, R.J., 1963, The Quantum theory of optical coherence, *Phys. Rev.* **130** (6): 2529-2539.

Mandoli, D.F., and Briggs, W.R., 1982, Optical properties of etiolated plant tissues, *Proc. Natl. Acad. Sci. USA* **79**: 2902.

Mehta, C.L., and Sudarshan, E.C.G., 1966, Time evolution of coherent states, *Physics Letters* **22**(5): 574-576.

Popp, F.A., Ruth, B., Bahr, W., Böhm, J., Graß, P., Grolig, G., Rattemeyer, M., Schmidt, H.G., and Wulle, P., 1981, Emission of visible and ultraviolet radiation by active biological systems, *Collective Phenomena (Gordon & Breach)*: 187-214.

Popp, F.A., 1992, Some remarks on biological consequences of a coherent biophoton field, in: *Recent Advances in Biophoton Research an its Applications*, F.A. Popp, K.H. Li, and Q. Gu, eds., World Scientific, Singapore - New Jersey - London - Hong Kong., pp. 357-373.

Popp, F.A., Li, K.-H.,1993, Hyperbolic relaxation as a sufficient condition of a fully coherent ergodic field, *International Journal of Theoretical Physics* **32** (9): 1573-1583.

Popp, F.A., Gu, Q., and Li, K.H., 1994, Biophoton emission: Experimental background and theoretical approaches. *Modern Physics Letters* **B8**: 1269-1296.

Popp, F.A., and Chang, J.J., 2000, Mechanism of interaction between electromagnetic fields and living organisms, *Science in China (Series C)* **43** (5): 507-518.

Popp, F.A., and Yan, Y., 2002, Delayed luminescence of biological systems in terms of coherent states, *Physics Letters A* **293**: 93-97.

Popp, F.A., 2004, Quantum Phenomena of Biological Systems as Documented by Biophotonics, in: *Quo Vadis Quantum Mechanics?,* A.C. Elitzur, S. Dolev and N. Kolenda, eds., Springer, Berlin-Heidelberg-New York, pp. 371-396.

Schamhart, D.H.J., and van Wijk, R., 1987, Photonemission and the degree of differentiation, in: *Photon Emission from Biological Systems*, B. Jezowska-Trzebiatowska, B. Kochel, J. Slawinski, eds.), World Scientific, Singapore, pp. 137-152.

Schlebusch, K.P., Maric-Oehler, W., and Popp, F.A., 2005, Biophotonics reveal meridian structure of the human body, *Journal of Alternative and Complementary Medicine (JACM)*, 11(1): 171-173.

Scholz, W., Staszkiewicz, U., Popp, F.A., and Nagl, W., 1988, Light stimulated ultraweak photon reemission of human amnion cells and Wish cells, *Cell Biophys.* **13**: 55.

Smith, H., 1982, Light piping by plant tissues, *Nature* **298**: 423.

Vogel, R., and Süßmuth, R., 1998, A model for the generation of low level chemiluminescence from microbiological growth medium an its depletion by bacterial cells, *Biochem. Bioenerge.* **45**: 93.

Thomas, R.N., 1961, Some aspects of non-equilibrium thermodynamics in the presence of a radiation field. University of Colorado Press, Boulder, Colorado.

Yan, Y., Popp, F.A., Sigrist, S., Schlesinger, D., Dolf, A., Yan, Z., Cohen, S. and Chotia, A., 2005, Further analysis of delayed luminescence of plants, *J. Photochem. Photobiol.*, **78**: 179-263.

3
QUANTUM SQUEEZED STATE DESCRIPTION OF SPECTRAL DECOMPOSITIONS OF A BIOPHOTON SIGNAL AND THE POSSIBILITY OF REMOTE INTERVENTION

R.P. Bajpai *

1. INTRODUCTION

Gurwitsch[1] hypothesized the existence of morphogenetic field of a living system and suggested some of its properties. The suggested properties are rephrased below:

1. Every living system has a morphogenetic field. The field is system and situation specific. The word "morphogenetic field" is used in generic sense and it includes both morphogenic and degrading radiation.
2. Morphogenetic field interacts with non-living matter. It is absorbed differently in different materials. It is nearly transparent to quartz. It propagates from one point to another in an open channel.
3. Morphogenetic field interacts with living systems connected by an open channel. The effects are observable after few hours of interaction in few systems only.
4. Gurwitsch suggested vector character of morphogenetic field and postulated that it is emitted from constellations of constituents. He also prescribed a geometrical method for calculating the influence of a morphogenetic field on living matter.

Gurvitsch inferred the above properties after performing many experiments. Some of his experiments have been repeated in other laboratories, e.g., onion root experiment and detection of morphogenetic field by yeast culture. The quality of detection of these experiments is considered poor because only a few living systems can detect some specific morphogenetic fields after many hours of interaction. Questions are raised about the efficacy of detection and source, nature and purpose of the field. Even then the existence of morphogenetic field is doubted. The detractors portray morphogenetic field as an extra ingredient introduced to explain some doubtful results of little relevance. A change in the attitude of detractors is occurring after the discovery of the phenomenon of biophoton emission[2], which is similar to morphogenetic field. Biophoton field is

* R.P. Bajpai, Sophisticated Analytical Instruments Centre, North Eastern Hill University, Shillong 793022, India, and International Institute of Biophysics, Raketenstationen, Insel Hombroich, Neuss D41472, Germany.

detectable by a sensitive non-living detector and its nature is electromagnetic. The source, origin, and purpose of the field are not known. The properties of morphogenetic and biophoton fields are too similar to ignore the possibility that two fields are identical. The possibility is confirmed by observing the biophoton signal of a living system interacting with another living system connected by only an open optical channel.

The semi-classical framework of photon emission fails to describe a biophoton signal. The failure occurs in the description of shape, situation specific nature, fluctuations, and broadband spectrum[3]. The shape of a biophoton signal has two parts; the one occurring earlier is small and decays too rapidly but the other occurring later is long and almost constant. Both parts do not have exponential decay. Both parts are sensitive to many environmental and physiological factors but the sensitivity is not easily measurable. Photon flux fluctuates non-classically in both parts. The spectrum is broadband in both parts. The different spectral decompositions of a biophoton signal obtained with different filters also show non-exponential shape and fluctuations of non-classical nature. We need a new framework to describe a biophoton signal. A phenomenological model based on quantum description provides such a framework[4]. The model assumes that biophoton signal is a quantum photon signal in a pure state. The signal acquires the shape from the evolution of its quantum state. The quantum state and its evolution are situation and system specific; so is the shape of biophoton signal. The fluctuations are attributed to repeat measurements made in different systems of an ensemble representing the quantum state. Fluctuations are, therefore, probabilistic and the set of probabilities of detecting different numbers of photons in a bin characterise the ensemble. The set is called photo count distribution. The set of probabilities is easy to measure in any portion of a signal of constant average signal strength. The model describes the average photon flux of a biophoton signal by four real positive parameters, called decay parameters. Fluctuations in photon flux of the non-decaying part are described by another four real parameters, called squeezed state parameters[5]. Both decay and squeezed state parameters are estimated in 21 different spectral decompositions of a biophoton signal emitted by a lichen sample. Only three decay parameters have different values in different spectral decompositions; other five parameters have nearly same values in different spectral decompositions. It suggests that three decay parameters are extensive variables and the other five parameters are intensive variables. The values of the parameters encourage us to consider them as new measurable attributes of a living system. The new attributes are holistic in nature. The new attributes open up new planes of exploration, which may explain some phenomena hitherto considered bizarre. One such phenomenon involves remote intervention in living system through non-substantial channel of communication. The onion root experiment is probably an example of remote intervention where the roots of an inducer onion intervene in the physiological processes of the detector onion connected by an open channel. The phenomenon will be bizarre in the absence of biophoton field. It is suspected that remote intervention is more widespread and its influence is observable through changes in the values of new attributes in much shorter time. The suspicion was confirmed in an experiment by observing changes in the values of intensive attributes of a lichen sample after only 5m of remote interaction. The experiment was a variant of onion root experiment in which a psychic person from a distance induced the interaction.

2. THE MODEL

The model describes biophoton signal by a pure quantum state whose dynamics is given by a Hamiltonian H. Different forms of Hamiltonian give different versions of the model. The Hamiltonian in the simplest version of the model is

$$H = \frac{p^2}{(1+\lambda t)^2} + \frac{1}{2}(1+\lambda t)^2 \omega^2 q^2 , \tag{1}$$

where λ is the damping coefficient, ω is the mode frequency, t is time, and p, q are canonically conjugate momentum and position of the damped field. It is the Hamiltonian of a frequency stable damped harmonic oscillator with time dependent damping and mass terms. Popp and Li proposed this Hamiltonian to explain broad features of biophoton signals in a semi-classical framework. The Hamiltonian has exact classical and quantum solutions. The quantum solution[6] is described by time dependent quasi particle annihilation and creation operators b(t) and $b^+(t)$. The quasi particle operators are related to photon creation and annihilation operators a^+ and a by following unitary transformation:

$$b(t) = (\mu(t)\, \mu_0 + \nu(t)\, \nu^*_0)\, a + (\mu(t)\, \nu_0 + \nu(t)\, \mu^*_0)\, a^+. \tag{2}$$

The constants μ_0 and ν_0 are fixed by the state of the signal at t = 0. The time dependent coefficients $\mu(t)$ and $\nu(t)$ are given[4,7] by

$$\mu(t) = \cos(\omega t) + \frac{\lambda}{2\omega}\frac{\lambda t}{2\omega(1+\lambda t)}\sin(\omega t) + i\frac{\lambda}{2\omega}\frac{\lambda t}{(1+\lambda t)}\cos(\omega t) - i\left\{1 + \frac{\lambda^2}{2\omega^2(1+\lambda t)}\right\}\sin(\omega t) \tag{3}$$

$$\nu(t) = \frac{\lambda}{2\omega}\left\{1 + \frac{1}{(1+\lambda t)}\right\}\sin(\omega t) + i\frac{\lambda}{2\omega}\frac{\lambda t}{(1+\lambda t)}\cos(\omega t) - i\frac{\lambda^2}{2\omega^2(1+\lambda t)}\sin(\omega t). \tag{4}$$

The model proposes that biophoton signal is an eigen state of b(t). This state is also a squeezed state of photons. It is represented in the usual notation[8] by $|\alpha, \xi(t)\rangle$, where α is the complex eigen value of b(t) and $\xi(t)$ is a complex parameter specifying the unitary transformation. The state is connected to vacuum state $|0\rangle$ by $|\alpha, \xi(t)\rangle = S(\xi(t))D(\alpha)|0\rangle$, where $D(\alpha)$ and $S(\xi(t))$ are usual displacement and squeezing operators. The time dependent photon flux n(t) specifies the shape. It is given by the expectation value of photon number operator $< a^+a >$ in the quantum state of the signal averaged over mode frequency ω. The time dependence of n(t) has the following simple form:

$$n(t) = \int_{t-\frac{\pi}{\omega}}^{t+\frac{\pi}{\omega}} dt' \left\langle \alpha, \xi(t') \,|\, a^{+}a \,|\, \alpha, \xi(t') \right\rangle = B_0 + \frac{B_1}{(t_0 + t)} + \frac{B_2}{(t_0 + t)^2} . \quad (5)$$

The real positive coefficients B_i's ($i = 0,1,2$) are short forms of three long algebraic expressions of μ_0, ν_0, λ, ω, and α, and $t_0 = \lambda^{-1}$. The signal acquires situation specific nature from the dependence of B_i's on μ_0 and ν_0. Equation (5) provides the framework for describing non-decaying and decaying parts of a biophoton signal by four parameters only. These parameters are called decay parameters. The photo counts detected in a bin of size Δ starting from (T-Δ) to T second after stimulation are obtained by integrating Eq. (5) and are given by

$$\int_{T-\Delta}^{T} n(t)dt = B_0\Delta + B_1 \ln\left[\frac{(T + t_0)}{(T + t_0 - \Delta)}\right] + B_2 \frac{\Delta}{(T + t_0)(T + t_0 - \Delta)} \quad (6)$$

The non-decaying part is given by B_0 and it includes the contribution of background noise. The decaying part is given by B_1, B_2, and t_0. The usual concepts of strength and shape are no longer valid in the new framework. It is, however, possible to specify quantities portraying these concepts. The specification of signal strength is different in the two parts. It is specified by B_0 in the non-decaying part where it measures the average photon flux. Its specification in the decaying part is problematic as this part has two independent components with different time dependencies. The strengths of the two components are measured by the coefficients B_1 and B_2. A single measure of strength does not exist. The number of photons detected in the first bin just after exposure to light is used as a pragmatic single measure of strength of the decaying part. It has limitations and it leads to contradictions in some signals. The specification of decay shape is also problematic for it depends on B_1/B_2 and t_0. B_1/B_2 is usually more sensitive than t_0 and can be used as another pragmatic measure of decay shape.

The model uses the following approximation to calculate the probability $P_{cal}(n)$ of detecting n photons in the non-decaying part[9]:

$$P_{cal}(n) = \mathrm{Lim}_{t\to\infty} \int_{t-\frac{\pi}{\omega}}^{t+\frac{\pi}{\omega}} dt' \left|\left\langle n \,|\, \alpha(t'), \xi(t') \right\rangle\right|^2 = \left|\left\langle n \,|\, \alpha_{eff}, \xi_{eff} \right\rangle\right|^2 \quad (7)$$

where $|n\rangle$ is n photons state and $|\alpha_{eff}, \xi_{eff}\rangle$ is an effective squeezed state with parameters α_{eff} and ξ_{eff}. The complex parameters α_{eff} and ξ_{eff} are expressed in the polar form as

$\alpha_{eff} = |\alpha| \exp(i\phi)$ and $\xi_{eff} = r\exp(i\theta)$ by four real parameters $|\alpha|$, ϕ, r and θ. The standard calculations give

$$\left\langle n \middle| \alpha_{eff}, \xi_{eff} \right\rangle = \frac{1}{\sqrt{n! \cosh r}} \left[\frac{1}{2}\exp(i\theta)\tanh r\right]^{\frac{n}{2}} \exp\left[-\frac{1}{2}\left(|\alpha|^2 + \alpha^{*2}\exp(i\theta)\tanh r\right)\right] \times$$

$$\times H_n\left[\frac{\alpha+\alpha^* \exp(i\theta)\tanh r}{(2\exp(i\theta)\tanh r)^{\frac{1}{2}}}\right] \quad (8)$$

where H_n is Hermite polynomial of degree n. The squeezed state parameters $|\alpha|$, r, ϕ, and θ are estimated by minimising the function F ($|\alpha|$,r, θ,ϕ) defined by

$$F(|\alpha|, r, \theta, \phi) = \sum_n \left(P_{obs}(n) - P_{cal}(n)\right)^2 . \quad (9)$$

P_{obs} (n) is the observed probability of detecting n photons in the squeezed state and summation is over those values of n for which P_{obs} (n) is non-zero. The parameters $|\alpha|$ and r determine the calculated photon flux

$$k_{cal} = \left\langle a^+ a \right\rangle = \sinh^2 r + |\alpha|^2 . \quad (10)$$

The observed photon flux k should be more than the calculated photon flux as k also includes the background noise. An estimate of background noise is provided by the difference

$$k_{noise} = k - k_{cal} \quad (11)$$

The quality of estimation is judged by value of F ($|\alpha|$,r, θ,ϕ) at its minimum. Background noise increases the minimum value of the function and flattens the minimum.

3. MATERIALS AND METHODS

The samples of lichens of species *Parmelina wallichiana* were collected in the first week of July 2003 from a forest around Shillong, India, but measurements were made in the last week of August 2003 at the laboratory in Germany. The collected samples along with the substrates were stored in polythene bags. The measurements were made without detaching the substrates. The measuring set up has been described many times in the literature[10]. It consists of a sample chamber, a mercury lamp, a rotating filter wheel, and a broadband photo multiplier detector. The sample chamber is dark and isolated from external light. The chamber has an opening for placing and removing samples in quartz cuvettes. The chamber has two quartz windows with shutters, one each on its two adjacent walls. One window is for light from the mercury lamp to enter in the chamber for stimulating a sample placed in it, and the other window is for photons emitted by the sample to come out for detection by a broadband photo multiplier detector operating in single photon detection mode. The detector is sensitive in the range (300-800) nm. The rotating wheel inserts a filter in the path of photons just before detection. The wheel can be rotated to 21 positions. The first position of the wheel is empty, and no filter is inserted in this position. It is meant for measuring unfiltered signals. The wheel in the 21st position inserts a ground glass (Trocken). The wheel in other positions inserts either interference or long pass filter. Every interference filter has peak transmission of ~30% at a specific wavelength and half width of ~20nm. An interference filter is identified by its wavelength of peak transmission in nm. The spectral decomposition obtained from an interference filter is the nearest approach to single mode photon field. The wavelengths of peak transmission of various interference filters fixed on the wheel cover the entire visible range. The transmission of a long pass filter is ~90% for wavelengths longer than a specific wavelength; the wavelength in nm is suffixed to the name of the filter. Two positions of the wheel are marked defective; the spectral decompositions from these positions give anomalous results.

Biophoton signals for 200s were repeatedly determined in many samples for ten days. A study of these signals confirmed two well-known[11] results in samples of this species: 1. A sample emits nearly same biophoton signal after every exposure to light at least for a week. 2. Biophoton signals emitted by different samples have similar shapes but different strengths. The first result allowed the measurement of spectral decompositions of a biophoton signal in repeat measurements after inserting filters. The second result suggested the existence of species-specific features in a biophoton signal. These features are identified and measured in the quantum framework of the model.

Samples became dry due to storage, but measurements were made with wet samples. A sample was selected, it was put in a quartz cuvette, and then it was soaked with distilled water a few hours before a measurement. The strength of biophoton signal begins to increase just after soaking and stabilizes to a much higher value within an hour. A sample soaked with water and emitting stable biophoton signal is called wet sample. The wet sample was placed in the chamber for half an hour to eliminate photon emission arising from exposure to laboratory illumination. The temperature of the chamber was maintained at 25°C during measurements. The measurements had three goals that were accomplished with the following protocol:

1. **Identification of decaying and non-decaying regions of a biophoton signal:** Photon flux emitted by a wet sample after 5s exposure to white light was measured for 11111s in 5000 consecutive bins. The bins were of five sizes namely 1ms, 10ms, 100ms, 1s, and 10s. The bin size was increased after 1000 bins.

2. **Measurements of shape and fluctuations in spectral decompositions:** A different wet sample was used in the following four parts of measurements:
   a. The prestimulation photon flux emitted by the sample was measured by detecting photons in 1000 contiguous bins of 50ms. The measured photon flux included contributions of background noise.
   b. The sample was then stimulated by exposure to white light for 5s and the emitted flux was measured in 1000 contiguous bins of 50ms. It was first measurement of decay. A second measurement of decay was similarly made after 5m.
   c. The sample was left for 10m after the second measurement of decay to make its biophoton signal non-decaying. The photon flux in the non-decaying region was measured in 10000 contiguous bins of 250ms. The measurements determined average photon flux and probabilities of detecting various numbers of photons in a bin. The

steps (a-c) constituted a set of measurements at one position of the filter wheel and took nearly an hour to complete.

d. The filter wheel was rotated to its next position, and measurements of steps (a-c) were made in the new spectral decomposition. Measurements in the first ten positions were made one after the other, then no measurements were made for 12h and finally measurements in the remaining positions were made. It is pointed out that the pre stimulation flux measured in step 1 is the flux of non-decaying part of the signal emitted nearly 52m after previous stimulation.

3. **Remote intervention by a psychic person:** The experiment originated from a chance comment by a participant of the summer school on Biophotons in 2003. The participant claimed that he could energise a sick person by his will power. He was persuaded to demonstrate his capabilities by energising a lichen sample in the night. The person is a professor of Philosophy in a respectable university and is referred to here as psychic person. The protocol was devised to compare decay shapes of biophoton signals emitted before and after energisation. The experiment was performed in the night. The laboratory illumination was dim. The temperature of the measuring chamber in this experiment was 20°C. A wet sample was placed for 2m in chamber and then 1000 measurements of photo counts in bins of 50ms were made to obtain prestimulation flux. The sample was then stimulated by 5s exposure to white light, and again 1000 measurements of photo counts in bins of 50ms were made to obtain the shape of decay called Stim1. A second decay shape called Stim2 was obtained after a pause of 5m by following the same procedure. The sample was then taken out. A psychic person who claims to energise living object, by his will power energised the sample by placing his hands 5cm away from the closed quartz cuvette containing the sample. He energised the sample for 5m and then pre-stimulation flux, Stim1, and Stim2 were measured by following the same procedure.

4. RESULTS AND DISCUSSION

Figure1 depicts photo counts in bins of different size measured in a biophoton signal in a double logarithmic plot. Photon flux in the figure decreases nearly thousand-folds and attains an almost constant average value in less than 500s. The 99% of the decrease occurs in the first 50s and photon flux at 50s is nearly ten times of its constant value. Photon flux depicted in the figure includes contribution of background noise. A constant average background noise implies its contribution is smaller than the average photon flux of non-decaying region and is negligible in the first 50s of decay. The figure also depicts a fit of the data based on Eq. (6). The values of parameters are B_1=6475, B_2=0.02, B_0=11.8, and t_0=0.369. The four parameters explain the decrease of average photon flux by four orders of magnitude in time varying by eight orders of magnitude. The contribution of B_2 to photon flux is very small and can be ignored without

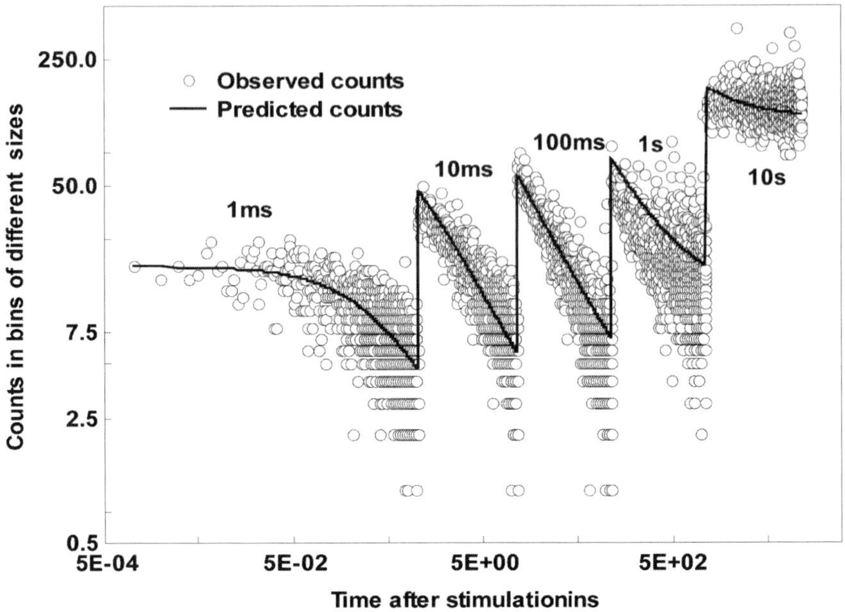

Figure 1. Light induced biophoton signal of a sample of *Parmelinia wallichiana*: Photo counts observed in contiguous bins of different sizes are plotted against time after stimulation. The size of the bin is indicated on the figure. The prediction of the model is depicted as continuous line.

affecting the fit. The figure depicts the occurrence of fluctuations at every bin size.

There were two sets of measurements of first 50s of decay region. Each set contains decaying signals in 21 spectral decompositions. The decay parameters were estimated in both measurements. The errors of estimation were also found for these parameters. The decay parameters and errors of estimations in a spectral decomposition have nearly same values in both sets. The values obtained in the first measurement are given in Table 1. The error of estimation is an indication of the quality of its estimation. The error is different in parameters indicating differences in the quality of estimations of parameters. The table indicates that B_2 is ill determined, t_0 is poorly determined, B_0 is reasonably determined, and B_1 is well determined. B_2 is ill determined because its contribution is negligible. The term containing B_2 can be ignored without affecting the fit of a signal. The estimated value of t_0 is small and not much different in different spectral decompositions. The error in its estimation is usually much larger than its estimated value, which suggests the possibility of a common value of t_0. The average and standard deviation of its estimated values in different decompositions are 0.118 and 0.079 in the first set of measurements. It is expressed as $<t_0> = 0.118 \pm 0.079$. The second set of measurements gives $<t_0> = 0.122 \pm 0.053$ and the two combined sets give $<t_0> = 0.120 \pm 0.067$. The estimated value of t_0 shows a little variation among its determinations in different spectral decompositions, but its average value is stable and gives the common value.

Table 1. Decay parameters of spectral decompositions of a biophoton signal: The decay parameters B_1, t_0, B_2, and B_0 estimated from the initial 50s portion of the decay are given for spectral decompositions obtained from different filters. The unit of B_0 is counts/50ms, of B_1 is counts.s, of t_0 is s, and of B_2 is counts.s^2. The errors in estimation of these parameters are given in columns of EB1, Et0, EB2, and EB0

Filter	B1	EB1	t0	Et0	B2	EB2	B0	EB0
Empty	23147	112	0.11	0.002	1860	96	0.0	0.8
227.9	35	7	0.03	14.63	0.01	509	0.8	0.1
850	62	10	0.03	4.50	0.01	277	4.4	0.1
WG320	19491	71	0.08	0.33	0.00	6335	8.3	0.8
RG715	7393	38	0.08	4.19	0.10	30958	4.3	0.4
RG665	17570	119	0.08	0.87	2.07	15283	8.3	0.9
RG610	17664	79	0.08	2.72	1.03	48109	11.5	0.9
GG495	18450	76	0.09	1.25	0.05	22988	9.0	0.9
WG280	951	16	0.09	3.17	0.10	3024	1.4	0.1
GG375	17299	74	0.08	5.18	0.16	89555	9.6	0.8
753	1089	22	0.10	11.12	0.07	12108	4.1	0.2
695	802	17	0.08	8.73	0.01	6994	4.0	0.1
652	80	12	0.07	5.12	0.03	412	3.2	0.1
595	0.0053	29	0.29	0.09	50	34	3.9	0.1
550	24	20	0.16	0.05	21	15	3.0	0.1
496	11	32	0.36	0.16	46	47	2.5	0.1
422	208	12	0.06	5.89	0.02	1219	2.9	0.1
278.9	63	11	0.16	0.05	11	10	0.7	0.1
246.6	16	14	0.18	0.15	6	14	0.7	0.1
375.5	15785	80	0.10	0.01	224	167	2.9	0.6
Trocken	4580	42	0.15	0.01	377	52	0.00	0.25

the common value of t_0 is, perhaps, an intensive attribute of the system. The common value of t_0 preserves the integrity of spectral composition in decaying biophoton signals. B_0 is the non-decaying part of a signal. It is very small in the first 50s of decay region but its estimates in different spectral decompositions are reasonable, which is an indication of the validity of the model. B_1 is well determined in most spectral decompositions, where its contribution is dominant. It is not well determined and does not give dominant contribution in relatively weak spectral decompositions of the signal. The background noise is probably not negligible and is responsible for poor estimations in these decompositions. The values of B_0 and B_1 do not appear correlated in different spectral decompositions.

B_1 can serve as a measure of decay in signals in which its contribution is dominant. Photo counts in the first bin have been used as a pragmatic measure of strength of a biophoton signal. The contributions of B_0, B_1, and B_2 to photo counts in the first bin are separately calculated for examining the connection between two measures. The separated contributions, observed photo counts in the first bin and prestimulation photon flux are depicted in Fig. 2 for various

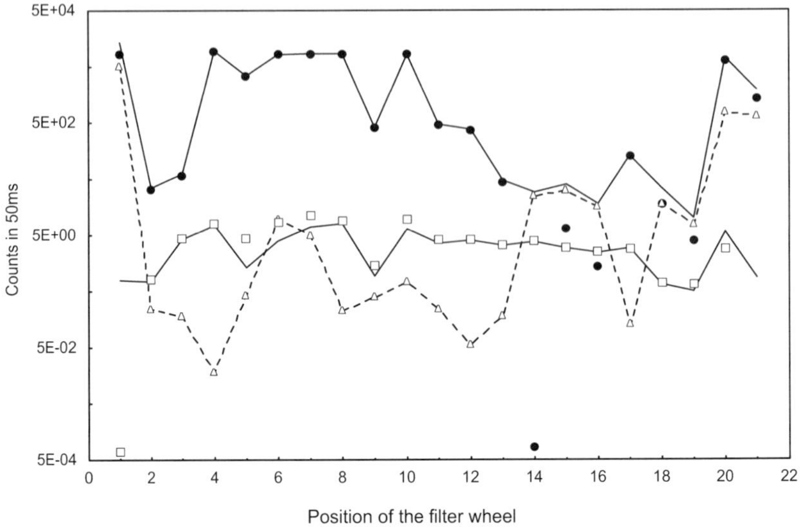

Figure 2. The contributions of B_0, B_1, and B_2 to photo counts in the first bin, observed photo counts and prestimulation flux are depicted in spectral decompositions obtained at 21 positions of the filter wheel. The contributions of B_0, B_1, and B_2 are represented respectively by □, and Δ. Connecting the values by solid lines depicts the observed photo counts and prestimulation photon flux. The points representing B_2 are joined by dashed line to improve clarity.

spectral decompositions. The observed photo counts and the contribution of B_1 differ only marginally, so that two measures are equivalent. The photo counts in the first bin may not be a good measure of strength in situations where the contribution of B_1 is not dominant and may lead to incorrect conclusions.

The values of B_1 in spectral decompositions obtained from interference filters are comparable and so are the values in spectral decompositions obtained from long pass filters. The values in decompositions from interference filters are smaller than the values in decompositions from long pass filters, which implies that strengths of decaying part in decompositions of interference filters are comparable and the spectrum of decaying part is broadband. The small strength arises from small bandwidth. The spectral decompositions from two defective positions yield anomalous value of B_1. The filters inserted in these positions are WG280 (long pass filter) and 375.5 (interference filter). The decomposition by WG280 yields a small value of B_1, and the decomposition by 375.5 yields a large value of B_1. The values of other parameters in these decompositions are not anomalous. The prestimulation flux is not much different from the contribution of B_0. Both give the flux of non-decaying part. The estimate of B_0 is obtained from measurements made just after the stimulation of sample and pre stimulation flux was measured 52m after the stimulation of sample. The measurements of photo count distribution provide another determination of the flux of non-decaying part. It is given as k in Table 2 in counts/250ms. It corresponds to measurements from 11m to 52m after the stimulation of sample. The three determinations agree qualitatively; quantitative agreement is not found probably, because of fluctuations. The values of B_0 in spectral decompositions obtained from interference filters are higher than its values in spectral decompositions obtained from long pass filters. It is an expected result and is opposite to the behaviour of the values of B_1. It implies either different spectral profiles of decaying or not-decaying parts or an artefact. A possible artefact is the behaviour of filters. A filter responds differently to weak and strong signals because of the re-emission of absorbed light. Interference filters absorb more light and hence re-emit more light. The effect should be pronounced in signals comparable in strength to re-emitted light, which seems to occur in the non-decaying part. The average signal strength and the probability of detecting photons in a bin were determined in the non-decaying part of 20 spectral decompositions. The signal strength varied from 2.94 counts/250ms to 21.9 counts/250ms, and number of non-zero probabilities varied from 12 to 67 in different spectral distributions. The squeezed state parameters were estimated from measured probabilities by minimizing the function $F(|\alpha|, r, \theta, \phi)$. The minimum value of $F(|\alpha|, r, \theta, \phi)$ is different in different spectral decompositions. Its lowest value was 7.8×10^{-5}, which was obtained with 59 non-zero probabilities in the spectral decomposition from the interference filter 850. Its highest value was 7.0×10^{-4}, which was obtained with 37 non-zero probabilities in the spectral decomposition from the interference filter 496. The estimated values of squeezed state parameters, observed photon flux k and noise k_{noise} are given in Table 2. The serial number in this table gives the position of filter wheel. The observed and calculated fluxes are higher in spectral decompositions of interference filters indicating that a part of the re-emission of absorbed light also occurs in the quantum state. The estimated values of r, θ, and ϕ in different spectral decompositions do not vary much. The estimated value of $|\alpha|$ varies considerably in different spectral decompositions but $|\alpha|^2/k$ is almost constant. The average and standard deviation of the estimated values in spectral decompositions give $|\alpha|^2/k = 0.88 \pm 0.05$, $r = 0.34 \pm 0.05$, $\theta = 0.099 \pm 0.002$, and $\phi = 1.31 \pm 0.36$. The standard deviation of ϕ is a little large but the dependence of $F(|\alpha|, r, \theta, \phi)$ on ϕ is weak. Photo count distributions of different spectral decompositions of a biophoton signal are specified by the same set of values of $|\alpha|^2/k$, r, θ, and ϕ. The parameters are new attributes of a biophoton signal and of its emitting system. The attributes are measurable by a broadband detector.

The new attributes specify a quantum state and hence their nature is quantum. The state is of a biophoton signal that remains intact for macroscopic time. It implies that the constituents of a living system emitting a biophoton signal are in a quantum state and the processes involved in the emission are of quantum nature. The essential ingredients of a living system comprise a quantum state that is macroscopic in space and time. Such a scenario is also visualised by the conjecture that living systems transfer information and make selections in fundamental biological processes with optimal efficiency. As a physical system operates with optimal efficiency in a quantum state, the conjecture implies that the constituents of a living system participating in fundamental biological processes are in a quantum state and remain in a quantum state during the life of the living system. A spontaneous fundamental biological process has to emit energy, and if the process is quantum then it will emit the balance of energy in a pure quantum state. It is suspected that biophoton signal originates from quantum processes of fundamental relevance to biology and the signal reflects biological

Table 2. Squeezed state parameters of spectral decompositions of a biophoton signal: The values of squeezed state parameters $|\alpha|$, r, θ and ϕ estimated from the observed photo count distribution along with the average observed photon flux k and estimated k_{noise} (noise) in counts/250ms are given for various spectral decompositions. The units of θ and ϕ are in radians, r and $|\alpha|$ are dimensionless

Filter	B1	EB1	t0	Et0	B2	EB2	B0	EB0
Empty	23147	112	0.11	0.002	1860	96	0.0	0.8
227.9	35	7	0.03	14.63	0.01	509	0.8	0.1
850	62	10	0.03	4.50	0.01	277	4.4	0.1
WG320	19491	71	0.08	0.33	0.00	6335	8.3	0.8
RG715	7393	38	0.08	4.19	0.10	30958	4.3	0.4
RG665	17570	119	0.08	0.87	2.07	15283	8.3	0.9
RG610	17664	79	0.08	2.72	1.03	48109	11.5	0.9
GG495	18450	76	0.09	1.25	0.05	22988	9.0	0.9
WG280	951	16	0.09	3.17	0.10	3024	1.4	0.1
GG375	17299	74	0.08	5.18	0.16	89555	9.6	0.8
753	1089	22	0.10	11.12	0.07	12108	4.1	0.2
695	802	17	0.08	8.73	0.01	6994	4.0	0.1
652	80	12	0.07	5.12	0.03	412	3.2	0.1
595	0.0053	29	0.29	0.09	50	34	3.9	0.1
550	24	20	0.16	0.05	21	15	3.0	0.1
496	11	32	0.36	0.16	46	47	2.5	0.1
422	208	12	0.06	5.89	0.02	1219	2.9	0.1
278.9	63	11	0.16	0.05	11	10	0.7	0.1
246.6	16	14	0.18	0.15	6	14	0.7	0.1
375.5	15785	80	0.10	0.01	224	167	2.9	0.6
Tocken	4580	42	0.15	0.01	377	52	0.00	0.25

aspects of the emitting system. The four decay and four squeezed state parameters needed for the description of a biophoton signal should also be able to describe a living system and its metabolic activities. The new parameters are more appropriate for measuring holistic behaviour of living system indicated by biophoton signals. It is illustrated in the experiment that measures remote influence of a psychic person on a lichen sample. A lichen sample was placed in the measuring chamber after 10m pouring water. The prestimulation flux measured after 2m of the placement was 1.38 ± 2.2. The biophoton signals in two measurements - Stim1 and Stim2 - of the decay region were nearly identical indicating that the sample has attained its wet state. There was a gap of 5m between Stim1 and Stim2. The sample was then taken out and was energised for 5m and then placed in the measuring chamber. The prestimulation flux measured after 2m of the placement was 23.9 ± 6.9. The biophoton signals of Stim1 and Stim2 after energisation were again nearly identical but were different from the signals observed before energisation. The four signals were analysed to obtain the decay parameters. The values of B_1 were (18010 ± 79 and 20770 ± 77) and of B_0 were (19 ± 9 and 225 ± 10) in two measurements before energisation and (25799 ± 119 and 27849 ± 72) and (425 ± 16 and 381 ± 15) after energisation. The values of B_2 were (2903 ± 89 and 560 ± 128) and of t_0 were (0.17 ± 0.002 and 0.12 ± 0.004) before energisation. The values of B_0 and t_0 are highly ill determined with estimation errors many times the estimated values after energisation. The values of B_0 were (0.45 and 0.06) and of t_0 were (0.11 and 0.12). It appears that energisation decreases the contribution of B_2 and increases the contribution of B_0. Because B_2 gives the classical contribution while B_1 and B_0 give quantum correction in the model, energisation enhances quantum aspects of the system. The observed signals in the first and second stimulation are depicted in Fig. (3a-3b).The same sample was used in the measurements of spectral decompositions.

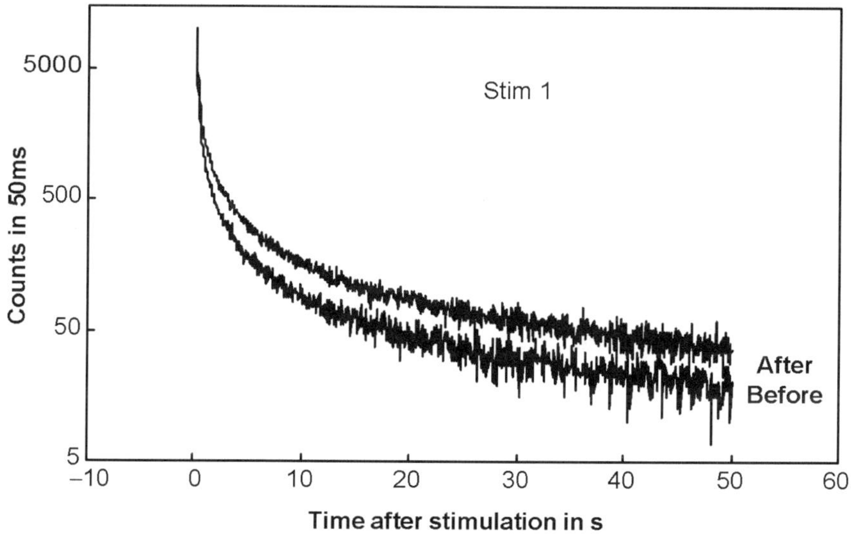

Figure 3a. Biophoton signals emitted by a sample before and after its energisation in the first stimulation.

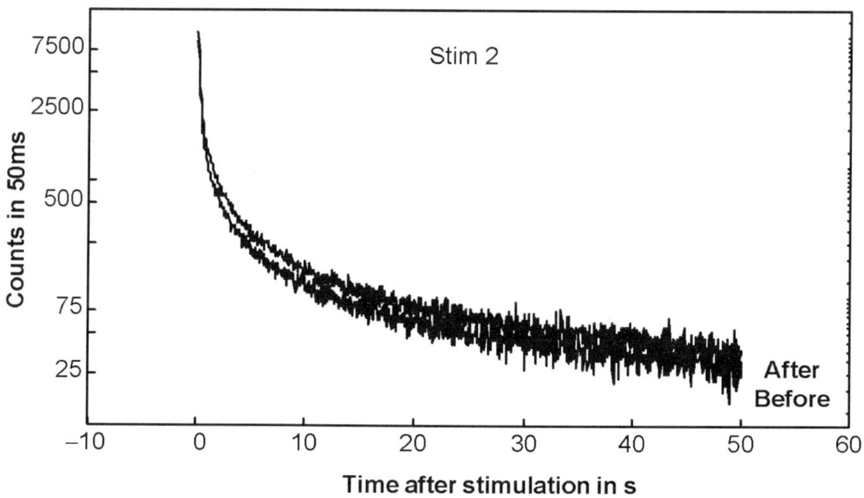

Figure 3b. Biophoton signals emitted by a sample before and after its energisation in the second stimulation.

The experiment was repeated with four psychic and a non-psychic persons. The results with psychic persons were similar; only the amount of changes differed. No change was observed with the non-psychic person. The effect persisted for a few hours and then the signal reverted back to its earlier shape.

REFERENCES

1. M. Lipkind, Alexander Gurwitsch and the concept of the biological field, *21st Century Science & Technology*, **11**, 34, 36 (1998).
2. F.A. Popp, in: *Recent Advances in Biophoton Research and its Applications*, edited by F.A. Popp, K.H. Li, and Q. Gu (World Scientific, Singapore, 1992), pp. 1-46.
3. R.P. Bajpai, in: *The Physical Basis of Life in Integrative Biophysics*, edited by F.A. Popp and L. Beloussov) (Kluwer Academic Publishers, Netherland, 2003), pp. 439-465.
4. R.P. Bajpai, S. Kumar, and V.A. Sivadasan, Biophoton Emission in the evolution of a squeezed state of a frequency stable damped oscillator, *Applied Math and Comp.* **93,** 277-288 (1998).
5. M. Orszag *Quantum Optics* (Springer Verlag Heidelberg, 2000), 44.
6. H.P. Yuen, Two photon coherent states of the radiation field, *Phys. Rev.* **A13**, 2226 (1976).
7. R.P. Bajpai, in: *Biophotons*, edited by J.J. Chang, J. Fish, and F.A. Popp (Kluwer Academic, Netherland, 1998), pp. 323-338.
8. D. F. Walls, and G.J. Milburn, *Quantum Optics* (Springer Verlag, Heidelberg, 1994), p. 29
9. R.P. Bajpai, Biophoton Emission in a Squeezed State from a Sample of *Parmelia tinctorum* Phys. Lett. A **322**, 131-136 (2004).
10. F.A. Popp, Biophoton & their theoretical implications *Ind. J. Exp. Bio.* **41**, 391(2003).
11. R.P. Bajpai, Biophoton Emission of a Lichen Species *Parmelia. tincotrum, Ind. J. Exp. Bio.* **41**, 403-410 (2003).

4
BIOLOGICAL STRUCTURE AS A CONVERTER OF COHERENT RADIATION

A. Budagovsky, O. Budagovskaya, and I. Budagovsky*

1. INTRODUCTION

For no less than forty years, the problem of coherence in the biological systems has been extensively discussed, becoming the area of contradictory views [1-3]. On one hand its solution may open quite new approaches permitting to understand the functioning of the living organisms. On the other hand, these investigations are aggravated by requiring complicated combinations of physical, biological, and mathematical methods. One of the particular tasks within this problem framework is the involvement of biochemoluminescence (BCL) in photoregulatorial processes. In the long run, it is reduced to the explanation of the distant cell interactions (DCI). Previously it was shown that the conditions required for involving endogenous irradiation in communication processes are indeed present in the living organisms [4, 5]. Due to its low intensity, that part of BCL that fulfills signal functions has to acquire the sufficient level of statistic ordering (coherence). (By its nature, BCL is non-homogeneous. Together with a stochastic component, induced by spontaneous process of peroxide oxidation of lipids [6], a coherent component determined by other mechanisms of photon generation may also exist.) This permits to suggest that the cells are able to generate a flow of coherent photons and to detect them on the background of considerably more intensive stochastic interference, i.e., natural illumination. The both properties has been theoretically substantiated and experimentally confirmed [4, 5, 7, 8]. Spreading of a coherent signal within optically non-uniform cellular structures inevitably will result in its stochastization, i.e., in a partial loss of a stochastic ordering of a photon collective. It is usually admitted that the coherence of the optic irradiation can be kept only in thin layers of biological tissues [9, 10]. According to V.S. Sinyakov and V.I. Khaidakov [10], irradiation of helium-neon laser (wavelength 632.8 nm) loses coherence and polarization by passing through samples of animal tissues of no less than 200 μm thickness.

* A. Budagovsky, I.V. Michurin All-Russian Research Institute of Fruit Crops Genetics and Breeding Michurinsk, Russia. O. Budagovskaya, I.V. Michurin All-Russian Research Institute of Fruit Crops Genetics and Breeding Michurinsk, Russia. I. Budagovsky, P.N. Lebedev Physical Institute Russian Academy of Sciences, Moscow, Russia

Meanwhile, if assuming so strong stochastization of a coherent wave in cell structures, it is difficult to explain the transmission of regulatory signals by means of endogenous irradiation. The contradiction includes the following questions:

1. Does an endogenous irradiation responsible for DCI really possess a certain degree of a stochastic ordering?
2. To what degree do the biological structures affect the transmission of coherent signals? Our paper attempts to answer these questions.

Two sets of experiments using different objects and methods of investigations have been performed. The first set of experiments was proposed by A.M. Kuzin and was aimed to confirm the coherence of a secondary biogenic irradiation (SBI). G.N. Surkenova, B.Y. Gurvitz, G.A. Gudi, and the authors of this paper took part in the experiments. Beforehand, a qualitative description of the obtained results has been given [11]. In this paper, we performed a mathematical modeling of the effect of destabilizing factors upon DCI and gave a quantitative estimation of reliability of the conclusion on statistic ordering (coherence) of SBI.

The aim of the second set of experiments was to establish the effect of biological medium (cellular tissue) on statistic properties of coherent waves that are spread into it. For this purpose, a method of laser analysis of tissue microstructure [12, 13], making it possible to determine the parameters of coherence of scattered light flow, has been used.

2. STOCHASTIC SCREEN WITHIN A COMMUNICATIVE CHANNEL OF DCI

2.1. Working Hypothesis

A proposed working hypothesis is that the manifestations of DCI effect will vary as depending upon the microstructure of a phase screen, installed in the communicative channel between a bioinductor and a detector. Space and ordering phase screen with constant angular phase along the wave front shouldn't affect the process of cell interaction. On the other hand, a stochastic phase screen disturbing phase correlation in photon collective will reduce the effect.

2.2. Materials and Methods

SBI induction was performed and registered according to A.M. Kuzin [14]. Fresh donor's blood of a healthy man was used as an inductor. It was stabilized by heparine (anticoagulator) and exposed to γ-irradiation by ^{60}Co in the radiation device RXM-γ-20 (Russia). Radish seed of the *cv. Zhara* placed on wet filter paper in Petri dishes (diameter 100 mm) were used as a biodetector. Proposed by A.M. Kuzin, growth index (*GI*) was used for estimating the effect of distance interaction. It was calculated as: $GI = (\Sigma \times 100)/N$, where Σ is cumulative length of all the seedlings in each set, N - total number of emerged and non-emerged seeds in the given set. Growth index was calculated in 48 h of cultivation in the dark at 28°C. Experiments were conducted with fresh donor's blood (1 h after selection) stored for 48 h at 4°C.

A specific feature of this experiment was that the irradiation of the same bioinductor reached the biodetectors through phase screens of a different structure. For this purpose, a standard photometric quartz cuvette with two opposite transparent sides and two mat sides has been used. All the sides were characterized by an equal weak light absorption, but the mat sides scattered the light, reducing a space coherence of irradiation. After filling the cuvette with γ-irradiated blood, its lateral walls become the phase screens (filters) for SBI. The cuvette was sealed and put in the middle of Petri dish separated into four sectors by non-transparent screens (Fig. 1). Opposite each wall of the cuvette a biodetector (wet reddish seeds) was placed within its own sector. Biogenic irradiation from an inductor to the detectors 1 and 3 was passed through stochastic (non-ordered) phase screen while to the detectors 2 and 4 through an ordered one.

Non-transparent screens between the sectors prevented other forms of optical interaction. Blood in cuvette non-subjected to γ-irradiation served as control. Control and experimental preparations (Petri dishes with seeds and cuvette with blood) were incubated in thermostat for 48 h. During this period, seed germination and distant field interactions between inductor and detector took place.

2.3. Results and Discussion

The experiments showed different effects of the ordered and non-ordered phase screens upon DCI (Fig. 2). Stimulation of functional activity of the seeds

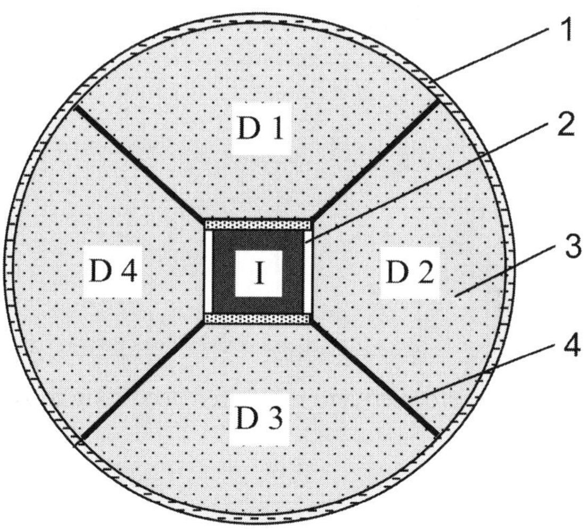

Figure 1. Scheme of the experiment on the effect of phase screens of the different types upon distant cell interactions. I - inductor, donor's radish seeds. D1...D4 - detectors, radish seeds. 1 - Petri dish, 2 - quartz cuvette with blood; 3 - wetted radish seeds on filter paper, 4 - non-transparent screens.

was considerably higher in the sectors with non-mat walls, i.e., where phase field distortions were of regular character. The observed effect was quantitatively evaluated by means of dispersion analysis. Index of seed growth of the detectors D1 and D3 which were separated from inductor did not show any significant differences from control values. A calculated Fisher criterion F was less than critical value F_c under confidential level $P = 0.95$ $(F = 4.32 < F_c = 5.99)$. The rate of the seed germination in the detectors D2 and D4, connected with inductor by means of an optic tract with ordered phase screen, reliably exceeded control values. The level of significance of zero hypothesis was $\alpha << 0.001$, $F = 180.8 >> F_c = 5.99$ at $P = 0.95$. In this case only 3.2% are considered to be the occasional dispersion factors while the distance interaction controls the result to 96.8%. In general the experiment has demonstrated the existence of a distant interaction between inductor and detector. However only ordered phase screen which retained the retaining initial statistics of irradiation provided a reliable. Therefore a secondary biogenic irradiation possesses rather high level of coherence, which is necessary for DCI.

It was of interest to know to what extent DCI depends upon the functional state of a bioinductor. For this purpose a fresh donor's blood and that incubated for 2 days at low temperature were used. In the second case, blood cells preserved viability but their activity declined. The experiment was planned and its results elaborated by the double factor dispersion analysis.

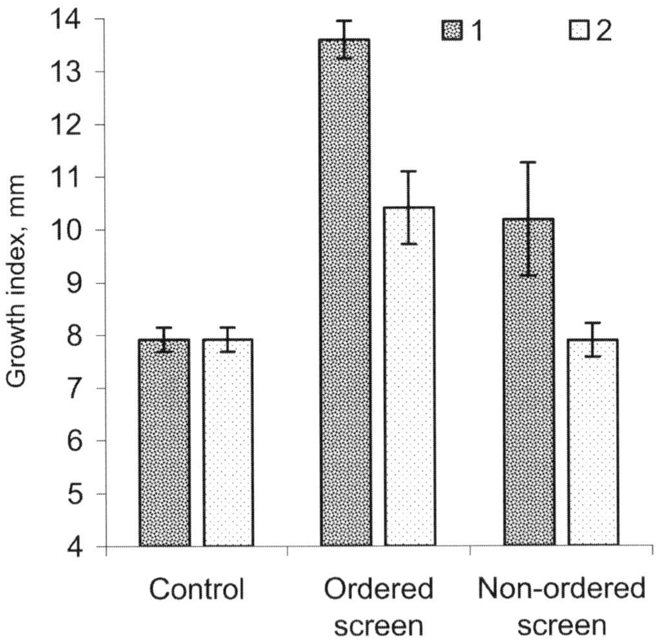

Figure 2. Germination of radish seeds exposed to SBI through ordered or non-ordered phase screens. Excited screens: 1 - fresh blood. 2 - blood incubated for 2 days at 4°C.

2.3.1. Analysis of the effects of destabilizing factors

In the given experiment, two factors affecting DCI activity were used:

- Factor A. Phase screen. Factor gradations: ordered screen (A_1) and non-ordered screen (A_2).
- Factor B. The state of inductor: functionally active state – fresh blood (B_1) and the state with reduced activity – blood after artificial aging (B_2).

The following experimental scheme was used:

A_1		A_2	
B_1	B_2	B_1	B_2

The results of dispersion analysis (Table 1) showed that both factors have a statistically significant effect upon DCI activity. Factors A and B have almost similar effect and their interaction can be out of consideration. Low level of significance of zero hypothesis confirms significant differences in biodetectors response both to the changes in phase screen type and in the functional state of bioinductors. Same was confirmed by a calculated Fisher criterion (F) exceeding critical value (F_c) under a confidential level $P = 0.99$. Therefore, the suppression of functional activity of cells and decline in coherence of their biogenic irradiation prevents DCI.

3. SCATTERING OF COHERENT LIGHT BY BIOLOGICAL STRUCTURES UNDER NORMAL CONDITIONS AND PATHOLOGICAL STATES

3.1. Working Hypothesis

The changes of a functional state of organism affects a microstructural organization of its tissues and therefore statistic parameters of spreading light wave. Thus a biological structure can affect the properties of communication DCI channel playing a role of a converter of coherent irradiation.

3.2. Materials and Methods

Leaf blades of spinach *(Spinacea oleracea L.)*, apple (*Malus domestica Borkh.)*, zisusia (*Cissus rhombifolia L.)*, ambrosia (*Ambrosia trifidia)*, as well as apple and cucumber fruits were used for investigation. Flat-parallel cuts of different thickness were prepared from the fruit tissues and the measurements of their optic characteristics immediately started. Spinach leaves were taken both from healthy plants (lacking any symptoms) and from those damaged to a different degree by *Perenospora spinaceae,* a parasite with a biotroph form of nutrition. Apple leaves were taken from one crown layer and similar light

Table 1. Dispersion analysis of experimental results

Source of variation	Statistic effect, %	Level of significance	Fisher criterion, F	Fisher criterion, F_c
Factor A	34.2	0.002	16.3	9.33
Factor B	39.8	0.001	19.03	9.33
Factor A and Factor B interaction	0.9	0.523	0.43	9.33
Occasional factors	25.1			

exposure, either healthy or damaged by scab shoots (*Venturia iaequalis*). Leaves taken from zisusia were either healthy or with the visible traits of chlorosis. Experiments on ambrosia were conducted in early autumn by using leaves being of different aging stages. The special computer device was used for measuring the intensity and degree of space coherence [12, 15]. Helium-neon (632.8 nm) or semiconductive (650 nm) lasers were used as a source of irradiation. Poorly spreading sounding beam was directed to the studied object perpendicularly to its surface (Fig. 3).

Characteristics of scattered irradiation were registered by means of a photosensitive detector (light-gathering displacement interferometer), installed with CCD-camera. Signals from camera were transferred to computer where they were processed by means of specially developed software. Graphs of intensity change I(t), degree of space coherence G(t) and radius of correlation R(t) of sounding beam scattered by biological object as well as all of these characteristics together were registered during one cycle of measurements at $t < t_0 = 3$ s and presented on the monitor screen in a real-time scale. Degree of coherence G(t) was determined at one and the same value of path difference $dr = |\boldsymbol{r}_1 - \boldsymbol{r}_2| = dr_1$ in the cross section of a beam. A certain subprogram was used for calculating a type of a normalized spatial correlation function of a field $G(dr) = 100 \times |\boldsymbol{\Gamma}^{(1)}(\boldsymbol{r}_1,\boldsymbol{r}_2) / (I_1\, I_2)^{1/2}|$, where $\boldsymbol{\Gamma}^{(1)}(\boldsymbol{r}_1,\boldsymbol{r}_2)$ is a function of space coherence of the first order, I_1; I_2 are the intensities of interfering rays. All information was stored in computer as table data.

Measurements were performed in two regimes: discrete and dynamical ones. Sounding beam had a relatively low intensity in a discrete regime (up to 200 W/m^2) and was acted to a biological object no more than for 3 sec. In this case, laser irradiation did not significantly affect the state of the investigated object. Each next measurement was made in the new zone of the same object by using another object (for more technical details see [12, 13]. The results of the experiments presented in Figs. 4 and 5 were obtained just by this method. In the dynamical regime, a sounding beam of a semiconductive laser was directed to the same zone of an object during the whole period of measurements having the intensity of 400W/m^2 (Figs. 6, 7, 9, 10). For observing a thermal photodestruction of leaf blade tissues, a local heater installed in the device was used (Fig. 8). Its working surface had a form of a circle 20 mm in diameter, with a hole in the center for a sounding beam passage. A power density in this case achieved 700 W/m^2.

Data obtained under discrete regime of measurements were averaged per 10…30 biological replications. Dynamical measurements were made from 5 to 10 times. Out of obtained dependences the most typical graphs were selected and presented in figures. Degree of coherence of scattered irradiation was calculated in percent and its intensity in nominal units corresponding to the device data. Radius of correlation also was given by nominal units – number of pixels of CCD-camera installed in the optic tract of interferometer at the distance of 80 mm from scattering object.

3.3. Results and Discussion

Experiments with different plant organisms showed that a stochastization of coherent light wave in cell structures is not so significant as supposed earlier. Spatial coherence of sounding laser beam passing through juicy tissues of fruit were reliably registered in 6 mm thick samples corresponding to hundreds of cell layers (Fig. 4). The degree of spatial coherence of irradiation scattered by leaf blades of different plants achieved 30….70%. This fact indicates that scattering elements of plant tissues are distributed with a certain order rather than

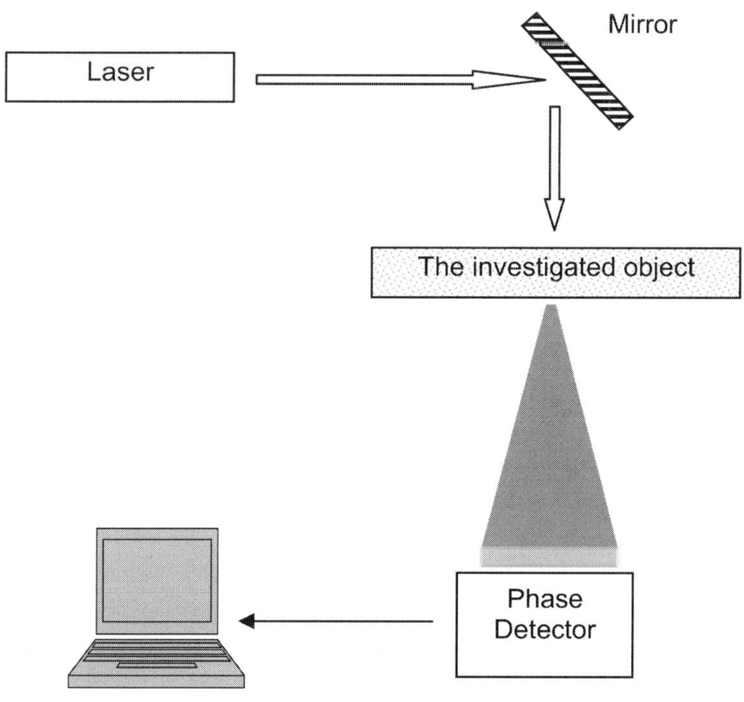

Figure 3. Functional scheme of measurements of space coherence of scattered laser beam.

chaotically. Changes in statistic characteristics of coherent irradiation scattering in optically heterogeneous environment is described by the following equation [16]:

$$\Gamma(dr) = 1 - \frac{dr}{a} + \frac{dr}{a}\int_{-\infty}^{+\infty} \omega(h)\exp(i2\pi d\mathrm{r}h/\lambda a)\,dh, \tag{1}$$

where Γ(*dr*) is a complex degree of a spatial coherence; *s* is a path difference in cross-section of sounding beam; *a* is a width of optical heterogeneity; *ω(h)* is the distribution function of optical heterogeneity according to height *h*; *λ* is a wave length of sounding irradiation.
The equation (1) shows that the degree of scattered irradiation is determined by the parameters of tissue microheterogeneity and can be considered as a measure of its structural organization. In the experiments mentioned below, a module of this function is G(*dr*) = *100* | **Γ**(*dr*) | expressed in percent was measured.

The effects of different destabilizing factors (physical, chemical, biotic) that disturb a normal functioning of an organism can be revealed by the decrease of a degree of a spatial coherence of scattered sounding beam [12, 13]. A typical picture could be observed under pathogenesis of plant tissue. With the development of infection, a spatial coherence was decreased while the intensity of irradiation passed through leaf blade increased (Fig. 5). In a healthy (intact) tissue (point 1), a coherent bundle was strongly absorbed but preserved a considerable part of its spatial ordering. Since the selected wavelength was 633 nm, a chlorophyll is considered to be the main pigment absorbing light quantum.

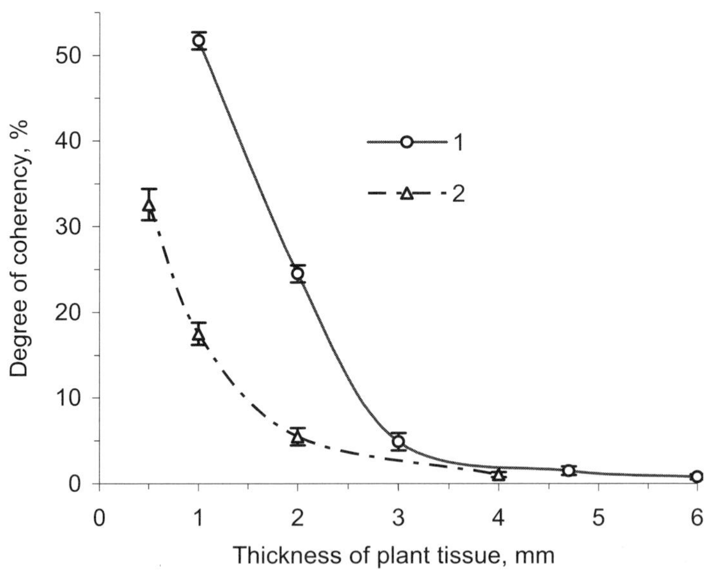

Figure 4. Variation of degree of spatial coherence of sounding laser beam scattered by fruit tissue of a different thickness. 1 - apple, 2 - cucumber.

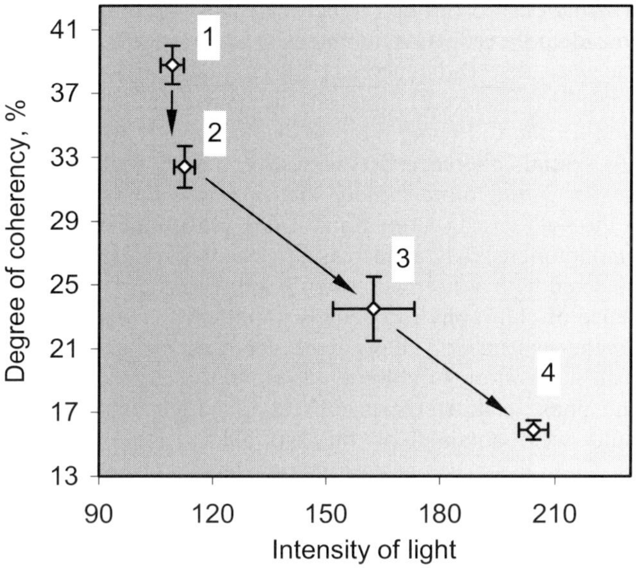

Figure 5. Parameters of coherent irradiation scattered by leaf blades of spinach at different pathogen (*Perenospora sp.*) injury (see text).

In the latent phase of infection (point 2), its concentration remained constant but the disturbance of tissue microstructure became evident. This can be illustrated by a significant decline in the degree of a spatial coherence of a scattered sounding beam. The appearance of the visible disease symptoms (point 3) is followed by further stochastization of light flow and the increase of its intensity. The reason is probably a breakdown of chlorophyll and protein complexes, which is confirmed by luminescence studies of photosystem 2 [13].

A severe infection (point 4) is accompanied by a 2-fold change of amplitude-phase characteristics of a scattered sounding bundle. Therefore the functional state of an organism considerably affects statistical characteristics of coherent irradiation passing through its tissues. This effect can be observed both in photosynthesizing and non-photosynthesizing organs under short time exposure (1….3 sec) of sounding beam. The corresponding measurements should be of a discrete, localized character for minimizing an effect of light upon the organism. A continuous irradiation taking place under the dynamic recording regime provides more complex and informative picture. In this case, the interaction between a light flow and the biological structures leading to both components chnages is taking place.

It has been established that monochromatic irradiation with intensity corresponding to natural illumination can cause relatively rapid (from tens up to hundreds sec) structure-functional rearrangement of photosynthesizing tissues of plant organisms. This effect was observed by measuring intensity and spatial coherence of sounding beam passing, for example, through a leaf blade. Already in 5….10 sec after its switching, a change of the amplitude and phase characteristics of scattered irradiation took place (Fig. 6).

Light flow that passed through intact leaf blade of an apple tree showed a sharp and considerable reduction of intensity I(t) (Fig. 6A). In 25 seconds its 2-fold decline was observed (curve 1). Leaves damaged by scab had a less pronounced response (curve 2). Intensity remained practically the same for 30 seconds and for the following 8 min it decreased only by 60% (curve 2). On the contrary, a spatial coherence G(t) of scattered beam increased, doing this in healthy leaves in 3-fold more rapidly than in those damaged by a pathogen (Fig. 6B). In the both cases (curve 1 and 2) the process developed more slowly and not so monotonously as a decrease in beam intensity (Fig. 6A). This is probably connected with a discrete cluster rearrangement of cellular structures.

The presence of chlorophyll in cells is important for this phenomenon to be expressed. In the absence of chlorophyll, for example in etiolated leaves or during thermal inactivation of chlorophyll-protein complex, the changes of the amplitude and phase characteristics of scattered light are not observed [15]. Obvious results were obtained on the leaf blades tissue with symptoms of chlorosis (Fig. 7). In this case, the relative time changes of registered data didn't exceed several percents (curve 2, Fig. 7A, B). Meanwhile, healthy leaves exhibited a pronounced dynamics of intensity and of a coherence degree of scattered sounding beam (curve 1, Fig. 7A, B).

I(t) and G(t) dynamics shown in Figs. 6 and 7 are of a universal character and can be observed in different photosynthesizing tissues. Amplitude and rate of changes of the registered optic indexes decreased together with a loss of their functional activity. Whatever being destabilizing processes affecting the organism, they always reveal the same tendencies.

A heat stress proved to be an appropriate object for investigation. A leaf blade of zisusia (*Cissus rhombifolia L.)* was exposed to intensive heating for 2 min. After achieving a certain temperature, a sharp (gradual) change in optic indexes of scattered beam (Fig. 8) was observed. A two fold decrease in the degree of a spatial coherence during 20 sec period indicates the disturbance of a native microstructure of a tissue. After switching off the heater, a relatively slow reparation process took place for 15 min. This is confirmed by the restoration of an initial level of a spatial coherence. However, more prolonged effect of a sounding beam with intensity of 700 W/m^2 resulted in a chlorophyll breakdown and the development of a new destructive process. The latter was characterized by increase in the intensity and decrease in the coherence of registered irradiation (Fig. 8 to the right).

A heat stress considerably decreased the threshold of a leaf blade resistance to the intensive light. In the absence of heating, much higher intensity and duration of irradiation was required for initiation of photodestruction. Therefore the effect of different factors can result either in reversible or irreversible structural rearrangements of plant tissue, which affect in their turn the value of a spatial coherence of a scattered beam.

The obtained results allow one to consider a plant tissue as a dynamic amplitude and phase screen, the properties of which depend upon the state of organism. Spreading of a coherent wave in such medium will be followed by decrease in its amplitude as well as by a decline in spatial coherence of a field. At the same time, the effect of interdependent changes in the microstructural organization of a tissue and the statistical ordering of scattered irradiation is taking place,

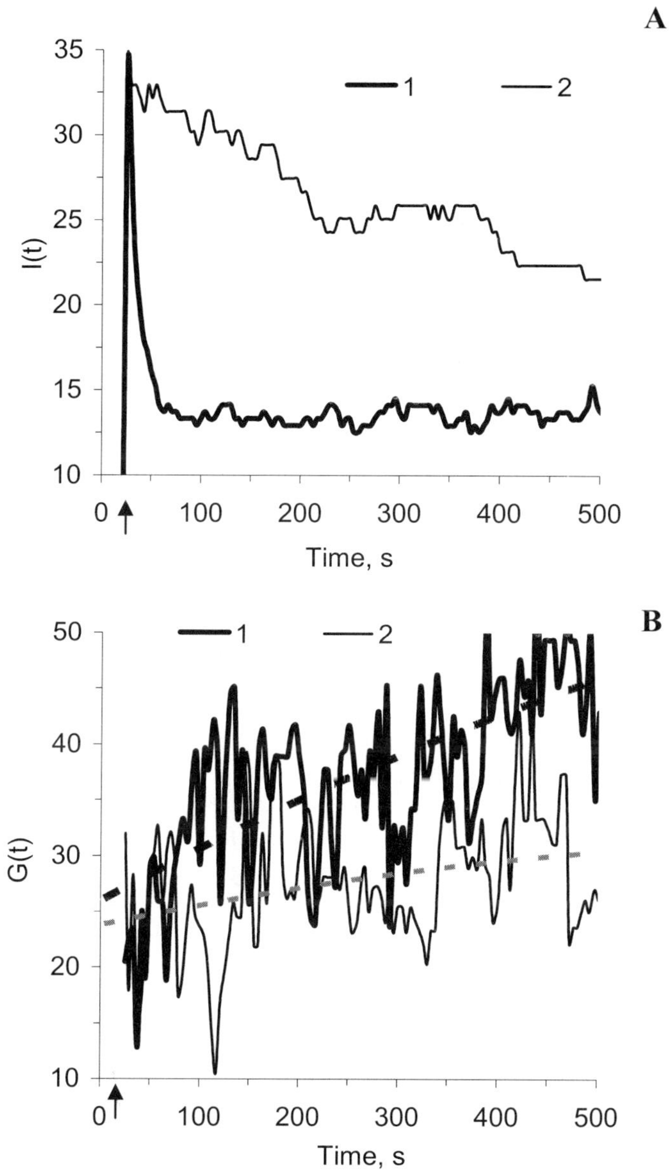

Figure 6. Temporary dependence of intensity I(t) and degree of spatial coherence G(t) of sounding laser beam scattered by healthy (1) and injured by scab (2) leaves of an apple tree (*Malus domestica Borkh.)*. Arrows indicate the moment of laser switching on. Dotted lines correspond to trend lines.

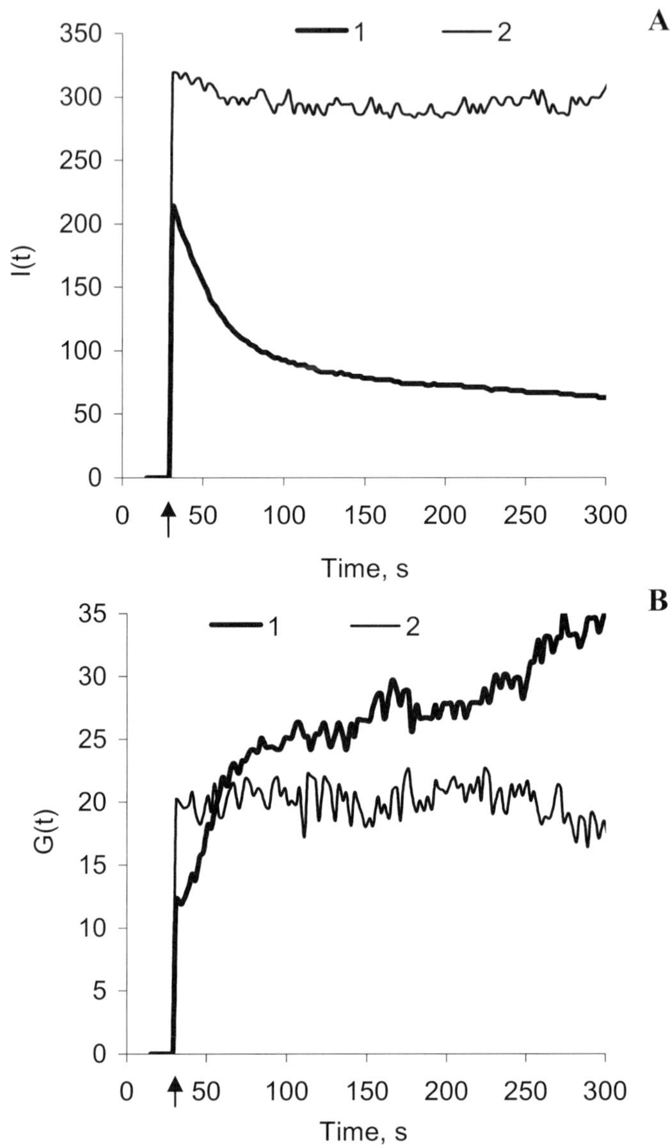

Figure 7. Intensity of irradiation I(t) and degree of spatial coherence G(t) of sounding beam scattered by leaf blades of zisusia (*Cissus rhombifolia L.*) in the different functional states: 1- normal leaf; 2- a leaf with symptoms of chlorosis.

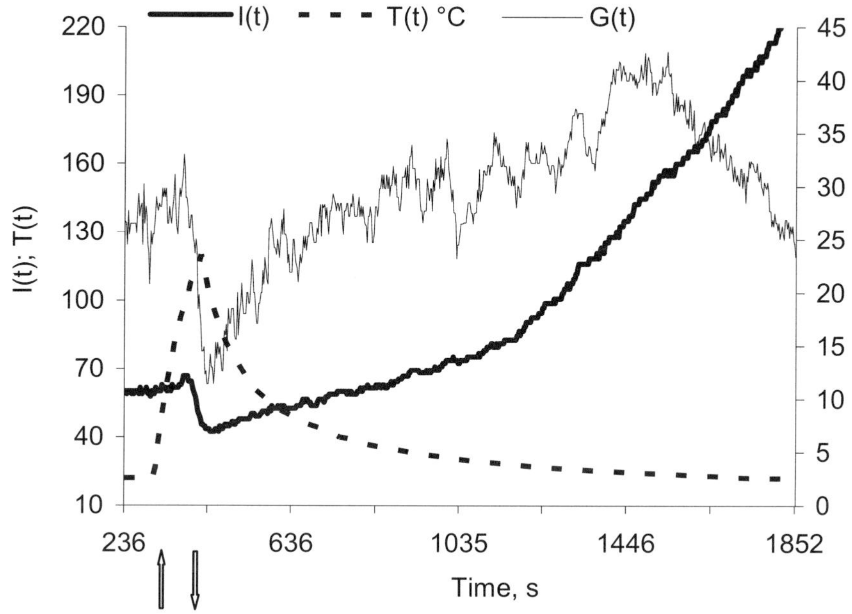

Figure 8. Changes in the parameters I(t) and G(t) of sounding beam, scattered by leaf blades of zisusia (*Cissus rhombifolia L.*) during short-term heating. T(t) - surface temperature of a heater. Arrows indicate the moments of switching the heater on and off.

pointing to an active response of an organism to irradiation. Analysis of a spatial transverse correlation function allows one to establish some regularities of such a complicated process.

In the experiment in which results are presented in Fig. 5, spinach leaves were analyzed by using radiation of one-mode helium-neon laser with a Gauss profile of a transverse distribution of intensity. Module of normalized correlation function of such bundle is described as [14]:

$$G\,(dr, z) = 100\ exp\left[-\frac{1}{2}\left(\frac{kad\,\mathrm{r}}{2z}\right)^2\right], \tag{2}$$

where a is the radius of sounding beam at object outlet; k is wave number, z is distance between object and image recording plane. Then the radius of correlation corresponding to a level 0.5 (50% decrease in the value of correlation function) is determined as $R_c = 0.37\lambda z/a$. The degree of a spatial coherence was measured at fixed means $s = s_1$ and $z = z_1$. The given expressions show that the decrease of $G(dr_1,z_1)$ from 39 up to 16% during development of pathogenic process was due to widening of the angular spectrum of a beam and 1.4-fold decrease of its correlation radius.

In this case, the measurement of $G(dr_1,z_1)$ was carried under short-term (1…3 sec) action of a sounding beam. As it has been already mentioned, more prolonged laser irradiation increased microstructural ordering, but did so only in the functionally active tissue. This is evident from the increase of degree of coherence of a scattered irradiation (Figs. 6B and 7B). In these cases, the evaluation made according to equation (2) demonstrates a rapid 30...40% increase in the correlation radius. It means that the biological structures interacting with a coherent wave can reduce their stochastization by means of their own rearrangement.

The results obtained by calculations are confirmed by direct measurements of a spatial correlation function (Fig. 9). In a healthy (functionally active) tissue, a loss of statistic ordering of a coherent optical signal is much smaller, what is confirmed by high values of correlation radius of scattered sounding irradiation (R_c was determined by 0.5 level, as shown in Fig. 9 by a dotted line). During a natural leaf aging, this index decreased 2.9-fold, while a volume of field coherence decreased 8.4-fold.

In the course of dynamical measurements, the general character of changes in the correlation radii corresponded to the time dependence of a coherence degree. Only in the functionally active tissue (Fig. 10, curves 1 and 2) R_c increase took place while lacking in the leaves with well-expressed aging features (curve 3). The increase in the value of statistic indexes G(t) and R_c(t) of scattered irradiation pointed to increase in the microstructure ordering due to passage of coherent waves.

Noteworthy, the observed dynamics of a scattered beam characteristics cannot be a result of a non-stable work of electronic devices or incorrectly functioning software. Similar characteristics of a tested object (glass scatter

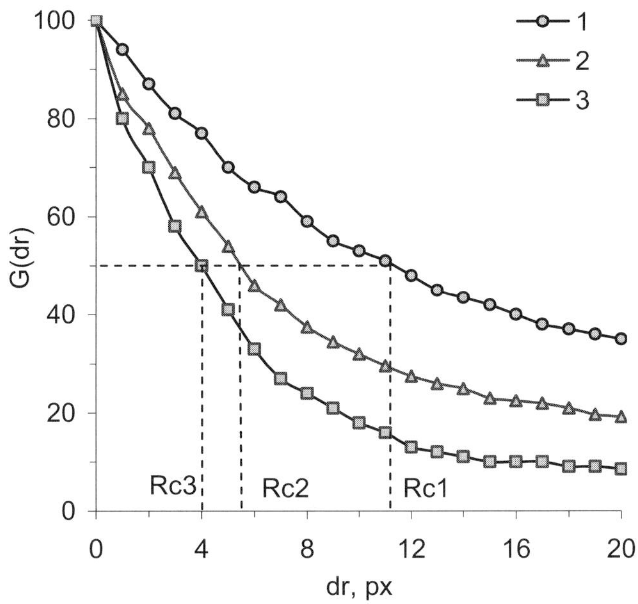

Figure 9. Space correlation functions G(*dr*) of a sounding beam, scattered by leaf blades of ambrosia at different functional states: 1 - onset of aging; 2 - development of aging process; 3 - expressed features of aging.

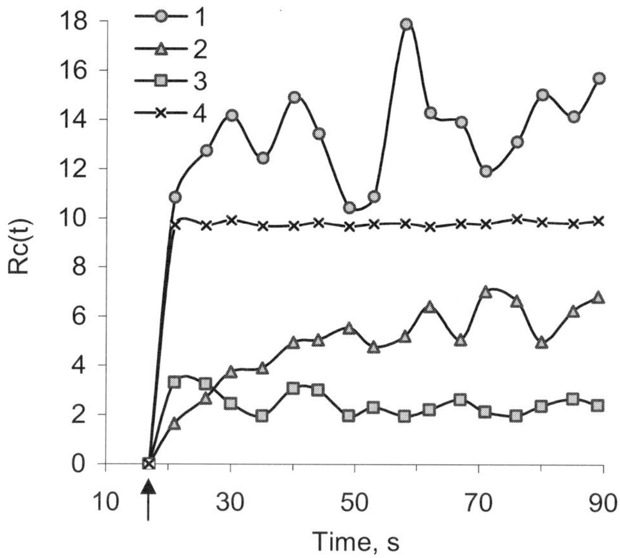

Figure 10. Temporal change of radius of correlation $R_c(t)$ of scattered irradiation at a continuous effect of sounding beam on leaf blades of ambrosia. The state of tissues (1, 2, 3) was the same as in Fig. 9; 4 - radius of correlation of glass scatter filter (tested object).

filter) remained constant during all the measurements. As an example, the fourth graph of time dependence of correlation radius can be taken.

4. CONCLUSIONS

The experiments performed had shown than the endogenous irradiation mediating DCI possesses a statistical ordering that is of principal importance for the functioning of the field communicative channel. A disturbance of cells functioning as well as an artificial stochastization of BCL light flow diminished the effects of optical interactions. Similar processes can proceed in the living organisms. Biological structures possessing a considerable heterogeneity are changing the statistics of coherent photons and therefore affect the properties of regulatory signals in the DCI channel. The investigation of optical properties of plant tissues allowed us to reveal some regularities important for understanding of the mechanism of the field (non-chemical) regulations in biosystems.

The effect of different destabilizing factors on the living organism results in the decline of a degree of coherence of scattered sounding beam. A connection between microstructural organization of a tissue and the statistics of a scattered light wave (formula 1) shows that the decrease of coherence is caused by the disturbance of the order on the cellular and sub-cellular levels. Therefore a degree of stochastization of a sounding beam can be used as a criteria of the functional state of an organism. The greater is a destructive process, the lower is a spatial coherence of irradiation. This is mostly evident under a short-term (several seconds) light exposure of any kind of plant tissues. More prolonged laser irradiation can cause photoinduced processes affecting a microstructural organization of a biological object.

Under the regime of dynamical measurements, a considerable (1.5-2.5-fold) and relatively rapid (10...100sec) change of intensity and of a coherence degree of scattered sounding beam has been observed. Such processes are typical only for phosynthesizing tissues. A decrease in chlorophyll concentration in the cells or any decline of a functional activity of an organism make this effect less pronounced. In this case, the amplitude and the rate of change in optical indexes decrease up to zero. Preliminary investigations showed that the discussed response of the organism depends upon the parameters of irradiation. A spectrum of the effect corresponded to the region of chlorophyll absorption, and the intensity can change from tens up to hundreds of W/m^2. Under low intensities, the effect decreases and becomes hardly observed, while under high intensities a photodestruction of chlorophyll-protein complex becomes evident.

A biological structure plays a role of a converter (transformer) of coherent irradiation, which properties depend upon a functional state of organism. In the damaged (sick) tissue, a coherence of sounding irradiation considerably declines and the angle spectrum of spatial frequencies enlarges even if the depth of penetration is small. As a result, the correlation radius and, correspondingly, the volume of field coherence decreases. One of the conditions of detection of a coherent optic signal, namely $R_c > D$ [4, 5] (D is a typical cell size), can be disturbed and the cells located behind a damaged region will stop to accept it as regulatory (stimulating) influence. Just the same effect took place under the artificial introduction of stochastic factor into a communicative DCI channel (Fig. 2).

In the normal (healthy) tissue, coherent irradiation is spread with a smaller decay of statistical ordering, so that a coherence volume sufficient for performing bioregulatory functions can be preserved along tens and hundreds of cell layers. If such an influence will prolong for some time, structure and functional reconstructions can take place affecting in their turn the scattering statistics of a light wave. Interrelated ordering of a tissue structure and of irradiation statistics is taking place. In this case, the distance of spreading of a coherent signal (depth of its penetration in cell layers) increases.

In no way a plant tissue is a passive participant of communication process. As depending upon its functional state, it can either limit or, on the contrary, enlarge the region of the distant cell interactions. Accordingly, one more kind of connections (namely, optical one) can be traced between the structure (in the given case microstructure) and the function of the biosystems. Its existence is confirmed by the experiments on holographic induction of morphogenesis [4, 5, 18].

REFERENCES

1. Dicke R. (1954) Coherence in spontaneous radiation processes. *Phys. Rev.* 93. Pp. 99-110.
2. Frohlich G. (1977) The coherent stimulation in biological system. *Biophysica.* V.22. Issue.4. Pp. 743-744. (Russ.).
3. Lobko V.V., Karu T.J., Letohov V.S. (1985) Is it impotant a coherency of a low-intensity laser light during its action upon biological objects?. *Biophysica.* V. 30. Issue. 2. Pp. 366-371. (Russ.).
4. Budagovsky A.V. (2000). On the physical nature of "Biological fields". In: *Biophotonics and coherent systems.* University Press. Moscow. Pp. 173-188.

5. Budagovsky A.V. (2004). Distant intercellular interaction. Technika. Moscow. 104 p. (Russ.).
6. Budagovsky A.V., Turovzeva N.I., Budagovsky I.A. (2001). Coherent electromagnetic irradiation in distant intercellular interaction *Biophysica*. V. 46. Issue. 5. Pp. 894-900. (Russ.).
7. Popp F.-A., Li K.H. (1993). Hyperbolic relaxation as a sufficient condition of a fully coherent ergodic field. *Int. J. Theor. Phys*.V. 32. No. 9. Pp. 1573-1583.
8. Bajpai R.P. (1999). Coherent nature of the radiation emitted in delayed luminescence of leaves. *J. Theor. Biol.* V. 198. No. 3. Pp. 287-299.
9. Krylov O.A. (1980) On the ways of studying the mechanism of laser irradiation. *Question of a health resort studying and physiotherapy and medical physical culture.* N.6. Pp. 1-5. (Russ.).
10. Sinyakov V.S., Haydakov M.I. (1983). In: Actual problems of disease and convalescence. Moscow. Pp. 154-156 (Russ.).
11. Kusin A.M., Surkenova G.N., Budagovsky A.V., Gudi G.A. (1997) Secondary biogenic radiation of a γ- irradiated human blood. *Radiobiology. Radioecology*. V. 37. Issue. 4. pp. 577-580. (Russ.)
12. Budagovsky A., Budagovskaya O., Lenz F., Mirovskaya A., Alkayed K. (1998) In: Puti povysheniya ustoichivosti sadovodstva. Michurinsk. Pp. 98-113. (Russ.).
13. Budagovsky A., Budagovskaya O., Lenz F., Keutgen A., Alkayed K. (2002). Analysis of the functional state of cultivated plants by means of interference of scattered light and chlorophyll fluorescence. *Journal of Applied Botany*. V. 76. Pp. 115-120.
14. Kusin A.M. (1997) Secondary biogenic radiation – rays of life. Pushino. 38 p. (Russ.).
15. Budagovsky A., Budagovskaya O., Budagovsky I. (2004). In: Mobilization of adaptation potency of fruit plants in dynamic environmental. Proceeding of International Conference. August 24-26. Moscow. pp. 92-99. (Russ.).
16. Asubov F.M. (1982) In: Diss. Ph. Physic-mathematic science. MGU. Moscow-143 p. (Russ.).
17. Ahmanov S.A. Djakonov U.E. Chirkin A.C. (1981) Introduction in statistical radiophysics and optics. Nauka. Moscow. 640 p. (Russ.).
18. Budagovsky A.V., Evseeva R.P. (1995) Experimental models of a distant transfer of the morphogenetic information in the system of two vegetative fabrics with different potencies for differentiation. In: Mechanisms of action of extra small dozes. 2nd Int. Symp. M. Pp. 124-125. (Russ.).

5
THE OSCILLATION BEHAVIOR OF THE DELAYED LUMINESCENCE OF PLANT LEAVES

Yu Yan*, Fritz-Albert Popp, Sibylle Sigrist, Daniel Schlesinger, Andreas Dolf, Zhongchen Yan, Sophie Cohen, Amodsen Chotia, and Daniel Busch

1. INTRODUCTION

Delayed luminescence (DL) is the afterglow of biological tissues after illumination with monochromatic or white light. In contrast to fluorescence or phosphorescence, this long term relaxation of light reemission lasts seconds up to even hours. DL was discovered in green plants by Strehler and Arnold 1951[1]. At the beginning, it was assigned to chlorophyll reactions. A new idea arose around 1970 at the University Marburg suggesting that DL originates from a coherent photon field in cells [2,3,4]. Theoretically, DL can be explained in terms of excited states of a coherent ground state[5,6]. However, further and more detailed analysis of DL is necessary in order to provide deeper insight into the physical problem, i.e., the mechanisms of DL. In particular, after exposure of plants to red and infrared light, not only continuous relaxation but also a "relative maximum" was observed in 1965 during the reemission of photons[7]. The most popular detailed explanation of the origin of DL traces it back to light emission of charge recombination during a back-flow of charges in the photosynthetic electron-transport chain PS II and PS I[7,8]. According to this hypothesis, the relative maximum could be assigned to the superposition of different decay curves of photon emission from PS II and PS I[7,8] or from different sub-populations of PS II[9]. In 1981, it was shown, the first time[3] that DL of plants and animal tissues follows a hyperbolic relaxation including hyperbolic oscillations while displaying the same kinetics for different wavelengths in the visible range. This result has been reproduced by several laboratories and was subjected to careful analysis[10-12]. It turned out that the "relative maximum" of DL belongs as a part just to this "hyperbolic" oscillation of the relaxation curve. Experimentally, as a result of limited sensitivity and resolution, it is not easy to show evidence of more than one or at the utmost two full oscillations during the hyperbolic delayed luminescence. Recently, correlations of delayed luminescence properties and whole-organism effects due to some chemicals, copper, cadmium, and zinc[13] have been reported. The dependency of the form of the first shoulder of DL on

* International Institute of Biophysics, Landesstiftung Hombroich, Kapellener Strasse o.N., D-41472 Neuss, Germany.

growth rate, cell density, and carbon content has been also discovered now[14]. Because the technical progress is of considerable importance for all these investigations, it is worthwhile to point to recent results of Tudisco et al.[15] and Niggli et al.[16] in developing and improving fast ultraweak luminescence analysis by means of Laser ultraviolet-A induced luminescence.

In the present work, the DL of green leaves, leaf homogenates, and isolated chloroplasts were investigated regarding their completeness, freshness, and tumor incidence in dependence on parameters of light exposure and external temperature. The DL relaxation curves were analyzed in terms of hyperbolic relaxation functions with oscillations as they were predicted by the coherent state theory of DL.

2. MATERIALS AND METHODS

2.1. Instrument of DL Measurements

All DL measurements were performed with the Photon-Measurement-System-2 (PMS-2) depicted in Figure 1, which was developed by our research group. This system is equipped with two photomultipliers (Thorn Emi, type 9558 QA) as photon detectors. For the purpose of DL measurement, one of the three light sources was employed in order to illuminate a leaf sample: an Argon-laser (Melles Griot, type 35MAP431-230, 457nm, 12mW), a stabilized He-Ne-laser (Melles Griot, type 05STP903, 632.8nm, 1mW) or five infrared LEDs (Roithner Lasertechnik, type ELD-780-514, 780nm, 11mW each).

2.2. Measurement of DL of Plant Leaves

Subjects of investigation were leaves of three kinds of indoor plant, i.e., *Crassula ovata*, *Anthurium*, and *Kalanchoe daigremontiana*, and leaves of corn salads (*Valeriana locusta*). The homogenate of corn salad leaves prepared in

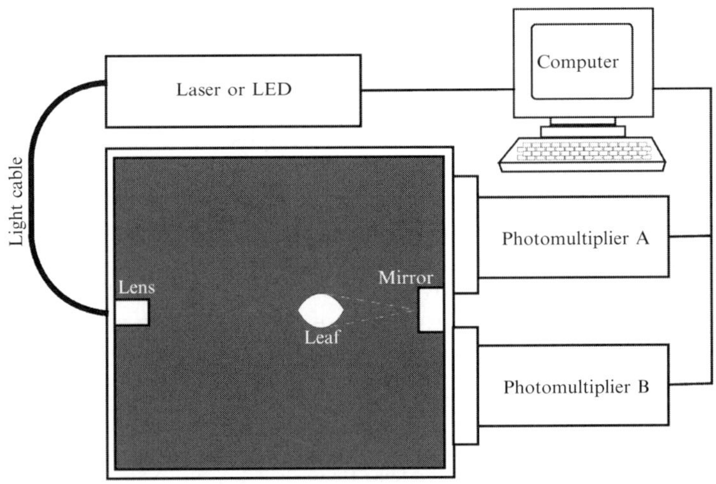

Figure 1. Photon-measurement-system-2 with two-channel photon counting system. (PMS-2).

0.4M sucrose buffer solution containing 60mM K_2HPO_4, 30mM KH_2PO_4, 5mM $MgCl_2$ and 35mM NaCl was also investigated. Furthermore, the chloroplasts were isolated from the homogenate through sucrose gradient centrifugation described by Richter[17] and investigated. In each measurement, a whole leaf, 10ml homogenate, or 10ml solution containing isolated chloroplasts was placed in the dark chamber of PMS-2, illuminated 30s by the LED or laser, and measured afterwards by the photomultipliers.

3. RESULTS AND DISCUSSION

3.1. DL of Plant Leaves and Leaf Fractions

After an infrared or red light illumination, the DL of a whole leaf was found to last over 10 minutes. The relaxation curve of DL shows an oscillation around a hyperbolic function (Fig. 2). The residues from the best fitting hyperbolic function can be well assigned to a "hyperbolic" oscillation function. Both these functions, the continuous and the oscillating one, were introduced in 1981[3] and are now derived from a quantum theoretical approach by Popp and Yan, in order to describe the coupling of coherent states [5].

For further investigation of the origins of DL and DL-oscillations, whole leaves, homogenates of leaves, isolated chloroplasts, and filtered (0.2μm pores) homogenates of corn salad were measured (Fig. 3). As expected from the coherent photon field under study, DL oscillation can only be observed in whole leaves. Homogenates exhibit DL with the same decay time as whole leaves, but without oscillation. Isolated chloroplasts display DL, too, but with much shorter relaxation times than the DL of homogenates. Again there is no further

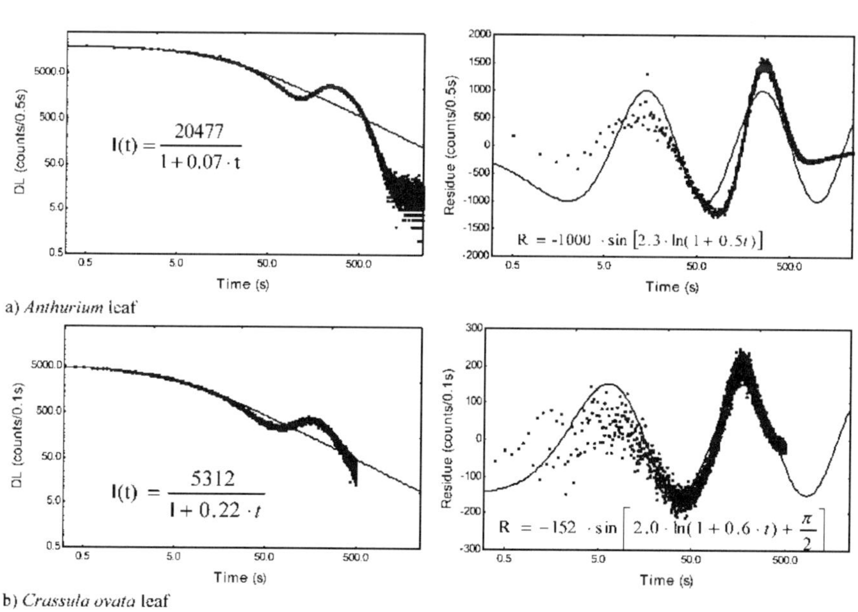

Figure 2. Two DL decay curves of leaves (illuminated by 780nm LEDs) from two plant species.

oscillation. Filtered homogenates show no DL at all, in spite of the fact that they contain a lot of chlorophyll and other molecular components of chloroplasts. But compared to the original homogenates, they contain no intact cells, chloroplasts, and other cell organelles.

The fact that DL oscillation exists only in whole leaves but not in leaf fractions indicates clearly the non-locality and coherence of DL oscillations in living tissues. Further experiments confirm that aging, where a leaf loses its cellular integrity gradually, affects correspondingly the DL pattern (Fig. 4). A fresh leaf shows a strong DL oscillation. During aging, its DL intensity even increases, but its DL oscillation diminishes gradually and disappears after three weeks.

The above results again show that DL of plant leaves is not a phenomenon of single molecules, but a property of tissues, cells, and cell organelles. Single light absorbers in plant leaves, e.g., chlorophyll, cannot store photons for such a long time (>10 minutes). Photon trapping is only possible in a living complex and ordered system with a large number of active light absorbers. The hypothesis that DL origins from PS I and PS II may explain the DL of chloroplasts, but it can hardly give an answer to the question of why a whole leaf has an oscillation pattern in its DL decay curve while a solution of chloroplasts displays no oscillation at all. The results again confirm that DL oscillation of a whole leaf has to be explained by non-local couplings of the light sources and sinks in a whole biological system. Only a system governed by a coherent field is able to build up such couplings. Theoretically, at least two coupled coherent states with different coupling frequencies are necessary to display a hyperbolic relaxation

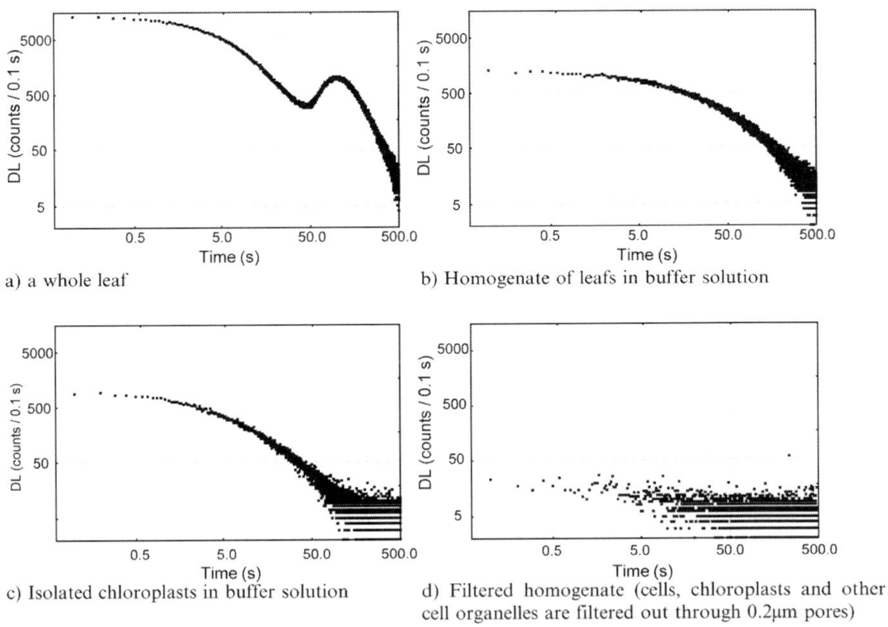

a) a whole leaf

b) Homogenate of leafs in buffer solution

c) Isolated chloroplasts in buffer solution

d) Filtered homogenate (cells, chloroplasts and other cell organelles are filtered out through 0.2μm pores)

Figure 3. DL-patterns of a whole leaf and leaf fractions of corn salad (illuminated by 780nm LEDs).

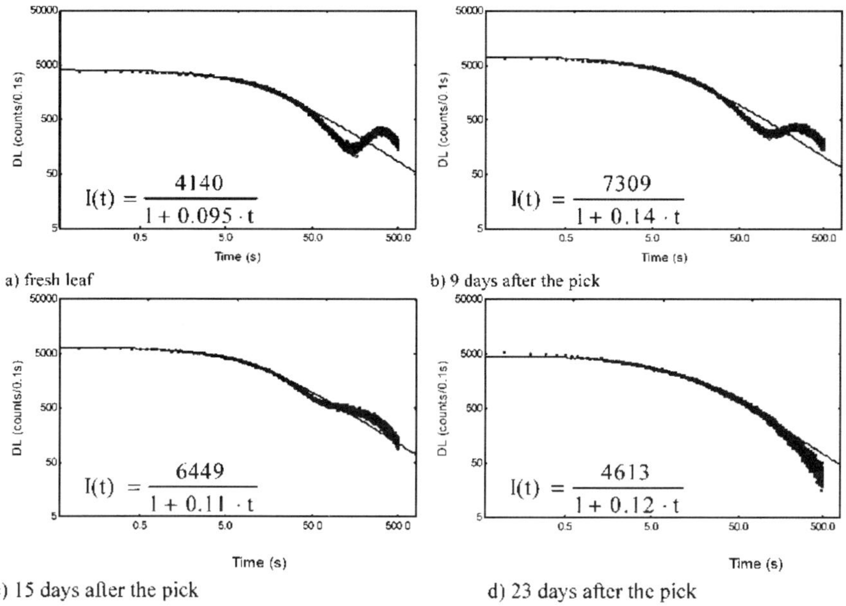

Figure 4. Changes of the DL pattern of an *Anthurium* leaf picked from the plant and stored at room temperature (illuminated by 780nm LEDs).

with an oscillatory pattern. Application of perturbation theory traces this kind of hyperbolic oscillation back to the following general mechanism. By means of switching on a coherent field, the energy of an ensemble of number states may split into at least two coherent states, where one has a lower energy than the former ground state; whereas the energy of the lowest new excited state is higher than before. This leads to a permanent slow oscillation of the occupation probability of photon(s) between the energetically lower new ground state and the higher excited state. This mechanism works like a photon trap, where the energy gap between the induced new lower-energy ground state and the new higher-energy excited states provides the extraordinary high storage capacity for photons. This mechanism of photon scattering reminds one of a kind of "photon sucking" or "light breathing" between the sucking phase by occupation of the new ground state and the emitting phase by occupation of the new excited state. This mechanism has been postulated also in order to explain the differentiation of cells and the connection to biological clocks[18]. It is worthwhile to note that this photon-breathing may be induced by exciplex states of DNA in permanent interaction with photon clouds.

On a leaf of *Kalanchoe daigremontanum,* tumor growth has been evoked by bacterial contamination in such a way that healthy, as well as tumor tissue on the same leaf becomes available for comparable investigation. In agreement with the results,[3] it turns out that the continuous part of the hyperbolic relaxation of DL, as well as the oscillatory part of tumor tissue, grow less distinct compared to that of the healthy tissue. It can be understood in terms of a clear loss of the degree of coherence as well as of the strength of coupling forces of the biophoton field.

3.2. DL of Plant Tumor

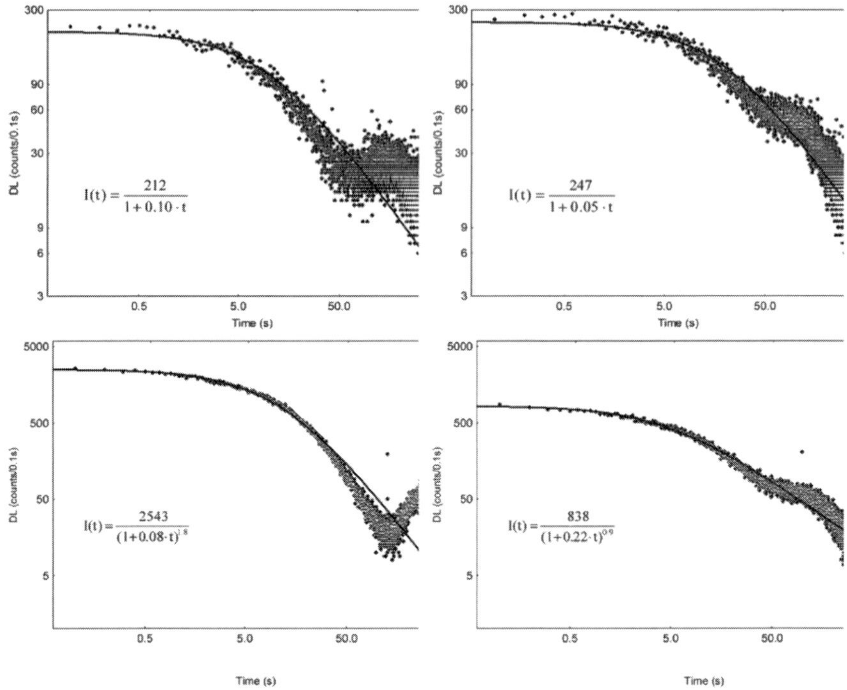

a) normal tissue b) tumor tissue

Figure 5. DL of the normal and tumor tissue on a leaf of *Kalanchoe daigremontiana* (illuminated by 780nm LEDs).

Figure 6 shows that increasing illumination time gives rise to higher amplitudes as well as more distinct oscillations of DL. However, after about 30s exposure, a saturation effect can be observed, that is, longer exposure time does not change the DL behavior any more. The DL can be reproduced several times without any quantitative change of its dynamics. These results can be explained in terms of an increase of the coherence time with increasing illumination time. It may explain the saturation effect at the same time.

Fig. 7 displays an example of the change of the continuous part of the relaxation function, as well as the oscillatory part, by changing the wavelengths of the illuminating light. The most distinct oscillation was observed in the DL after an illumination by infrared light arround 780nm. If we compare this observation to the spectrum of the DL (Fig. 8), we can find that this range is also the main part of the DL emission no matter what kind of illuminating light.

The DL-intensity of two arbitrarily selected *Anthurium* leaves have been measured in dependency on the external temperature and varied between 12 and 45° C. The Arrhenius-plot (Fig. 9) shows that the energy gap ("activation" energy) is of the order 0.5 eV, where the upper (blue) curve corresponds to 0.43 eV, the lower (red) one to 0.27 eV. The deviation from the straightness of the

3.3. Illumination Time and Wavelength

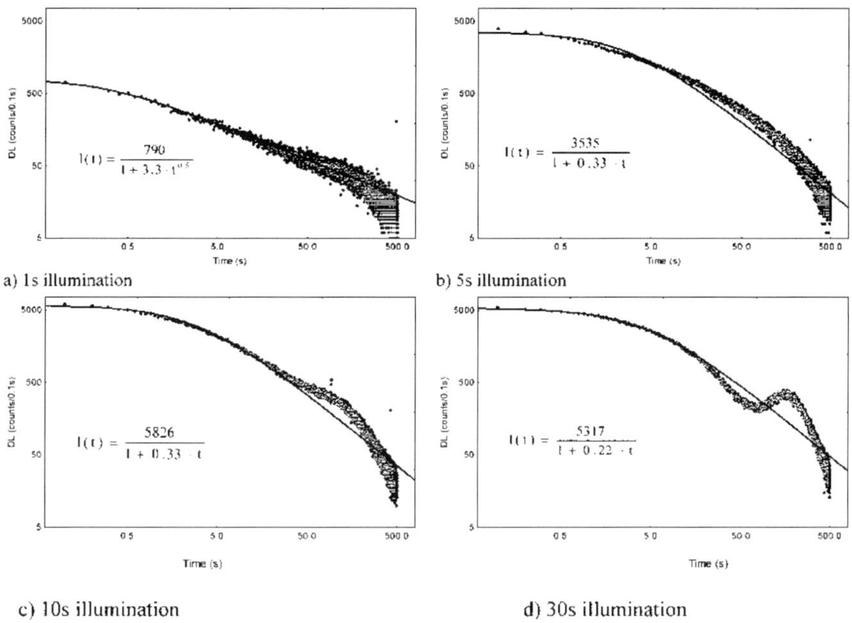

Figure 6. Relationship between DL-dynamics and illumination time (illuminated by 780nm LEDs).

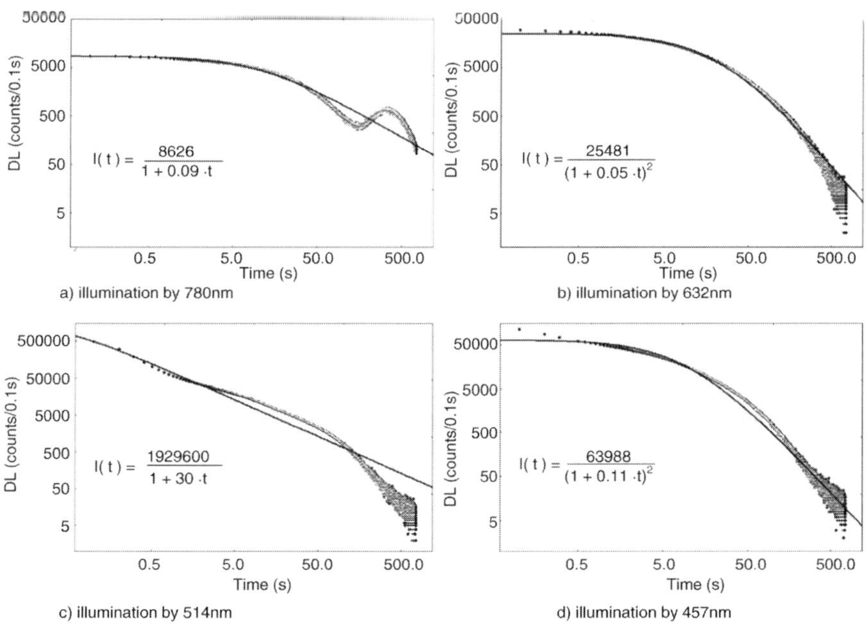

Figure 7. Relationship between DL-dynamics and illumination wavelength (780nm: LEDs, 632nm: He-Ne-laser, 457nm and 514nm: Argon-laser).

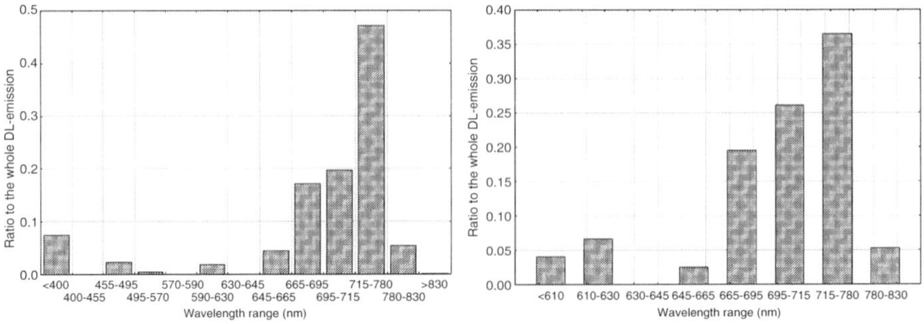

a) illuminated by 780 nm b) illuminated by 457 nm

Figure 8. Spectra of DL emitted by *Crassula ovata* leaves. The spectra were measured by use of long pass filters.

3.4. Temperature and DL of Plants

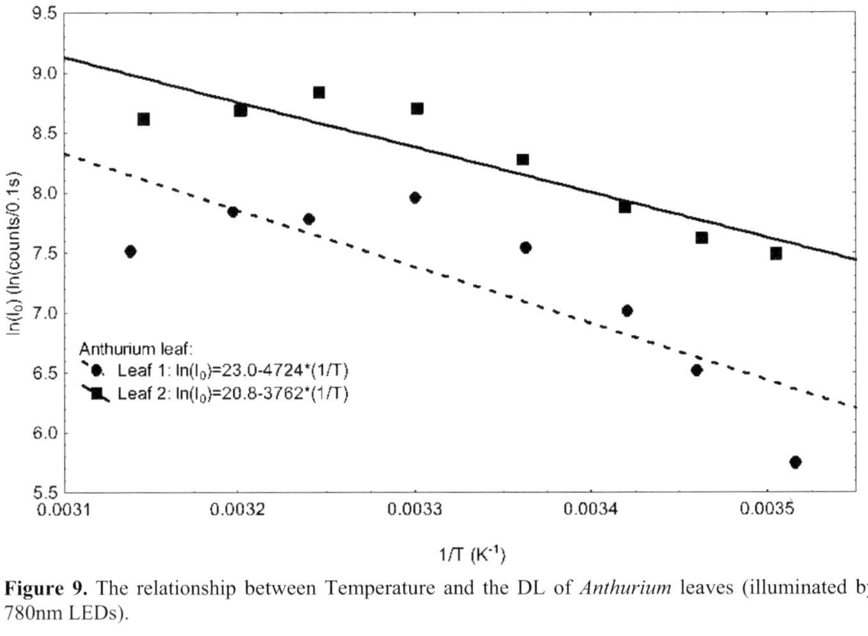

Figure 9. The relationship between Temperature and the DL of *Anthurium* leaves (illuminated by 780nm LEDs).

Arrhenius-curve shows that this "activation" energy cannot be assigned to molecular compounds. Rather, it corresponds to an energy gap of the photon traps of the whole plant tissue. The deviations from the perfect Arrhenius-law behavior concern non-linear temperature dependence and give rise to well known temperature hysteresis loops[19-20].

We may conclude that the "hyperbolic oscillations" in plant leaves show evidence of collective excitations which cannot be understood in terms of purely molecular interactions. They are based on coherent states of a photon field in living systems. Basically one is able to describe them by well-known laws of quantum optics. They express the capacity of living systems to trap light and to use it for organizational tasks.

ACKNOWLEDGEMENTS

We like to thank Prof. Dr. W. Klimek for the valuable discussions and his support. We also want to express our thanks to the German Ministry of Research (BMBF) and the Family-Ernst-Wendt Foundation for their financial supports.

REFERENCES

1. B.L. Strehler, W. Arnold, Light production by green plants, J. Gen. Physiol. 34 (1951) 809-820.
2. B. Ruth, experimental investigations on ultraweak photon emission, in: F.A. Popp, G. Becker, H. L. König and W. Peschka (Eds.), Electromagnetic Bio-Information, Urban & Schwarzenberg, Munich – Vienna – Baltimore, 1979, pp.107-122.
3. F.A. Popp, B. Ruth, J. Böhm, P. Grass, G. Grolig, M. Rattemeyer, H.G. Schmidt, P.Wulle, Emission of visible and ultraviolet radiation by active biological systems, Collective Phenomena (Gordon &Breach) 3 (1981) 187-214.
4. F.A. Popp, K.H. Li, Q. Gu (Eds.), Recent Advances in Biophoton Research and its Applications, World Scientific, Singapore-London-New Jersey, 1992.
5. F.A. Popp, Y. Yan, Delayed luminescence of biological systems in terms of coherent states, Physics Letters A 293 (2002) 93-97.
6. F.A. Popp, J.J. Chang, A. Herzog, Z. Yan, Y. Yan, Evidence of non-classical (squeezed) light in biological systems, Physics Letters A 293 (2002) 98-102.
7. W.F. Bertsch, J.R. Azzi, A relative maximum in the decay of long-term delayed light emission from the photosynthetic apparatus, Biochim Biophys Acta 94 (1965) 15-26.
8. W. Schmidt, H. Senger, Long-term delayed luminescence in Scenedesmus obliquus I. Spectral and kinetic properties, Biochimica et Biophysica Acta 890 (1987) 15-22.
9. E. Hideg, M. Kobayashi, H. Inaba, The far red induced slow component of delayed light from chloroplasts is emitted from Photosystem II, Photosynthesis Research 29 (1991) 107-112.
10. A. Scordino, A. Triglia, F. Musumeci, Analogous features of delayed luminescence from Acetabularia acetabulum and some solid state systems, J. Photochem. Photobiol. B 56 (2000) 181-186.
11. B. Chwirot, R.S. Dygdala, S. Chwirot, Quasi-monochromatic-light-induced photon emission from microsporocytes of larch showing oscillation decay behaviour predicted by an electromagnetic model of differentiation, Cytobios 47 (1986) 137-146.
12. D.V. Parkhomtchouk, M. Yamamoto, super-delayed luminescence from biological tissues, Journal of International Society of Life Information Science 18 (2000), 413-417.
13. L. Drinovec, D. Drobne, I. Jerman, A. Zrimec, Delayed fluorescence of *Lemna minor*: A Biomarker of the effects of copper, cadmium, and zinc, Bull. Environ. Contam. Toxicol. 72 (2004) 896-902.
14. M. Monti, A. Zrimec, A. Beran, M.B. Zrimec, L. Drinovec, G. Kosi, F. Tamberlich, Delayed luminescence of *Prorocentrum minimum* under controlled conditions, Harmful Algae 4: 643-650, 2005.
15. S. Tudisco, F. Musumeci, A. Scordino, G. Privitera, Advanced research equipment for fast ultraweak luminescence analysis, Rev. Sci. Inst. 74 (2003) 4485-4490.
16. H. Niggli, S. Tudisco, G. Privitera, A. Applegate, A. Scordino, F. Musumeci, Laser ultraviolet-A induced ultraweak photon emission in mammalian cells, Journal of Biomedical Optics (2004) in press.
17. Richter, G. Stoffwechselphysiologie der Pflanzen. Thieme (1998), Stuttgart.
18. W. Nagl, F.A. Popp, A physical (electromagnetic) model of differentiation, Cytobios 37 (1983) 45-62, 71-83.
19. F.A. Popp, K.H. Li, W. Nagl, A thermodynamic approach to the temperature response of biological systems as demonstrated by low level luminescence of cucumber seedlings, Z. Pflanzenphysiol. 111 (1984) 1-13.
20. J. Slawinski, F.A. Popp, Temperature hysteresis of low level luminescence from plants and its thermodynamical analysis, J.Plant Physiol. 130 (1987) 111-123.

6
HUMAN GENOME REALIZATION AT THE VIEWPOINT OF PHYSICS OF THE ALIVE

S.P. Sit`ko*

Quantum medicine rests upon the belief that understanding of the essence of the alive in its distinction from the non-alive must serve as a prerequisite for medical treatment, or "rendering aid to people".

Just this belief was introduced into quantum medicine by its theoretical basis, i.e., Physics of the Alive – a new trend of natural science that has turned biology and medicine from empirical into fundamental science. It is expedient to remind here that nowadays there exists a strict definition of the notion of fundamentality in natural sciences. They are the sciences in which the objects of investigation have discrete spectra of characteristic eigenfrequencies. Before the discovery of "manifestation of characteristic eigenfrequencies of a human organism" [1], that is, before the time when ideas of physics of the alive have been formed, there were three such sciences: nuclear, atomic, and molecular physics.

I believe Weisskopf [2] was the first one who has drawn attention of the scientific world community to the fact that just the principles of quantum mechanics, i.e., the principles of identity and discreteness, and also existence of characteristic eigenfrequencies related to them, ensure diverse stability of the world at nuclear, atomic, and molecular levels of the matter self-organization. Weisskopf introduced the notion about three stages of quantum organization of nature or, as it is often said, three steps of Weisskopf's Quantum Ladder. Guided by the well-known facts of the levels overlap in the energy spectra of manymolecules structures (due to the screening mechanism and close connections in solid bodies and liquids), Weisskopf has guessed that the third molecular level was the last level of quantum organization of nature, and molecular physics was the third and the last fundamental science, respectively.

At the same time, in nature, besides nuclea, atoms, and molecules, there is also at least one more class of objects that are characterized by diversified differential stability as well. There are the living beings. Life is not a substance that constantly varies its form and structure as "the ocean of life" in the well-known film "Solaris". The earth, water, air are inhabited by quite discrete representatives of flora and fauna. There are their species, genera, particular individuals. Their similarities and differences are stable in time: at any continent we distinguish cats, dogs, sparrows.

* Scientific Research Center of Quantum Medicine "VIDHUK", Kyiv, Ukraine, 01033 Volodymyrska str. 61-B. E-mail: sitko@i.kiev.ua

We, I mean the humans, are also much alike to each other, but each of us has individual features of appearance which remain unchanged so that we recognize ourselves (in a mirror) and our acquaintances when we see them. Thus there arises a temptation to explain diverse differential stability of the living by the same principles of quantum mechanics, i.e., the principles of identity and discreteness and, accordingly, to consider the living systems as the whole quantum-mechanical entities[1].

Microscopic dimensions do not serve as the necessary condition for quantum mechanics application. The presence of macroscopic quantum effects testifies to this fact: superfluidity, superconductivity, Josephson effect. Actually the necessary condition for quantum mechanics application is existence of the entire self-consistent potential in the system. The self-consistent potentials of the same type determine the existence of the objects which form the respective steps of quantum ladders.

In other words, **the necessary condition** for formation of the whole macroscopic quantum-mechanical entity is occurrence of the efficient long-range acting forces within a restricted energetic (frequency) range that would have created the coherent multimode fields of laser type in each entity.

The sufficient condition for existence of macroscopic quantum-mechanical entity at its own step of Weisskopf's quantum ladder is the availability of the mechanism of self-support of such types of fields, and of characteristic spectral composition defined by active centers, but certainly on condition of positive energy of their joining.

Such conditions are realized in the living systems.

Really so, as shown by Fröhlich [3], the frequencies of eigen-oscillations of cytoplasmic membranes of all living systems must lie within (10^{10}-10^{11}) Hz range. It means that this is the range where we can observe the effects of resonance amplification of selective modes related with the reaction to changes of spatial genome structures in the process of DNA replication, RNA transcription, proteins translation. In this context, of great importance is the existence of the so-called proton transport described by Mitchell [4], which consumes a considerable portion of metabolism energy of cells and which constantly maintains the great tension of electric field on cytoplasmic membranes (approximately 10^5 V/cm). Just this fact may (potentially) turn the cells (their membranes, to be more exact) into the active centers of formation and maintaining of coherent eigenfield of a body in millimeter range of electromagnetic waves.

However, with due regard for the fact that water prevails in chemical composition of human organism and this water intensively absorbs the mm-range electromagnetic radiation, so the necessary condition of occurrence of coherent modes generation is not sufficient as yet, though the favorable conditions exist ($h\nu << kT$). In this case, the relation of probability of induced transitions to spontaneous ones is much higher than unity ($P_{ind}/P_{spon} \sim kT/h\nu >> 1$) [7].

That is why the answer to the question whether the real situation in living organisms lies beyond the threshold of non-equilibrium phase transition to coherent state should be obtained by way of observation and research.

Such observations exist.

---

[1] We must do justice to Weisskopf: drawing schematically his quantum ladder, he has also drawn the fourth step with discrete energy levels – the level of life – as early as in 1972. With no comments, just as a foresight of a genius.

Several thousand years B.C., the Chinese men of wisdom, who have laid the foundation of what we call now the Ancient Chinese Medicine or acupuncture, were guided by the ideas that the internal organs of a man are intersected by the lines, the so-called meridians (channels), whose external tracks are situated at the surface of a body. There are 26 channels, twelve paired and two unpaired. The majority of biologically active points (BAP) or acupuncture points are situated just over them. These points are used for sticking the needles into them according to the needling technologies (by way of example 5).

Sceptical attitude of the official Western medicine toward the Ancient Chinese medicine in spite of undeniable achievements of the latter is related to the ideas concerned with the existence of a meridian network. The problem resides in the fact that channels are not observed at anatomic-morphological level, and the Western medicine based on the so-called chemical paradigm adheres to the visualization principle claiming that there actually exists and can be an object of scientific research in an organism only something that can he seen directly by an eye or with the help of a microscope. The origin of so primitive, at the first sight, ideas can be understood if we consider the history of development of Western science in general, and medicine and biology, in particular. The modern Western medicine had been forming in the middle ages staying under the pressure of religious dogmatism the canons of which in the struggle against heretics were defended by the Inquisition. The meticulous medical men were in a constant danger to be enlisted among the heretics. The most brave of them displayed their protest by spontaneous formation of primitive materialistic world outlook. In the struggle against official religious scholasticism, they shifted to positions of the extreme atheism, denying the very existence of God with the argument that "nobody saw him".

In my opinion, just this argument underlies the principle of visualization which has been considered the criterion of science in medicine and biology for many centuries.

During the same centuries, fundamental science studying the non-living nature expanded essentially our idea about it, in particular, due to the field concepts. And nowadays, even at domestic level nobody is surprised at the possibility to tune the radio or TV sets to a great number of stations or the possibility to chat by mobile telephone, though it is impossible in all these cases to "view" by an eye the information carriers.

As to the scientific notions, mankind enters the third millennium with strong realization of the idea that in the picture of the world a field and a substance are represented at the fundamental level as equal in rights.

It is worth noting that for several centuries, i.e., long before formation of quantum electrodynamics and physical vacuum concept, physics, being non-oppressed by ideological burden in contrast to medicine and biology, has been guided by the global principles that reflect the material unity of the world due to the existence of the effective long-range action and which underlie the laws of modern physics. I mean the principle of the least action (Maupertuis), the principle of the shortest optical path (Fermat), the least losses principle (for current), the principle of a system transfer to the lowest potential energy, etc.

It is difficult to imagine that not a single person in medicine and biology knew anything about it. Then a question arises: why is it considered the axiom that for a child birth nothing is needed apart from the union *in vitro* of a spermatozoid and an

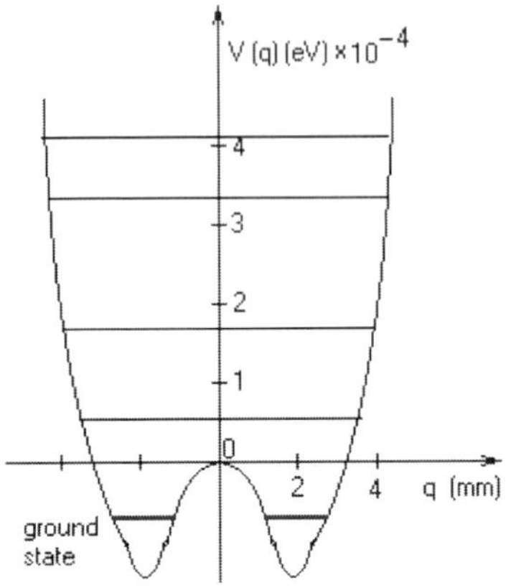

Figure 1a. Organism's ground state (health). Landau-Haken potential. $V(q)=kq^2/2+k_1q^4/4$ $(k<0,\ k_1>0)$

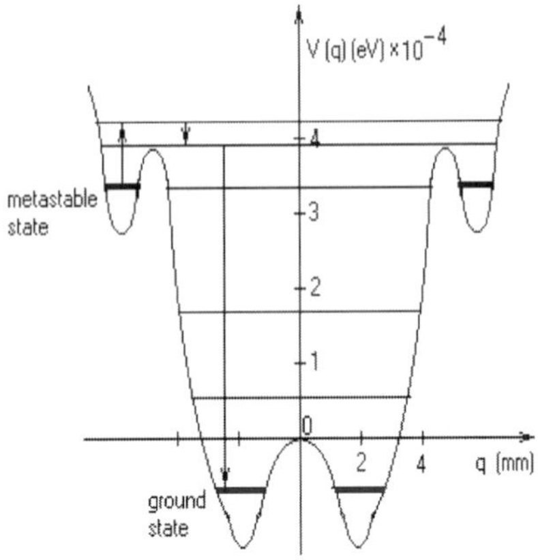

Figure 1b. Organism's metastable state (disease). Deformed Landau-Haken potential. The way out of metastable state is shown (treatment) with use of MRT.

ovum, for example? Or else, that it is necessary to look for the genes that are "responsible" for something? [6].

I am convinced that **the cause of such views is macroscopic dimensions of independently functioning living objects.**

Really, modern Western civilization was based on atomistic ideas of Democritus according to which cognition of nature must proceed by way of division of the macroscopic objects surrounding us into the smaller parts, up to the indivisible ones (atoms) and their study would give an answer to all questions. And though today our atoms are not the smallest objects of the microworld, the atomistic idea itself proved to be very fruitful and the achievements of Western civilization testify to this statement.

It should be recognized that the physicists who made revolution in natural science in the first decades of the past century have also contributed to consolidation of false idea that only in the microworld there occur the events having fundamental importance.

As it is known, the pretext for the above-mentioned revolution was impossibility to explain certain phenomena of the microworld by the laws of classical physics, and its consequence was the origin of quantum mechanics, the principles of which (identity and discreteness), as was noticed earlier, ensure the existence of three steps of Weisskopf's quantum ladder and, respectively, three fundamental sciences: nuclear, atomic, and molecular physics. It means that manymolecules objects having no discrete energy levels cannot carry the fundamental information.

In this way, beyond the interest of the fundamental science (with its field notions, virtual particles and photons, quantum transitions and metastable states, volume and length of coherence, etc.) there was left not only the whole macroscopic physics but the entire living world. It means that according to the standard notions, the integral living beings (the humans inclusive) must be studied within the scope of classical physics solely, painted with chemical reactions, and the phenomenon of life itself is a singularity that stays outside the science.

Thus, we are the witnesses of the absurd situation: all people have no doubts that the living differs from the dead, life from death, but dozens of biological and medical sciences the task of which must have been the support of life in its opposition to death were not imbued with the phenomenon of life, studying only its fragmentary signs.

Physics of the alive and quantum medicine have radically changed the situation. It became clear that an organism displayed all signs of the whole quantum-mechanical system, the ground state of which is health and metastable state - disease.

And really so, transition from the metastable to the ground state similarly to the three preceding steps of the quantum ladder is realized in a body during medical treatment by mm-range electromagnetic quanta the energy of which stimulates transition of the system from metastable state to such an excited state from where a cascade transition into the ground state (health) goes by the selected rules with the higher probability than the return to the metastable state (Fig. 1).

As is generally known [7, 8], the basic technology of quantum medicine is microwave resonance therapy (MRT), which makes use of the flows with spectral density (10^{-21}-10^{-20}) W/Hz·cm^2 in medical practice. This density corresponds to quite a few mm-range quanta.

In this way, the physician of quantum medicine working with superlow flows of mm-range electromagnetic radiation tries to do his best to implement the conditions depicted in Fig. 1b. At one of these "resonances" or therapeutic frequencies, the electromagnetic framework of a human returns to the ground state and as far as the framework is self-consistent with anatomic-morphological structure of a body, so such transition induces the process of adjustment of the anatomic-morphological structure of the restored framework, i.e., the process of cure starts. Taking into account that the organs and other morphological structures of a body cannot get reconstructed in a moment, there remains probability that with the lapse of time (several hours) the organism will return to metastable state though deformation of potential decreases and the state approaches to the one depicted in Fig. 1a. It should be noted that according to variation of the potential form, therapeutic frequencies may be changed in the following days of treatment, so the resonance "tuning" is necessary at each session. It is easy to notice that in a healthy organism there are no therapeutic frequencies and this fully corresponds to the practice of the MRT application.

Let us return to the question about formation of the coherent field of a body, existence of which in accordance with genome allows to perceive an organism as a whole quantum-mechanical entity.

The investigations showed that the maximal MRT efficiency is observed in those cases when the action of the source of the mm-range electromagnetic radiation is directed to biologically active points (BAP), of a body, which correspond to acupuncture points, and are located mostly, as it was noticed above, over the external tracks of the channels painted on sculptural images of a man by Chinese men of wisdom more than 5000 years ago.

I have already written that trajectories of the meridians do not have morphological peculiarities, i.e., they cannot be seen by eyes, that is why Western medicine denies their real existence in the belief that their only destination may be to help the physicians-needling therapeutists to find BAP on a human body.

We managed "to see" the channels [9].

They actually exist and really connect the fingertips of hands and legs with the internal organs, but not within a visible range seen by the eyes but just within mm-range of electromagnetic framework, coherent eigenfield of an organism, due to which there exist efficient long-range acting forces of an organism ensuring its quantum-mechanical entirety.

According to the ideas of physics of the alive, formation of a meridian system of a body begins during 14th week of embryo development. At this time cartilages harden and are turned into the bones, this is accompanied by spontaneous breaking of symmetry at fundamental level: the running waves are reflected from the nails thus forming dynamic interferential picture such as standing waves. This can be observed as a papilar picture at the fingertips of one's hands and legs [10]. The meridian system in the form of dynamic waveguides is formed due to reflection of the running waves from the bones, on the one hand, and, on the other hand, from the inside skin surface in the area of BAPs, positions of which on the surface of skin are defined with the places of falling of the running waves at angle of the complete internal reflection [10]. Stability of the meridian system during functioning of the joints is ensured by the obligatory presence of BAP in the center of flexions of each joint of the limbs.

The measurements carried out with the help of specially designed radiometric system with the level of the inherent noises $\sim 5 \cdot 10^{-23}$ W/Hz·cm^2 [8, 11] gave the possibility to obtain the important characteristics of the channels and BAP.

1. The channels have diameter (3-5) mm, at least at the spots of their nearing the surface in acupuncture points.

2. The refraction index inside the channel is the same as in atmosphere, that is $n = 1$, but not 5-6 as in the body outside the channel areas.

3. In case of functional disorders related to the concrete channel, at density of the external flux within the range of $(10^{-21}\text{-}10^{-20})$ W/Hz·cm^2, the respective acupuncture point completely absorbs this radiation, that is, the black-body mode is realized with the absence of reflection.

4. With the current density increasing up to 10^{-19} W/Hz·cm^2 and more, the situation changes in a triggering way – BAP completely reflects the external mm-radiation. (It can be suggested that just in this way life on the planet is preserved under condition of technogenic electromagnetic pollution of the environment within the life range, which is, in natural conditions, devoid of the sun effects due to the intensive absorption of mm-range electromagnetic waves by the atmosphere.)

The above-stated properties of the channels actually allow to consider them as dynamic waveguides along which light-exitons are running ensuring the coherence of the entire electromagnetic framework of a body [12]. Such an interpretation gives good reason to apply the electrodynamics laws in the attempts to understand the peculiarities of the metric scale used in ancient Chinese medicine.

It is known that the distance between acupuncture points along the external tracks of the channels in the ancient Chinese medicine is measured in the specific length unity – *cun*. One *cun* length is different with different people, because it is defined by anatomic characteristics of a particular organism. As a rule, in monographs concerned with acupuncture [5], to determine the *cun* length it is recommended to give due regard for certain anatomic peculiarities of a hand. Generally speaking, one *cun* is approximately the width of a thumb in the plane of a nail in a joint, that is, this value for the grown-up person with common anatomic proportions constitutes approximately 2.5 cm.

Let us turn our attention to the mechanism of formation of the field inside the channel considering the latter as dynamic cylindrical waveguide with diameters d = (3-5) mm and refraction index equal to unity (n = 1), i.e., the same as in the air (Fig. 2).

In the process of the standing wave formation along the waveguide (Z), at first the running wave is in motion. Axially symmetric problem is solved in cylindrical coordinates (p, z). Write down the wave equation:

$$\nabla^2 E - 1/c^2 \cdot \partial^2 E/\partial t^2 = 0. \qquad (1)$$

Its solution will be found as:

$$E = E_0 j_0 (k_\rho \cdot \rho) \exp (j(\omega t - k_z z)), \qquad (2)$$

where k_ρ and k_z are components of a wave vector in channel-waveguide along radius R and channel waveguide z, respectively, and $j_0 (k_\rho)$ is cylindrical Bessel function of the first degree of zero order (Fig. 3). Let us use the first root approximation: $k_\rho \cdot \rho_0 = 2{,}4$.

Taking into account that Bessel function is eigenfunction of Laplace equation, after substitution of (2) into (1) we obtain:

$$- k_\rho^2 - k_Z^2 + \omega^2/c^2 = 0; \qquad (3)$$

$$k_Z = \sqrt{(\omega/c)^2 - k_\rho^2}; \qquad (4)$$

$$k_Z = \sqrt{(\omega/c)^2 - (2{,}4/\rho_0)^2} \qquad (5)$$

where λ_0 is wave length in atmosphere, $\omega/c = k = 2\pi/\lambda_0$ is wave vector k in atmosphere.

By definition: $k_Z = 2\pi/\lambda_Z$, where λ_Z is wave length in a waveguide. Hence:

$$\lambda_Z = \lambda_0/[1-(\lambda_0/1{,}3 \cdot d)^2]^{1/2}.$$

For $\lambda_0 = (5\text{-}6)$ mm; $\lambda_Z = (5\text{-}6)$ cm.

In a standing wave formation, the distance between the maxima (and minima) equals the half of the wave length, i.e. (2.5-3) cm.

Certainly, the values of λ_Z given by formula (6) are very sensitive to the relation λ_0/d. But formation and support of the meridian proper in a body, in accordance with

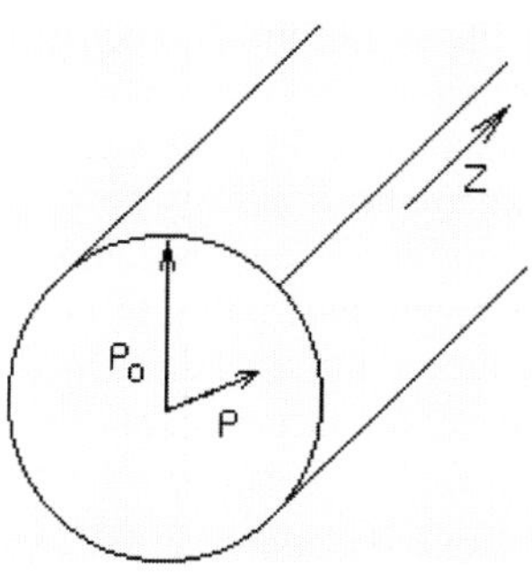

Figure 2. Scheme of the channel as dynamic waveguide $\rho_0 = d/2$.

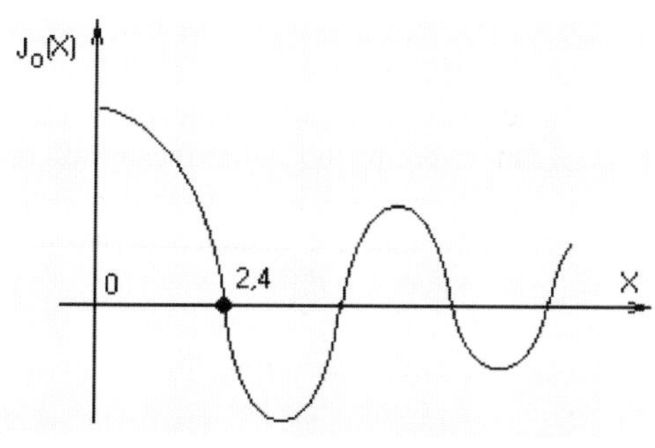

Figure 3. Cylindrical Bessel function of the first degree of zero order.

genome, represents a self-consistent process when depending on the state of an organism, the meridian diameter *d cun* "breath" by fractions of millimeter, preserving constant distance in *cun* between atomic-morphological structures; if we take into consideration the internal tracks of the channel, then, perhaps, it controls also the structure and form of the internal organs.

In this way, besides anatomic-morphological structures of a body, which we can see by eyes, there exists (actually exists, because it can be measured [11]) something that it is impossible to see – the so-called electromagnetic framework of a man or, to be more exact in scientific sense, the coherent eigenfield of a human in mm-range of electromagnetic waves. This field is formed owing to electromagnetic activity of each cell of a body, but having been formed, it coordinates, synchronizes, and directs the functioning of each organ, each structure of a body in a mother's womb and after the child birth during the whole life. Taking into account that genome of each somatic cell of a particular organism is the same, so just by way of formation and functioning of this coherent field, this electromagnetic framework, the genome is realized but not, as it was believed earlier, exclusively by way of chemical transformations within the cells (by cell division and proteins generation).

Apart from maintaining the organism's growth, these processes (DNA replication, RNA transcription, protein translation, etc.) are also realized for the vital requirements, i.e., for the case when an organism's coherent field does not match its anatomic-morphological structure. This happens in two cases.

The first case is related to the situation when the external factors (blows, falls, injury, etc.) break an organism's morphological structure and form mismatch between electromagnetic framework and its realization in a particular spot (for example, in a wound).

The second case is realized in the situation when under some extremely strong external stimuli, the coherent field gets deformed. It no longer corresponds to the genome and gradually imposes its deformation to the anatomic-morphological structure, which cannot be removed by the methods of medicamental therapy. In this way, chronical diseases arise.

The stated approach gives us the new attitude to solution of many well-known problems of biology. By way of example, let us consider two of them: the "garbage genes" problem and the wounds healing problem.

Existence of nearly 98% genes as if not participating in the hereditary information transfer is considered as one of the most painful paradoxes at the modern stage of biology development. This problem does not exist for Physics of the Alive, since it considers that all 100% of the genome's genes participate in formation of a body's coherent eigenfield in mm-range of electromagnetic waves (the electromagnetic framework). In conformity with the quantum mechanics laws, the potential wells of Landau-Haken type along the meridians, which at this approach are considered as Poincare's limit cycle, are filled with the energy levels. Transitions between these levels, in accordance with selection rules, form the spectrum of characteristic eigenfrequencies of a particular organism. The genome's hereditary information is retranslated just in this spectrum. It is (the spectrum) the universal passport of an organism and, as mentioned before, it is visualized in the form of papilar patterns on the soft flesh of fingers of hands and legs, which are (proceeding from positions of physics of the alive) nothing else but the dynamic interferential images (on concave screens) of the direct and reflected from the nails eigen-waves of an organism [10].

In a healthy organism whose quantum system is devoid of metastable states, the electromagnetic framework is self-consistent with anatomic-morphological structure. For maintaining of such a state, biochemical mechanisms of cell division and protein generation must switch on from time to time and in the definite spots of an organism, namely, in those where between the structure of a coherent field and its morphological realizations mismatch begins to exceed the definite threshold. This happens in the situations when even in natural conditions, life time of certain cells or tissues is restricted, for example, for epithelium tissues or erythrocytes. Let us remind that lifetime of the human erythrocytes constitutes 120 days and nearly 2.5 mln of them die and are generated again in a spleen and liver each second. For constant maintenance of these processes, the coherent field of an organism actually utilizes only very insignificant portion of the genome.

Quite a different picture must be realized during embryonal development (morphogenesis, forms creativity) and in post-natal period, in case of the damages of electromagnetic framework or injuries of anatomic-morphological structure of an organism. In all these situations, much greater part of the genome, up to one hundred percent[2], proves to be effectively actuated depending on the specific conditions in the chain "genome — coherent eigenfield — anatomic-morphological structure".

By way of example, consider a prosaic situation that can take place with everyone. You have cut a finger. Why is the wound healing? Why just that kind of tissue is formed that is required and in the quantity that is required and in such a way that if the cut is not very deep, then in some days even trace of it will disappear?

Despite the seeming simplicity of these questions, the answers on them are related with solution of one of the fundamental problems of biology – the problem of morphgenesis, form creativity, and differentiation of tissues. Within the boundaries of classical biology and linear physics, there were no answers to these questions, moreover, it was unclear how to get closer to their solution.

At the late years of the past century when the revolution in natural sciences has taken place, the situation changed. It was due to recognition of the importance of non-linearity and openness in formation of stable self-organized systems far from thermodynamic equilibrium. That is, the conditions of local entropy decrease became clear. Implementation of these ideas resulted in origin of two new sciences: synergetics and theory of dissipative structures.

Undoubtedly, B. M. Belintsev [13] was the one who applied the means of self-organization theory for development of the foundations of biological formativity and solution of the related problems in the most professional and fruitful way. Unfortunately, he remained within a chemical paradigm; he believed that the carrier of long-range forces were the so-called morphogenes, chemical substances formed by some elements of future organism during form creativity and absorbed by the others. This approach did not allow him to make a step toward understanding of the

---

[2] The real situation is much more complex. I understood this, preparing for publication the unpublished proceedings of my father, Prof. Sit`ko Panteleimon Onufrievich, Doctor of biological sciences, genetics scientist, on occasion of his 100 years birthday anniversary (b. 1906). To all appearances, he was the first to pay attention to so-called polygenomity of heredity, i.e., that there should be inherited not only the genome connected to DNA, but also division mechanisms (among them occurrence of division spindle, ensuring divergence of chromosomes in mitosis and meiosis), formation and functioning of mitochondria as energetic pool of cells, etc. So the development of exclusively chromosome heredity theory is of rather fragmentary, initial character.

living as the whole quantum-mechanical entity which is situated at the fourth step of the quantum ladder when self-consistent potential is formed in accordance with genome as coherent eigenfield of a body within millimeter range of electromagnetic waves. Just this definition lies at the basis of physics of the alive.

From these positions, the phenomenon of healing of the injured (cut) finger finds its **schematic explanation**. In the wound area, a certain number of cells were destroyed, but electromagnetic framework — coherent eigenfield of an organism remained – since it was created by billions and billions cells of an organism carrying the same hereditary information. The mismatch between the structure of a coherent field of a body (realized owing to the spectrum of its characteristic eigenfrequencies and which describes by the universal electromagnetic language all the details of a body structure and its functioning) and the deformed morphology at the injured spot initiates the standard and well-known mechanisms of cells division and generation of the particular proteins just at the injured spot (DNA replication, RNA transcription, the proteins translation). These processes must proceed under control of the electromagnetic framework until the mismatch between a framework (which gives what is necessary) and morphological structure at the injured spot becomes less than sensitivity threshold of the system realizing this mechanism of communication.

The expression "schematic explanation" was underlined earlier because I do not actually have a claim on description of the details of formative mechanisms. It is just a scheme as yet but the real scheme based on the modern scientific ideas [3, 9-25], medical-biological and physical measurements [26-31], the impressive clinical results [32-41] obtained during 20 years in the process of curing of hundreds of thousands of patients in many countries of the world.

So I hope that the above-stated, the new in principle, ideas about the nature of life which form physics of the alive and quantum medicine, will enable biology (and medicine as well) to overcome the prejudice the historical roots of which I described in this paper and which essentially hamper the development of the relevant sciences. I hope as well that planning of the further research in biology and medicine will be carried out with due regard for that stated above.

We discussed the process of healing, that is self-cure. But what can be done if the disease becomes chronic and is not cured by itself, or with the efforts of surgery and medicamental therapy. It was mentioned that this corresponds to the situation of disorder (deformation) of the electromagnetic framework itself.[3] Quantum medicine (and, respectively, its basic technology – microwave resonance therapy – MRT) are aimed at restoration of electromagnetic framework of a human. The patented technologies of diagnostics and quantum medicine therapy [42] allow for determining of disorders in these or that channels and for eliminating them.

As a rule, the course of treatment consists of 10-12 sessions, 45-70 minutes each. During this time the metastable state of the framework decreases so much that not a single self-organization level can be formed here. In other words, the framework of an organism is constantly in the ground potential well.

---

[3] I have no answer to the question why this occurs. The most natural is to suppose that the reason consists in the presence of strong external influence: blows, falls, operations, leaving the scars on a body through which the channels cannot pass [9], supercooling, starvation, stresses. However, under the same conditions in other cases, the framework is not disturbed, and the criterion for distinguishing of these conditions is not clear to me as yet.

The express-diagnostics methods [29, 30, 33, 40] used by us permit us to monitor the dynamics of treatment and to make the adequate corrections, if necessary.

The most impressive (even fantastic from the point of view of Western medicamental therapy) results are observed at the first session. At the moment of resonance which ensures returning of quantum system from metastable state to the ground state, the patient feels that his pain disappears practically instantly, the feeling of lightness, of imponderability, complete delight arises, in the closed eyes there appear dark blue, light blue, violet, green colours or bright white radiance.

Let me remind that MRT is monotherepy, i.e., its application envisages the complete rejection of any drugs or medicaments several days before the first session. Thus our patients pass to the new, non-chemical medicine.

It is important to underline that the procedures of BAP stimulation are realized by the powers carrying a few quanta (10^{-21}-10^{-20}) W/Hz·cm^2. The points of action are located, as a rule, in a distal way, on fingertips of hands and legs, sometimes in other places. To apply stimuli directly to the wound or painful area is prohibited categorically. The first and very important MRT rule is formulated as: never exert influence upon a focus of a disease.

It was found that subjective sensations of a patient are more reliable and efficient method of "tuning" to resonance. The thing is that human sensations have been formed as a result of action of millions of self-consistent structures of an organism, which ensure its functioning. The most reliable value herewith is the painful sensation. Adjustment to resonance aimed at removing the pain proved to be the important and obligatory prerequisite of successful treatment. The modern quantum medicine technologies (Sit`ko–MRT) ensure practically complete removing of pain even in case of the grave oncological patients when anesthetic drugs fail to help.

Unfortunately, this does not mean that in all cases of rendering help to very grave and "incurable" patients, we can save them from their disease and also from the consequences of their treatment by well-recognized methods of Western medicine: surgical operations, chemical therapy, irradiation. But almost always we manage to improve their quality of life: to prolong it maximally without taking drugs and other chemical preparations; to remove the pain allowing the patient to keep his dignity and to associate with his relatives and fellow men up to the last day. In this way, the objective reasons for discussion about euthanasia disappear.

As the treatment proceeds, and the depth of metastable well of an organism's self-consistent potential decreases, the probability of an organism's residence in this well also decreases, the averaged sensations during the session become less acute. In case of the complex potential restoration which corresponds to the healthy organism criterion according to our ideas, the "resonance sensations" disappear, which gives the reason to speak about experimental definition of a healthy person as such who does not respond in any way to the external mm-range electromagnetic radiation of superlow intensity.

As we see, even the first steps of the new sciences – physics of the alive and quantum medicine, based on the comprehension that life is the fourth fundamental level of quantum organization of nature, make it possible to approach the solution of global problems of biology and medicine in a new way. So it is a natural hope that practical medicine would take into account, as soon as possible, the new ideas about the nature of life and basing on these positions it would reconsider the available treatment technologies in order that the declared slogan: Medicine of the third millennium – "Life without pain" would be realized in the forthcoming ten years.

To conclude the paper, I would like to pay attention to philosophical aspect of the physics of the alive concepts.

I have underlined more than once that all the living beings are macroscopic quantum-mechanical entities that obey the laws of quantum mechanics (beginning with its principles).

At the same time, we get used to treat ourselves and other living beings surrounding us (people, dogs, cats, birds, etc.) as the common macroscopic entities that obey the laws of classical mechanics. According to these laws, we are moving, the forces of gravitation and inertia affect us, in a free state we are positioned with the minimum of the potential energy, our extremities and jaw bones work by the law of levers. Moreover, millions of chemical reactions in body structures take place in accordance with the laws of chemical transformations, just those, which can be observed and re-created outside a body. And what is the living organism – the quantum-mechanical entity, the object of classical mechanics or the extremely complex computer that defines the sequence of chemical transformations, generation of the adequate ingredients, etc.?

The first, the second, and the third, all at once. And something else above it. Under the words "something else above it" I imply "the special point" around which there occur the events inside a mother's womb related to formation of the electromagnetic framework. From mathematical viewpoint, this corresponds to Poincare solution of nonlinear differential equations as the limit cycles on the phase plane. During embryonic period and further on, during the whole life, there have been developed and sustained synergetic scenarios [43] which are called the dissipative structures hierarchies. And at different intersections of the cognition planes, they characterize the living as a whole quantum entity as well as a complex computerized factory and also as an object of classical mechanics.

"The special point" is and, I am convinced, will always be beyond the cognitive possibilities of science. Mechanical and chemical aspects of life are studied by the existing medical-biological sciences. Physics of the Alive and Quantum Medicine investigate the fundamental quantum-mechanical level of the living.

REFERENCES

1. Andreev Ye. A., Bely M. U., Sit`ko S. P. Manifestation of the Characteristic Eigenfrequvencies of Human Organism, *Dopovidi AN UkrSSR, B,* (1984), № 10, P. 56-59 (in Ukrainian).
2. Weisskopf V. F., 1972, *Physics in the Twentieth Century: Selected Esseys*; The MIT Press, Cambridge, Massachusetts, and London, England.
3. Fröhlich H. (Ed.), *Biological Coherence and Response to External Stimuli*, (Springer-Verlag, Berlin, Heidelberg, New York, London, Paris, Tokyo; 1988), p. 268.
4. Mitchell P., *Chemiosmotic Coupling and Energy Transduction*, (1968).
5. Tabejeva D. Ì. *The Handbook on Acupuncture*, (Moscow, Medicine, 1980) (in Russian).
6. Sit`ko S. P. "The Gene Responsible for ..." – Anthropomorphism or a Tribute to Primitivism?, *Physics of the Alive*, V. 11(1), 2003; P. 9-12.
7. Sit`ko S. P., Martchian L. N. *Introduction to Quantum Medicine*, (Kiev, "Pattern", 1994), P. 127.
8. Sit`ko S. P., Skripnik Yu. A., Yanenko A. F. *The Hardware of Modern Quantum Medicine Technologies*, (Kiev, "FADA, Ltd", 1999), P. 200 (in Russian).
9. Sit`ko S. P., Andreev E. A., Dobronravova I. S., The Whole as a Result of Self-Organization, *Journal of Biological Physics* v. 16 (1988), p. 71-73.
10. Sit`ko S. P., Gizko V. V., Towards a Quantum Physics of the Living State, *Journal of Biological Physics*, v. 18, №1 (1991), p. 1-10.
11. *Physics of the Alive* v. 6, №1 (1998).

12. Sit`ko S. P., Tsviliy V. P., Electrodynamic Model of the Human Organism's Electromagnetic Frame, *Physics of the Alive*, v.5, №1 (1997), p. 5-8.
13. Belintsev B. N. *Physical Fundamentals of Biological Formacreation*, (Moscow, "Nauka", 1991), P. 252.
14. Ñhaly A. V., Non-Equilibrium Phase Transitions in Living Systems Affected by Low-Intensive Microwave Radiation, *Physics of the Alive*, v.1, №1 (1993) p. 81-92.
15. Chaly A. V., Dobronravova I. S., Sit`ko S. P. Synergetics and Phase Trensitions: Mounting the Quantum Ladder of Nature, *Physics of the Alive*, v.2, №1 (1994), p. 5-11.
16. Sit`ko S. P., Tsvitiy V. P., "Space-Time Structures" of Synergetics in Physical Terms of Quantum Mechanics, *Physics of the Alive*, v.7, №1 (1999), p. 5-11.
17. Ermakov V. N., Ponezha E. A., Modeling of Microwave Radiation Action on Alive Systems by Nonlinear Resonant Tunneling, *Physics of the Alive*, v.10, №1 (2002), p. 16-25.
18. Brizhik L., Davydov Solitons and Physics of the Alive, *Physics of the Alive*, v.10, №2 (2002), p. 6-30.
19. Serikov A. A., On the Role of Molecular Ensembles in Primary Reception of Micro-wave Radiation by Biosystem, *Physics of the Alive*, v.1, №1 (1993) p. 62-71.
20. Lisitsa M. P., Sit`ko S. P. One More Mystery About Relict Radiation?, *Ukr. Phys. Journal*, V. 39(9,10) (1994), P. 973-975 (in Ukrainian).
21. Alipov Le. D., Belyaev I. Ya. et al., Experimental Justification for Generality of Resonant Response of Procaryotic end Eucaryotic Cells to Mm Waves of Superlow Intensity, *Physics of the Alive*, v.1, №1, (1993), p. 72-80.
22. Sit`ko S. P., Suhakov V. P. The Role of the Spine States of Proteins Molecules, *Dopovidi AN Ukraine*, À, № 6, (1984), P. 63-64 (in Ukrainian).
23. Sit`ko S. P., The Physical Sense of Schrodinger Equation in the Context of the Synergetics Conseption, *Dopovidi AN Ukraine*, À, 10 (1993), p. 98-101.
24. Sit`ko S.P., Life as a Fourth level of Quantum Organization of Nature, *Proceeding of the International Workhop: Energy and Information Transfer in Biological Systems*, Acireale, Catania, Italy, 18-22 September 2002, (World Scientific, New Jersey, London - Singapore - Hong Kong, 2002), p. 293-307.
25. Brizhik L. S., Eremko A. A. Soliton Induced Electromagnetic Radiation and Selfregulation of Metabolic Procesess, *Physics of the Alive*, V. 9, № 1 (2001), p. 5-11 (in Ukrainian).
26. Kozakova L. G., Svetlova S. U., Subbotina T. I., Yashyn A. A. Morphological and Biological Analysis of Marrow Blood Creation for Rats Under Low Intensive Electromagnetic SHF-Radiation, *Bulletin of New Medical Technologies*, (Toula, 1999), V. VI, N 3-4, p. 39-41 (in Russian).
27. Skripnik Yu. A. et al. *Microwave Resonance Radiometric of Physical and Biological Objects* (Edited by professor Yu. A. Skripnik), (Zhytomyr, "Volyn", 2003), p. 406 (in Russian).
28. Yanenko A. F. Microwave Radiometry as Instrumental Basis of Physics of the Alive and Quantum Medicine, *Physics of the Alive*, V. 7, № 1 (1999), p. 12-18 (in Russian).
29. Ivanchenko I. A. et al. Application of Differential SHF-Reflectometry Method for Research Polarizing Properties Acupuncturic Points, *Physics of the Alive*, V. 8, № 2 (2000), p. 52-62 (in Russian).
30. Phlinkel L. S., Sit`ko S. P. Statistical Approach to the Representation of Clinically Observed Organism States as Observable of the Heisenberg Quantum - Mechanical Formalism, *Physics of the Alive*, v.1, №1 (1993), p. 132-140.
31. Ponezha G. V., Ponezha S. G., Nizhelskaya A. I. Physical Aspects of Measurements of the Microwave Emission From Human Body, *Physics of the Alive*, V. 9, № 2 (2001), p. 33-54 (in Russian).
32. Binyashevsky Ye. V. et al. *Collection of Methodological Recommendations and Legal Acts of Microwave Resonance Therapy*, (Kyiv, "Oberig", 1992), ISBN 5-87168-014-3 (in Russian).
33. *The Cytobiophisical Methods of Estimating the Condition of Humen Organism Using in the Practice of Microwave Resonance Therapy* (Methodical Recommendations), Approved by the Ministry of Public Health of Ukraine 18.05.2001, Kyiv 2001, page 12, MH of Ukraine, Devised by Kharkiv National University and "SRC" of Quantum Medicine "Vidhuk" MH of Ukraine (in Ukrainian).
34. Grubnik B. P., Chayalo P. P. Clinical Criteria of Evaluation of Microwave Resonance Therapy Efficiency, *Physics of the Alive*, V. 11, № 2 (2003), p. 95-100 (in Ukrainian).
35. Moskalenko V. F., Sit`ko S. P., Gorban' Ye. M., Grubnyk B. P. Yanenko O. P. Quantum Medicine: from Fundamental Theories to Practical Use, Ukrainian Medical Bulletin, (2002), N2, p. 106-109 (in Ukrainian).
36. Sit`ko S. P. (consultant), Mkrtchian L. N. et al., "Physics of the Alive" in Medico-Biological Aspect, *Physics of the Alive*, v.1, №1 (1993), p. 110-131.
37. Grubnik B. P., Sit`ko S. P., Shalimov A. A. Experience of Using Sit`ko-MRT Technology for Rehabilitation of III-IV Stage Oncologic Patients, *Physics of the Alive*, V. 5, № 1 (1997), p. 90-95 (in Russian).
38. Teppone M. V. *SHF-Puncture (Super High Frequency Puncture*, (Moscow, "Logos", 1997), 308 p. (in Russian).
39. Gayko G. V. et al. *Variety Estimation of the Application of Sit`ko-MRT (Microwave Resonance Therapy) Technologies in Complex Treatment of the Osteomyelitis* in the book *"Renewal Surgery of Destructive Forms of the Bone-Articulate Tuberculosis and Osteomyelitis and its After-Effects*", V. II, part 9, p. 295-306, (Kyiv, Knyga plus, 2002) (in Ukrainian).
40. Bundyuk L. S., Grubnik B. P., Nikishina N. G., Sit`ko S. P., Shakhbazov V. G. Clinical Meaning of Intracellular Microelectrophoresis in Thechnology of Microwave Resonance Therapy, *Physics of the Alive*, V. 9, № 1 (2001), p. 58-66 (in Ukrainian).
41. *Microwave Resonance Therapy in Family Doctor Practice. (Medical Recommendations)*, (Kyiv 2004), MH of Ukraine (in Ukrainian).
42. *Sit`ko S. P. "Microwave* Resonance Therapy Method by Sit`ko S. P." Ukraine Patent N2615 dated 15.03.1994 (in Ukrainian); Russian Federation Patent N2053757 dated 10.02.1996 (in Russian); Sit`ko S. P. Microwave Resonance Therapy, US Patent N 5.507.791, Apr. 16, 1996.
43. Dobronravova I. S. Physics of the Alive as a Phenomenon of Postnoneclassical Science, *Physics of the Alive*, V. 9, № 1 (2001), p. 85-95 (in Russian).

7
FUNDAMENTAL ROLE OF WATER IN BIOENERGETICS

Vladimir L. Voeikov*

Abstract: Water plays the key role in generation, transformation, and utilization of energy for the realization of biological functions. Its direct involvement in hydrolytic processes in which primary "fuels" are produced, in ATP synthesis and energy gain due to ATP hydrolysis is well known, but not appreciated. Recently, water became known to be one of the major sources of high grade energy – energy of electronic excitation (EEE) generated in the reactions in which active oxygen participates. Due to quazi-polymeric properties of interfacial water, it may transform low grade into high grade energy. Besides, singlet oxygen may directly oxidize water, in structured environment providing for the generation of EEE, which may "spark" other energy donating processes. EEE may also be used locally for the performance of different kinds of chemical and physical work; it may accumulate and pool in aqueous systems and migrate within them without dissipation on macroscopic distances. Slow combustion in water and combustion of water is capable to self-organization in space and time expressed in the development of oscillatory-wave regimes of these processes serving as time-keepers of other biochemical processes dependent on them as well as sensitive antennas for external oscillatory signals.

1. INTRODUCTION

A human being may survive for 40 days without food, no more than for 5 days without water, and only for few minutes without breathing. Thus the role of two simple substances, water and oxygen for life support, is much more important than of thousands of complex molecules that food contains. Certainly full life is impossible without the latter, but still it is based on water and oxygen.

It seems clear why overwhelming majority of living organisms critically depend on oxygen: it burns food to get energy for the performance of all vital

* Faculty of Biology, Lomonosov Moscow State University, Moscow, Russia, 119234.
E-mail: vvl@soil.msu.ru

processes. It is less clear why we so depend on water taking into consideration that any organism contains plenty of it. It is usually said that water content of an adult person is around 70% by weight. But in terms of chemistry this figure is inappropriate – chemistry deals not with kilograms, but with moles. Molarity of water in water is 55.6 M. In the most "liquid" tissue – blood, sodiumchloride is the next most abundant substance subsequent to water. Its molarity in plasma is only 0.15 M. Molarity of the most important low molecular weight organic compound in blood, glucose, does not exceed few millimolar and of the most abundant protein, albumin, is approximately 0.00005 M. Even in erythrocyte that is packed with haemoglobin, concentration of the latter does not exceed 0.01 M. Thus contribution of water into chemical composition of blood exceeds 99%. If water was just a solvent for other biomolecules, loss of few percent of it should not practically change its overall concentration. Nevertheless, it is well-known how dangerous even mild dehydration for the state of health of a person is [1]. On the other hand concentration of the majority of specific bioorganic molecules may vary and do vary manifold in the course of normal metabolism. Crucial dependence of an organism on the constancy of water contents in it indicates that water should be no less important for all vital activities than all other biomolecules that it embraces.

Albert Szent-Gyorgyi has noted long ago, "The cell is a machine driven by energy. It can thus be approached by studying matter or by studying energy" [2]. Present-day biology focuses on profound study of the tiniest details of material particles representing admixtures in the organismal water. Much less attention is paid to studying energy that drives incessant and directed transformation and renewal of all material particles of which a living matter consist. And even more sorrowful that practically no attention has been paid until recently to the basic matrix in which bioorganic and bioinorganic "solid" particles are imbedded – to water whereas water plays the primary role in metabolism (literally – "exchange of substances") – a necessary condition of living state. When metabolism comes to an end, a living organism turns into a dead body. Water participates in both indivisible aspects of metabolism – in exchange and transformation of substances, including water itself, and in energy supply and transduction for the persistent flow of former. Here we'll reason that both structural and dynamical aspects of life are based on unique water virtues and that studying water is a necessary condition for the better understanding a living cell, an organism, as a peculiar dynamic material organisation driven by energy.

2. WATER – A DYNAMICAL STRUCTURE

There is growing understanding that water can not be regarded as some unstructured "liquid gas". Many models of water structures are put forward [3, 4, 5, 6, for more references see 7]. In "real" water, structuring is expressed much more than in ideal ultra-pure boundless water because of the contribution of multiple interfaces. They include an interface between bulk water and walls of a vessel, a water/air interface, interfaces with gas bubbles and other substances dissolved and suspended in water. Vicinal water with special properties may extend very far from the interface which it solvates [8]. For example, many layers of structured water extend beyond the initial monolayer hydrating a protein surface, and induced protein conformational change modifies the extent of non-ideally behaved water [9]. Several resilient water molecular layers close to the surface of a solid material immersed in water were detected using atomic force microscope [10]. It was shown by subfemtosecond x-ray absorption spectroscopy that liquid water in a first coordination shell of ice consists of structures with two strong hydrogen bonds of each molecule to its neighbours, resulting in water chains and rings [11]. Potentiality of water to form linear polymer-like associations has remarkable consequences for both biological shaping (not to be considered here) and for its participation in bioenergetics – to be discussed later. But in general ability of water to form bulky rather monolayer structures around the surfaces it hydrates necessitates to reconsider many long-term concepts of biochemistry based on statistical laws of classical chemistry (e.g., free 3-dimentional diffusion of substances in living cells) and to acknowledge that physical interactions between interfacial water, macromolecules and low molecular weight organic and inorganic compounds could play a primordial role in vital processes [12].

Only recently understanding has come that high concentration of macromolecules in cells and intercellular matrix which was given a special term "macromolecular crowding" may drastically influence kinetics and even specificity of biochemical reactions in comparison to that observed in diluted *in vitro* reaction systems [13]. A significant part of water in a cell is to different extent "bound" and unavailable for classical diffusion of biomolecules due to crowding. On the other hand it is well-known that the rates of biochemical reactions in a cell may exceed those *in vitro* hundreds and thousands fold and many of them proceed simultaneously in a tiny volume of a cell. How to resolve this contradiction between "structure" and "dynamics"?

Actually there is no such contradiction if to recall that there exist dynamical structures such as a vortex or a tornado that possess their special form and stability only due to flow of a substance, in particular of water through space. The tornado example is especially impressive, as its power, the power of a structure existsing only due to unrestricted input and output of material particles through space, is so powerful that it easily destroys seemingly robust stationary structures. This analogy could be considered artificial in relation to living structures if recently it has not been shown using molecular dynamics simulation that coherent patterns of water (nano-vortexes) may persist in locales defined as ''site-dipole field'' much longer than orientation memory of individual molecules persists [14].

Thus, when the notion of water structure is mentioned, one should consider that these structures may be extremely versatile. Rapid development of studies of water structures using computer simulation as well as sophisticated experimental techniques calls for the necessity of the emergence of classification of water structures important for the performance of particular vital processes. This classification is still lacking but intuitively it becomes clear that water of living things may exist and coexist in at least few alternative forms, for example layered and fibril water [e.g., 8, 11] vs. "globular" water [e.g., 5], "high density" water vs. "low density" water [15] and superposition of these forms. All these forms participate in metabolism including quite a few alternative and complementary vital processes: catabolism (degradation of a substance) and anabolism (synthesis of new material particles); energy gain and storage vs. energy utilization.

3. WATER PARTICIPATION IN CLASSICAL METABOLISM

Water participation in major biochemical processes is well-known but until recently it has not been appreciated to its true worth. In fact different types of processes need different waters. For example food digestion and innumerous biochemical reactions in an organism represent hydrolysis of macromolecular and low molecular weight substances. For hydrolysis to be realized, a water molecule should split in two parts. Thus efficient hydrolysis depends not only on chemical structures of hydrolyzed molecules, on the efficiency of enzymes that catalyze it, but on the availability of a particular water form that is appropriate for the hydrolysis.

Hydrolysis among other purposes provides building bricks for the synthesis of new molecules. New biopolymer synthesis proceeds as the reaction of polycondensation: when the new brick sticks to a growing polymer chain, a water molecule is released into the surrounding. This chemical reaction is opposite to hydrolysis. The question how this process may be realized in a medium where water molecules represent the overwhelming majority of any other molecules is usually omitted by biochemical manuals though it should seem not too advantageous to push a newly appeared water molecule into surrounding water unless if most part of water in a locale where polymers are synthesized is not free but somehow bound. Alternatively, water should have much higher chemical activity (be more free) in a locale where hydrolysis occurs. To our knowledge this idea was initially suggested by P.M. Wiggins [16].

Energy supply for all vital processes also needs immediate participation of water. For example, synthesis of generally recognized universal "energy currency" of any living cell – an ATP molecule from ADP and orthophosphate is accompanied by a water molecule release, while its splitting at which energy is gained is in fact a reaction of hydrolysis. It is logical to suppose that water organization in locales where ATP is synthesized and where it is utilized is different.

On the other hand how exactly energy value of this "currency" is yielded is still unclear, though credible models suggesting participation of water organized by specific biopolymer surfaces in this mechanism has been suggested long ago [17, see also 18]. The role of aqueous environment in both ATP synthesis and utilization with yield of energy is considered also by P.T. Wiggins who suggests that water may exist in two states – "high density" and "low density" state [19]. She experimentally proved [20] that ATP is synthesized from ADP and Pi in dense cellulose acetate membranes in the presence of KCl, but not NaCl – under the conditions favoring low density water state. Further details of this surprising mechanism and of biological significance of two states of water may be found in [21].

Thus water plays a key role in already well-known metabolic processes related both to substance transformation and energy transduction. However, energy transduction process related to ATP synthesis and energy yield from it is by no means the only one significant for general bioenergetics of living organisms. There exist more deep levels of energy acquisition and processing and water actively participates in them.

4. PUTREFACTION AND COMBUSTION

According to the current concept of bioenergetics, the overwhelming majority of living organisms gain energy from food burning by oxygen. In a simplified form of this concept specific dehydrogenases abstract "hot" electrons (plus protons) from "fuel" (sugars and fats) and transfer them to NAD^+ and $NADP^+$. Reduced forms of these carriers donate electrons to the respiratory chain in mitochondria, where their energy is released stepwise while they pass downhill from one redox center to another and is used for the synthesis of ATP. Oxygen here is the final acceptor (a "trash box") of electrons that had exhausted most part of their redox potential. As energy portions released in mitochondrial oxidation are equivalent to quanta of middle-far IR-photons (≤0.5 eV), this process is analogous to SMOLDERING COMBUSTION.

An alternative form of energy gain from oxygen-dependent oxidation is genuine COMBUSTION when direct one-electron oxygen reduction occurs, and quanta of energy equivalent to energy of visible and even UV-photons (>1 eV) are generated. One of the classical examples of combustion is direct oxygenation of hydrogen resulting in water production and at which high density energy is released. Combustion, in particular combustion of hydrogen is not commonly considered as relevant for bioenergetics. However, a lot of evidence argues that it should be taken into account as one of the most fundamental processes ensuring vital activity with high grade and well ordered energy.

Generally energy may be characterized by quantity and by qualities (forms, levels, and orderliness). Biomedical community is very much preoccupied with energy quantity rather than with its qualities while usefulness of energy for the performance of particular work is in the first place determined by its quality, in particular by the level to which it belongs. Levels of energy are subdivided into translational (energy associated with the motion of a molecule in space), rotational and vibrational energy of parts of di- and many-atomic molecules. The highest level of energy relevant to further discussion is energy of electronic excitation (EEE). As energy may be in principle transformed from one form into another, for example, mechanical energy may be transformed into electrical energy and vice versa, it is convenient to compare different levels of energy as belonging to different regions of electromagnetic spectra. Low level translational, rotational and vibrational energy belong to IR-part of electromagnetic spectrum (from $\sim 10^{13}$ Hz to $\sim 10^{14}$ Hz), while energy of electronic excitation covers the range of visible light – from 3.8×10^{14} (deep red) to 7.9×10^{14} Hz (violet) and transcends into UV-light region having higher frequencies. The higher is the frequency of oscillations the higher is energy density. High density energy is the energy of higher "quality" – it is easier transmitted to long distances, easier transformed into different forms of energy, more different types of work may be performed by the same quantity of high density energy that of low density energy. By the way progress of human civilization may be traced by the ability of human beings utilize energy of higher and higher density. Distinguished biologists Alexander Gurwitsch and Albert Szent-Giorgyi many decades ago already insisted that energy of electronic excitation should play a noteworthy role in bioenergetics.

It turns out that water in not less significant for realization of this form of energy generation and utilization as in classical bioenergetic processes. A. Szent-Gyorgyi was probably the first to argue that "bioenergetics is but a special aspect of water chemistry" and that "... water arranges an indivisible system with the structure elements (of a cell) making possible electronic excitations which otherwise are highly improbable... in structured water electronic excitation may be surprisingly long-living, and this may be of a paramount importance for the biological energy transfer" [22]. However, the idea of the importance of EEE for bioenergetics besides specialized biological functions such as photosynthesis and vision is not still sufficiently absorbed by biological community probably because the link between water properties, energy of electronic excitation and particular biochemical processes has not been yet clearly formulated.

The first question that needs to be answered is about the source of energy of electronic excitation, of quanta of energy equivalent to photons of visible and UV-light under mild conditions of living systems (temperature, pressure, pH, aqueous milieu). In fact, A. Gurwitsch has discovered more than 70 years ago that all living systems are the source of such hot photons as photons of UV-light (although of ultra-weak intensity), and suggested as early as in 1930s that this energy may be provided by free radical reactions. During the past half century, thousands of papers were devoted to the studies of "chemiluminescent" reactions of oxygen free radicals under conditions characteristic for living organisms. Still, most biologists consider such reactions abnormal in comparison to "normal" enzyme catalyzed biochemical reactions. Common opinion is that free radicals arise from malfunctioning of mitochondrial electron transport chain or such derogative reactions as lipid peroxidation.

But the situation is rapidly changing. It becomes more and more clear that oxygen free radicals and other reactive oxygen species are regular participants of normal metabolism; moreover, they turn out to be key participants of multiple bioregulatory processes. A lot of enzymes that specifically catalyze ROS generation are ubiquitously present in living cells, there are quite a few biochemical reactions in course of which ROS are inevitably generated (see below). However, the role of water in ROS production and generation of energy of electronic excitation in the course of their reactions has been considered only very recently.

5. ROS GENERATION IS AN INTRINSIC PROPERTY OF WATER

5.1. Water – A Transformer of Low Density Energy into High Density Energy

Above it has been argued that significant part of organismal water is interfacial and that such form of water has features of polymeric substance, in particular of linear polymers. Therefore mechanochemical phenomena are expected to take place in it. Polymers can undergo chemical transformations under the action of mechanical impacts, freezing-thawing and fast temperature variations, action of audible sound and ultrasound, and other low density energy forces that are too weak to induce chemical reactions in monomers or short oligomers. If macromolecules in polymers or their solutions are reluctant to shift along each other due to weak but multiple intermolecular bonds they may accumulate and concentrate mechanical energy to densities that comprise energy quanta enough to excite and break down internal covalent bonds in polymers. That means unpairing of electrons and appearance of a pair of free radicals followed with multiple chemical and physical consequences [23].

Basing on the presumption that liquid water contains quazi-polymeric structures, the team of Russian physicists headed by G.A. Domrachev started more then 15 years ago to investigate the effects of low density energy physical factors on homolytic water dissociation (H—O—H → HO• + •H, cf. ionic water dissociation: H—O—H → H^+ + OH^-). They estimated augmentation of hydrogen peroxide concentration in water because the most probable explanation for appearance *de novo* of H_2O_2 is recombination of HO• radicals arising in homolytic water dissociation (HO• + •OH → H_2O_2). It was shown that water freezing-thawing, evaporation-condensation, sonication even with audible sound, filtration through narrow capillaries resulted in an increase of H_2O_2 even in ultra-pure and carefully degassed water. Efficiency of water splitting resulting from evaporation/condensation and freezing/thawing is ~10 times as effective, sonolysis ~70 times and water filtration through narrow capillaries – more than 100 times as effective as its photodissociation with far UV-light [24, 25]. Yield of H_2O_2 in water containing common ions and dissolved oxygen was much higher. What is notable, H_2O_2 concentration continued to grow for some time after resumption of any treatment. About 3% of all energy used for viscous flow of water through capillaries with diameter of 0.2 mkm was used for water splitting.

Japanese authors who were looking for the new way to produce hydrogen by water splitting have shown that powders of NiO, Cu_2O, Fe_3O_4 suspended in distilled water by magnetic stirring, catalytically decompose it into H_2 and O_2. Efficiency of the mechanical-to-chemical energy conversion under these very mild conditions exceeded 4% [26]. Here water splits to the final products because presumably metal oxides instantaneously decompose intermediate peroxides.

In case if a water molecule has dissociated as a mechanically excited polymeric entity:

$$(H_2O)_n(H_2O...H\text{-}|\text{-}OH)\,(H_2O)_m \rightarrow (H_2O)_{n+1}(H\downarrow) + (\uparrow OH)\,(H_2O)_m \quad (1)$$

the initial products of water splitting are free radicals H↓ and ↑OH (here we symbolize a given electron as ↑ or ↓ to stress their alternative spin states). Indeed, if water is in an apparent rest this singlet pair of radicals readily recombines back to water:

$$H\downarrow + \uparrow OH \rightarrow H_2O \quad (2)$$

However, even in such a case this is not just a reverse, equilibrium reaction because water splitting has been achieved under the action of mechanical forces while back recombination of radicals gains an energy quantum of 5.2 eV. In condensed and organized media (such as water), long-range energy transfer of electronic and vibrational excitation has been demonstrated already in the 1930s and 1940s by J. Perrin, S. Vavilov, Th. Foerster, and others. This phenomenon was confirmed with new techniques recently [27].

The probability of radicals to move away of each other significantly increases in "real" water, in which dissolved gases and other molecules and particles are present, especially in cases when multiple layers of water are organized by surfaces which it hydrates and when these layers move along each other with different rates (consider a vortex as an example). This is proved by aforecited data on of the appearance of H_2O_2 in water filtered through narrow capillaries and H_2 and O_2 in water stirred in the presence of metal oxides. Here the following reactions may proceed:

$$HO\uparrow + HO\downarrow \rightarrow H_2O_2 \quad (3)$$

$$H\uparrow + \downarrow H \rightarrow H_2; \quad (4)$$

$$H\bullet + O_2 \rightarrow HO_2\bullet; \quad (5)$$

$$HO_2\uparrow + HO_2\downarrow \rightarrow H_2O_2 + O_2; \quad (6)$$

$$2H_2O_2 \rightarrow 2H_2O + O_2 \quad (7)$$

Most important of them are the reactions #6 and #7 in which oxygen molecules are released. It should be reminded that O_2 is unique among other molecules because in its ground state its two electrons are unpaired [$O_2(\uparrow\downarrow)_2\uparrow\uparrow$ or $O_2(\uparrow\downarrow)_2\downarrow\downarrow$] (besides, an oxygen atom also has two unpaired electrons). Thus, oxygen molecule is a bi-radical (in fact it is a tetra-radical) and it represents a vast store of energy. But it is stable because the laws of quantum physics forbid direct reactions of bi-radicals (they are called also particles in a triplet state) with molecules in which all electrons are paired (singlet state particles). That is why oxygen needs to be initially activated to release its energy reserve.

There are few ways for O_2 activation. It may be excited by an appropriate energy quantum (≥ 1 eV) and turn into a highly reactive singlet oxygen ($O_2(\uparrow\downarrow)$, its another symbol, $^1O_2$). A peculiar feature of O_2 is that electronically excited singlet oxygen may relax only to triplet state because oxygen unlike other substances does not have ground single state. As soon as singlet-triplet transition is "forbidden" by quantum physics laws, lifetime of excited singlet oxygen is usually much longer than that of any other molecules in an excited singlet state. Triplet O_2 may be also activated with transition metals because in their field its spin state is changed. Finally, triplet oxygen easily reacts with free radicals – atoms and molecular particles possessing an odd number of electrons on their valence orbital. In these reactions, oxygen gains or loses an electron, turns into a free radical that can easily take new electrons releasing large portions of energy at each consecutive step of one-electron reduction. Another peculiar feature of free radical reactions in which oxygen participates is that they may easily turn into branching (or run-away) process [28], and concentration of free radicals in a reaction mixture grows up exponentially until the rates of their production and annihilation equalize. That is why elevation of H_2O_2 yield in water equilibrated with air under the conditions favorable for its splitting occurs faster, continues longer after initial perturbation, and reaches higher levels than in degassed water.

Thereupon it is interesting to speculate that an outcome of water splitting may be significantly influenced by external magnetic fields. There are a lot of reports on the long-lasting effects of even a brief treatment of water with magnetic fields, though these effects are not easily reproduced. In principle, magnetic fields may modulate the outcome of free radical reactions. Initial radicals, as mentioned, emerge in a singlet form (H↑ + ↓OH) and they may easily recombine back into water. Under the action of a magnetic field singlet-triplet transition (H↑ + ↓OH → H↓ + ↓OH) may occur. This prohibits recombination of the radicals favoring the development of the array of reactions 3-7 and others. Development of branching chain reactions in aqueous systems containing oxygen and some other admixtures significantly changes their properties, but as free radical reactions, especially branching chain reactions are highly non-linear, the overall effect should depend drastically upon slight variations of initial conditions.

5.2. Burning of Water and Burning in Water

As it is mentioned above, singlet oxygen belongs to the family of ROS. Recently it was discovered that besides being a source of O_2, water may be directly oxidized with it. This reaction is readily catalyzed *in vitro* by antibodies (immunoglobulins) provided that energy of activation for excitation of molecular oxygen to its singlet state was supplied by dim light illumination of an antibody solution [29]. In other words, antibodies promote water "burning". Catalysts do not "invent" reactions that cannot go without them. They organize the reactants in space (and time) so that thermodynamically favorable processes go much faster. Quantum chemical calculations have shown that if two or more water molecules are arranged in space in particular disposition in relation to singlet oxygen and to each other, energy of activation for oxidation of a water molecule with singlet oxygen diminishes to reasonable values and such exotic peroxides as HOOOH, HOOOOH, HOO-HOOO may be produced under mild conditions as intermediates on the way to a more stable H_2O_2 [30]. Water oxidation goes on very fast in a solution of antibody because its active center provides for the optimal arrangement of water molecules for the process.

It is reasonable to suggest that if water is organized in a favorable way by some other means, if singlet oxygen is supplied, for example by the reactions #6 and #7, water oxidation may proceed in aqueous solutions in which water splitting had been initiated. We observed that in the course of branching chain reaction of slow oxidation of amino acids in aqueous solutions initiated with H_2O_2, concentration of H_2O_2 increases to the levels that can be explained only by water oxidation with O_2 [31]. Recently it has been shown that in water containing carbonates and phosphates [32] or noble gases, such as argon [33] concentration of H_2O_2 spontaneously increases and its augmentation goes on faster in case of water stirring. Using chemiluminescent methods we also found that such process goes on in aerated mineral waters from natural sources [34].

Thus, water – the most abundant substance in any living system, should regularly produce oxygen free radical and other forms of ROS under mild physiological conditions. The fact that a substantial part of organismal water is interfacial and dynamically structured increases the probability of its splitting and oxidation with all the above-listed consequences.

6. BIOLOGICAL SIGNIFICANCE OF ROS METABOLISM

Controversy related to biological significance of reactive oxygen species including free radicals in normal and pathological physiology and biochemistry is reaching its climax. Vast literature is devoted to ROS production especially under *in vitro* or cell culture conditions and detrimental chemical properties of ROS due to which they induce multiple lesions in important bioorganic molecules.

On the other hand during the past decade, more and more evidence appear that ROS play an important and probably even ubiquitous regulatory role in all living organisms – from microbes to higher animals, and the number of works devoted to the bio-regulatory role of ROS grows now exponentially. It turned out that adequate reaction of cells upon the action of hormones, neurotransmitters, cytokines, upon physical stimuli (light, temperature, mechanical stimulation) requires certain ROS levels in the environment. Some reactions, from the reversible activation or inhibition of certain enzymatic chains to genome activity regulation, which are provoked in cells by chemical signals, may be induced in them by ROS addition [35]. ROS are shown to have wholesome effects: they promote differentiation of cultured malignant cells into their benign counterparts [36], improve properties of taken out blood [37], and exercise significant therapeutic effects [38].

Besides, it turned out that common belief that under normal physiological conditions only a few percent of oxygen goes on ROS production is wrong. Due to ubiquitous presence of enzymes belonging to NADPH-oxidase family, to other means of direct oxygen reduction even under resting conditions up to 20% of all oxygen consumed is directly reduced and go on ROS production [39]; this share may increase up to 70% when physiological activity is enhanced [40].

Although a substantial part of inhaled oxygen is used for ROS generation, stationary levels of $O_2^{-}\bullet$ in cells and tissues do not exceed 10^{-10}-10^{-11} M [41], while that of H_2O_2 in a cell cytoplasm is estimated as 10^{-7}- 10^{-9} M [42]. ROS are kept at such low levels due to their nearly immediate elimination by the powerful antioxidant system. This seems to be puzzling: an organism converts a substantial share of oxygen into ROS and immediately eliminates these particles. How to explain such apparent squandering? And how can these particles, which are so short-lived and practically devoid of *chemical specificity*, exercise specific bioregulatory actions?

7. BIOENERGETIC FUNCTIONS OF ELECTRON EXCITED STATES

We suppose that difficulties in comprehension of the real role of ROS in vital activity are related to the attitude to them only as to chemical particles, while they should be considered as participants of continuous flux of oxygen reduction to water: $O_2 + 2H_2 \rightarrow 2H_2O$. This reaction consists of several steps, and in order to unveil its intimate details we suggest the following notation:

$$4(O_2 + e^- + H^+) \rightarrow 4HO_2\bullet \tag{8}$$
$$2(HO_2\bullet + HO_2\bullet) \rightarrow 2H_2O_2 + 2O_2 \tag{9}$$
$$H_2O_2 + H_2O_2 \rightarrow 2H_2O + O_2 \tag{10}$$

$$4O_2 + 4e^- + 4H^+ \rightarrow 2H_2O + 3O_2 \tag{11}$$

From such a notation of oxygen reduction, several important conclusions follow. First, if oxygen excess over the electrons that reduce it is less than 4-fold, combustion does not go to a final point, and intermediate ROS accumulate, which may initiate chain reactions with bioorganic molecules. Thus, an adequate supply of oxygen is necessary for maintaining low stationary level of ROS and other free radical particles. Second, all these reactions imply recombination of unpaired electrons. This applies also to the reaction #10 where one H_2O_2 molecule may be considered an electron donor and another an electron acceptor. Third, all these reactions are sources of energy quanta equivalent to electronic excitation energy. Energy yield in the reaction of dismutation of two superoxide radicals is ~22 kcal/mol, equal to the energy gap between triplet and excited singlet states of oxygen and equivalent to a near IR-photon (λ~1269 nm). When two singlet oxygen particles transit to triplet state simultaneously, EEE may be "pooled" and a doubled quantum of energy (equivalent to λ~635 nm, red light) is released [43]. Decomposition of two molecules of H_2O_2 donates an energy equivalent of 2 eV or λ<610 nm. When dismutation of $HO_2\bullet$ (reaction #9) is catalyzed by SOD or decomposition of H_2O_2 (reaction #10) is performed by catalase, quanta of high density energy should be generated with some megahertz frequencies due to very high turnover numbers of these two enzymes. This prevents energy from its immediate dissipation into heat and is favorable for energy pooling to even higher quanta.

A key role of EEE and related photon emission in the regulation of vital processes was discovered 80 years ago by A.G. Gurwitsch in the form of the so-called "mitogenetic radiation" – ultra-weak photon emission in the UV-range of EM-spectrum responsible for triggering cell division [44]. This radiation is emitted not only by living cells and tissues, but also by enzymatic (hydrolytic and glycolytic) and chemical reactions including gel-sol transitions in aqueous media. Water splitting and accessibility of active oxygen is a prerequisite condition for the emergence of this radiation [45]. Ultra-weak photon emission in the range from UV to near IR of electromagnetic spectrum from living cells and chemical reactions in aqueous media [46] affect activity of enzymes [47], activity and morphology of cells and tissues [48], regulate locomotion and mutual orientation of cultured cells [49]. Back reflected photons emitted during respiratory burst in human blood affect the intensity of this immune reaction by a feed-back mechanism [50].

In our opinion, regulatory role of ROS is provided by the unique feature of reactions with their participation – generation of electronic excitation energy (EEE) that continuously pumps biophotonic fields of living systems. But if reactions with ROS participation play such a versatile role, they should proceed in all living things including those that are considered to be anaerobic. Indeed, even obligate anaerobic bacteria are equipped with SOD [51] indicating that ROS appear even when molecular O2 concentration in water is negligibly low.

However, the intrinsic property of water to produce oxygen radicals due to its splitting makes their appearance in liquid water practically inevitable.

8. OSCILLATORY NATURE OF ACTIVE OXYGEN DEPENDENT REACTIONS IN AQUEOUS SYSTEMS

Besides serving a role of a source of the highest grade of energy, processes going on in aqueous systems in which EEE are generated may automatically acquire oscillatory character and are capable to serve as pacemakers for biochemical reactions dependent on them. Arousal of ROS in reactions going by in water and generation of EEE provides for the involving of other substances such as nitrogen and carbon dioxide into the process. They may beget amine and carbonyl compounds, and when concentrations of the latter exceed certain thresholds amino-carbonyl (Maillard) reaction develops. In this reaction, biologically significant heterocyclic, aromatic, polymeric substances appear [52]. Some of them activate oxygen resulting in ROS production and generation of EEE [28]. We found that profound oscillations of photon emission [53] and redox potential [54] emerge in Maillard reaction. Oscillations last for many hours and even days and their periods extend from fractions of minutes to tens of minutes. Amplitudes of redox potential variations may reach 0.3 V (from −0.2 V to −0.5 V).

High redox potential differences between different parts of the system can not be explained only from uneven distribution of reduced and oxidized forms of organic components because of their low concentrations (few tens of millimolar). It is interesting to speculate that these differences reflect gross changes in reduction and oxidation state of aqueous medium itself.

What is the primary cause of the development of oscillations of ROS production and oscillations of EEE generation? Our experimental data indicate that generation of EEE in reactions with ROS participation is prerequisite for self-organization observed as these processes develop. Initial building up of EEE fosters oxidation and oxygenation of available substrates resulting in an exhaust of dissolved oxygen and accumulation of reducing (easily oxidizable) equivalents. Oxygen continues to diffuse into the system from the air and when its concentration and concentration of reducers reach optimal ratio, a new wave of burning appear followed with oxygen depletion until the concentration of diffusing oxygen reaches a threshold value again. Thus, oscillatory behavior naturally emerges in such systems.

It is notable that oxygen consumption in single neutrophils and other cells that reduce it to ROS using NADPH-oxidase exhibits multimode oscillatory patterns of ROS generation [55]. Some hormones influence the amplitude of these oscillations, other affect their frequency. In other words, both deepness of respiration of single cells and its rate are related to their functional activity. Respiration rate and deepness (especially in case when oxygen consumption is realized through it one-electron reduction) define in their turn downstream regulatory processes.

Oscillatory behavior is characteristic not only of single cells but of their populations as well. We observed pronounced oscillations of photon emission from neutrophil suspensions containing hundreds of thousands of cells and even in whole blood, indication of a collective behavior of these big groups of cells related to metabolism of ROS in them [50].

Amino-carbonyl reaction proceeding in aqueous systems in which oscillations and waves spontaneously emerge is, in our opinion, the simplest model of arousal and performance of the respiratory process. Such conditions for the emergence of oscillations of EEE are common to all cells. A steep oxygen gradient between a metabolizing cell and its environment exists. Oxygen is poorly soluble in water, and what is more important, its delivery to a cell may be regulated by interfacial water at a cell-environment boundary. Reducing equivalents (e.g., NAD(P)H) accumulate in cells due to their metabolic activity. When the ratio of these equivalents to incoming O_2 reaches threshold values energy discharge primarily in the form of EEE occurs. Oxygen is rapidly exhausted, and released energy is directed for metabolic needs. Indeed, recently it has been experimentally demonstrated that oxygen is taken by single cells in an oscillatory mode [56]. Oscillations of EEE may play the role of pacemakers for the processes going on different levels of biological organization. On the other hand oscillatory nature of all these processes provides them the properties of sensible receptors for external electromagnetic and other physical fields.

9. RESPIRATION CYCLE OF WATER: A HYPOTHESIS

Respiration as we know it is a cyclic process. Though it is not so obvious that respiration at a level of a single cell should also be cyclic, experimental evidence supports this conclusion. It can be suggested that cyclic nature of respiration emerges on the one hand from the spatial relationship of oxygen consuming system and its environment and on the other from the orderliness of energy fluxes and high density of energy (EEE) that is generated in the course of oxygen-dependent processes in which ROS participate. Taking into consideration that all the aforementioned phenomena occurred in aqueous systems and that ROS generation is the intrinsic property of water, we suggest a hypothesis of the existence of the "respiratory cycle of water". Splitting of water molecules under the action of low density energy (mechano-chemical or mechano-catalytic water decomposition) results in the appearance of oxygen and hydrogen in aqueous systems:

$$8H_2O \rightarrow 4O_2 + 16H\bullet \qquad (12)$$

Four hydrogen atoms ($H\bullet$) are needed for complete reduction of one oxygen molecule, the rest of the hydrogen atoms recombine to H_2 molecules: $12H\bullet \rightarrow 6H_2 \Uparrow + n(h\nu)$. EEE released may be used, for example, for excitation of oxygen with the appearance of singlet oxygen, for sustaining of an aqueous system in a non-equilibrium, excited state, etc. This sequence of events may be by convention defined as the "exhale" stage because water splitting is accompanied with gas (hydrogen) release.

What may follow afterwards is analogous to the "inhale" stage, as oxygen is consumed here. We remind that for the complete reduction of oxygen molecule, a 4-fold excess of oxygen is needed:

$$4O_2 + 4H\bullet \rightarrow 2H_2O + 3O_2 + m(h\nu); \qquad (13)$$

Energy released in the course of the reactions #12 and #13 is enough to excite oxygen to a singlet state, and under appropriate conditions $^1O_2$ may go on water oxidation:

$$3O_2{}^* + 6H_2O \rightarrow 6H_2O_2 \qquad (14)$$

"Respiration cycle of water" allows to transform low density energy (freezing-thawing, evaporation-condensation, energy of sound, energy of shearing forces of water filtration or its vortexing) into a high density one; at least some part of the latter may accumulate in water in the form of metastable substances such as H_2O_2 and other peroxides as well as in long-living water excitation making it an active physical medium.

As other gases and substances that are present in "real" water should get involved in the process, respiration cycle should be considered not as a closed loop, but rather as a single convolution of an untwisting helix. Real processes proceeding in water should significantly depend upon the presence of positive and negative catalysts of particular reactions, of substances affecting water structure, upon the nature of interfaces that it solvates, upon the action of external physical factors and fields. Studies of phenomena related to water may help in solving many practical problems of medicine, agriculture, environmental problems, in providing people with healthy drinking water, in optimization of technologies in which water is important.

REFERENCES

1. S. Kleiner, Water: An essential but overlooked nutrient, *J. Amer. Diet. Assoc.* **99,** 200-206, (1999).
2. A. Szent-Gyorgyi, *Bioelectronics: A Study in Cellular Regulations, Defense, and Cancer* (Academic Press, New York, 1968), p. 4.
3. N. A. Bulionkov, Periodic dissipative-module structures of "bound water" – possible constructions defining biopolymer conformations in structures of their hydrates, *Krystallografia (Moscow).* **35**, 155-159, (1988).
4. S. V. Zenin, Hydrophobic model of water molecules associates, *Zhurnal Fizicheskoi Himii.* **68**, 634-641, (1994).
5. M. F. Chaplin, A proposal for the structuring of water, *Biophys. Chem.* **83**, 211-221, (2000).
6. S. Maheshwary, N. Patel, N. Sathyamurthy, A. D. Kulkarni, and S. R. Gadre, Structure and stability of water clusters $(H_2O)_n$, n = 8-20, An Ab Initio investigation, *J. Phys. Chem.- A.* **105**, 10525-10537, (2001).
7. M. Chaplin; http://www.martin.chaplin.btinternet.co.uk/index.html
8. G. N. Ling, A new theoretical foundation for the polarized-oriented multilayer theory of cell water and for inanimate systems demonstrating long-range dynamic structuring of water, *Physiol. Chem. Phys. & Med. NMR.* **35**, 91-130, (2003).
9. I. L. Cameron, K. M. Kanal, C. R. Keener, and G. D. Fullerton, A mechanistic view of the non-ideal osmotic and motional behavior of intracellular water, *Cell. Biol. Int.* **21**, 99-113, (1997).
10. S. P. Jarvis, T. Uchihashi, T. Ishida, and H. Tokumoto, Local solvation shell measurement in water using a carbon nanotube probe, *J. Phys. Chem. B.* **104**, 6091-6094, (2000).
11. P. Wernet, D. Nordlund, U. Bergmann, M. Cavalleri, M. Odelius, H. Ogasawara, L. A. Naslund, T. K. Hirsch, L. Ojamae, P. L. Glatzel, G. M. Pettersson, and A. Nilsson, The structure of the first coordination shell in liquid water, *Science.* **304**, 995-999, (2004)
12. P. Mentre, An introduction to "water in the cell": tamed hydra? *Cell. Mol. Biol.* **47**, 709-715, (2001).
13. D. Hall, and A. P. Minton, Macromolecular crowding, qualitative and semiquantitative successes, quantitative challenges. *BBA-Proteins Proteomics.* **1649**, 127–139, (2003).
14. J. Higo, M. Sasai, H. Shirai, H. Nakamura, T. Kugimiya, Large vortex-like structure of dipole field in computer models of liquid water and dipole-bridge between biomolecules, *Proc. Nat. Acad. Sci USA.* **98**, 5961–5964, (2001).
15. P. M. Wiggins, Enzyme reactions and two-state water, *J. Biol. Phys. Chem.* 2002, **2,** 25-37, (1990).
16. P. M. Wiggins, Role of water in some biological processes, *Microbiol. Rev.* **54**, 432-449.

17. G. N. Ling *A physical theory of living state: the Association-Induction Hypothesis.* (Blaisdell Publ. Co., Waitham, MA, 1962).
18. G. H. Pollack, *Cells, Gels and the Engines of Life: A New, Unifying Approach to Cell Function.* (Ebner & Sons, Seattle, WA, 20010.
19. P. M. Wiggins, Two states of water found in hydrophobic clefts: their possible contribution to mechanism of cation pumps and other enzymes. *Internat. Rev. Cytol.* **108**, 249-303, (1987) .
20. P. M. Wiggins, Water structure in polymer membranes, *Progr. Polym. Sci.* 1988, **13**, 1-35.
21. P. M. Wiggins, Life depends upon two kinds of water. 64 pp; http://www.lsbu.ac.uk/water/monograph200904pw.pdf
22. A. Szent-Gyorgyi, *Bioenergetics.* (Academic Press. New York, 1957). [Back translation from a Russian edition of the book: (GIZ Fiz-Mat. Literature, Moscow, 1960), p. 54-56].
23. N. K. Baramboim, *Mechanochemistry of High Molecular Weight Compounds.* 9Chimiya, Moscow, 1971).
24. A. Domrachev, G. A. Roldigin, and D. A. Selivanovsky, Role of sound and liquid water as dynamically unstable polymeric system in mechano-chemically activated processes of oxygen production on Earth, *J. Phys. Chem.* **66**, 851-855, (1992).
25. G. A. Domrachev, G. A. Roldigin, and D. A. Selivanovsky, Mechano-chemically activated water dissociation in a liquid phase, *Proc. Russ. Acad. Sci.* **329**, 258-265, (1993) .
26. S. Ikeda, T. Takata, M. Komoda, M. Hara, J. N. Kondo, K. Domen, A. Tanaka, H. Hosono, and H. Kawazoe, Mechano-catalysis -- a novel method for overall water splitting, *Phys. Chem. Chem. Phys.* **1**, 4485-4491, (1999).
27. S. Woutersen, and H. J. Bakker Resonant intermolecular transfer of vibrational energy in liquid water, *Nature.* **402**, 507 – 509, (1999).
28. V. L. Voeikov, and V. I. Naletov, Weak photon emission of non-linear chemical reactions of amino acids and sugars in aqueous solutions, *Biophoton,* edited by J.-J. Chang, J. Fisch, and F.-A. Popp (Kluwer Academic Publishers, Dortrecht, The Netherlands, 1998), pp. 93-108.
29. A. D. Wentworth, L. H. Jones, P. Wentworth, Jr, K. D. Janda, and R. A. Lerner, Antibodies have the intrinsic capacity to destroy antigens, *Proc. Nat. Acad. Sci. USA.* **97**, 10930-10935, (2000).
30. X. Xu, R. P. Muller, and W. A. Goddard 3rd. The gas phase reaction of singlet dioxygen with water: a water-catalyzed mechanism. *Proc. Nat. Acad. Sci. USA.* **99**, 3376-3381, (2002).
31. V. L. Voeikov, I. V. Baskakov, K. Kafkialias, and V. I. Naletov, Initiation of degenerate-branched chain reaction of glycin deamination with ultraweak UV irradiation or hydrogen peroxide, *Russ. J. Bioorg. Chem.* **22**, 35-42, (1996).
32. V. I. Bruskov, A. V. Chernikov, S. V. Gudkov, and Zh. K. Masalimov, Activation of reducing properties of anions in sea water under the action of heat, *Biofizika.* **48**, 1022-1029, (2003).
33. V. L. Voeikov, and M. V. Khimich, Amplification by argon of luminol-dependent chemiluminescence in aqueous NaCl/H2O2 solutions, *Biofizika.* **47**, 5-11, (2002).
34. V. L. Voeikov, R. Asfaramov, V. Koldunov, D. Kononov, C. Novikov, and N. Vilenskaya, Chemiluminescent analysis reveals spontaneous oxygen-dependent accumulation of high density energy in natural waters, *Clin. Lab.* **49**, 569, (2003).
35. V. L. Voeikov, Reactive oxygen species, water, photons, and life, *Rivista di Biologia/Biology Forum.* **94**, 193-214, (2001).
36. H. Sauer, M. Wartenberg, and J. Hescheler, Reactive Oxygen Species as Intracellular Messengers During Cell Growth and Differentiation, *Cell. Physiol. Biochem.* **11**, 173-186, (2001).
37. V. Bocci, Autohemotherapy after treatment of blood with ozone. A reappraisal, *J. Int. Med. Res.* **22**, 131-144, (1994).
38. C. F. Nathan, and Z. A. Cohn, Antitumor Effects of Hydrogen Peroxide in Vivo, *J. Exp. Med.* **154**, 1539-1550, (1981).
39. H. P. Souza, X. Liu, A. Samouilov, P. Kuppusamy, F. R. Laurindo, and J. L. Zweier, Quantitation of superoxide generation and substrate utilization by vascular NAD(P)H oxidase, *Am. J. Physiol. Heart Circ. Physiol.* **282**, H466-H474, (2002).
40. J. R. Trimarchi, L. Liu, D. M. Porterfield, P. J. Smith, and D. L. Keefe, Oxidative phosphorylation-dependent and -independent oxygen consumption by individual preimplantation mouse embryos, *Biol. Reprod.* **62**, 1866-1874, (2000).
41. V. Niviere, and M. Fontecave, Biological sources of reduced oxygen species, *Analysis of Free Radicals in Biological Systems,* edited by A. E. Favier, J. Cadet and B. Kalyanaraman (Birkhauser, Basel, Boston, Berlin, 1995), pp. 11-19.
42. D. D. Tyler, Polarographic assay and intracellular distribution of superoxide dismutase in rat liver, *Biochem. J.* **147**, 493-504, (1975).
43. E. Cadenas, and H. Sies, Low-level chemiluminescence as an indicator of singlet molecular oxygen in biological systems, *Methods Enzymol.* **105**, 221-231, (1984).
44. A. G. Gurwitsch, and L. D. Gurwitsch, Twenty years of mitogenetic radiation, emergence, development, and perspectives, *Usp. Sovrem. Biologii.* **16**, 305-334, (1943). (English translation: *21st Century Science and Technology.* Fall, 12, 41-53, (1999).)
45. V. L. Voeikov, Mitogenetic radiation, biophotons, and non-linear oxidative processes in aqueous media, *Integrative Biopysics. Biophotonics,* edited by F.-A. Popp and L. Beloussov (Kluwer Academic Publishers, Dordrecht 2003), pp. 331-360.

46. J. Slawinski, Luminescence research and its relation to ultraweak cell radiation, *Experientia.* **44**, 559-571, (1988).
47. G. Cilento, Photobiochemistry without light, *Experientia.* **44,** 572-576, (1988).
48. V. P. Galantsev, S. G. Kovalenko, A. A. Moltchanov, and V. I. Prutskov, Lipid peroxidation, low-level chemiluminescence and regulation of secretion in the mammary gland, *Experientia.* **49**, 870-875, (1993).
49. G. Albrect-Buehler, Changes of cell behavior by near-infrared signals, *Cell. Motil. Cytoskeleton.* **32**, 299-304, (1995).
50. V. L. Voeikov, R. R. Asfaramov, E. V. Bouravleva, C. N. Novikov, and N. D. Vilenskaya, Biophoton research in blood reveals its holistic properties, *Indian J. Exp. Biol.* **43**, 473-482, (2003).
51. J. Hewitt, and J. Morris, Superoxide dismutase in some obligately anaerobic bacteria, *FEBS Lett.* **50**, 315-318, (1975).
52. M. Namiki, T. Hayashi, S. and Kawakishi, Free radicals developed in the amino-carbonyl reaction of sugars with amino acids, *Agric. Biol. Chem.* **37**, 2935-2937, (1973).
53. V. L. Voeikov, V. V. Koldunov, and D. S. Kononov, Long-duration oscillations of chemi-luminescence during the amino-carbonyl reaction in aqueous solutions, *Russ. J. Phys. Chem.* **75**, 1443-1448, (2001).
54. V. L. Voeikov, V. V. Koldunov, and D. S. Kononov, New oscillatory process in aqueous solutions of compounds containing carbonyl and amino groups, *Kinetics and Catalysis (Moscow).* **42**, 606-609 (2001).
55. A. L. Kindzelskii, and H. R. Petty, Apparent role of traveling metabolic waves in oxidant release by living neutrophils, *Proc. Natl. Acad. Sci. USA.* **99**, 9207-9212, (2002).
56. D. M. Porterfield, R. F. Corkey, R. H. Sanger, K. Tornheim, P. J. S. Smith, and B. E. Corkey, Oxygen consumption oscillates in single clonal pancreatic - cells (HIT), *Diabetes.* **49**, 1511–1516, (2000).

8
THE HYDROPHOBIC-HYDROPHILIC BALANCE IN WATER SOLUTIONS OF PROTEINS AS THE POSSIBLE TARGET FOR EXTREMELY LOW FREQUENCY MAGNETIC FIELDS

Victor S. Martynyuk* and Yulia V. Tseyslyer†

1. INTRODUCTION

Alexander Gurwitsch was the first to demonstrate the presence of extremely low photon radiation (mitogenetic radiation) in the living organisms that plays a regulatory role. Together with his follower Gleb Frank, he attributed this radiation to the ultraviolet diapason of electromagnetic scale[1,2,3].

Numerous further studies showed meanwhile that the spectral range of low intensity radiation in the living organisms is significantly wider, including the visible and infrared waves. During past decades, the radiation of the living organism in the extremely high radiofrequency range (millimeter waves) was also revealed[4]. The interaction of these electromagnetic waves with living organisms seems to proceed via resonance mechanisms. Low intensive millimeter waves are biologically very active, therefore the millimeter waves are now widely used for therapeutic treatments in medicine[4].

A biological activity of the extremely low frequency magnetic fields has been also revealed at the end of past century[5]. The ideas about regulatory and informational role of extremely low frequency magnetic fields in biosphere were developed by B.Vladimirsky and N.Temuryants[6].

Although a high sensitivity of living organisms to the action of weak magnetic fields of the natural and artificial origin is reliably established, their primary physical-chemical mechanisms remain unclear. In this paper, we propose that the biological effects of electromagnetic fields are mediated by the changes in hydrophobic-hydrophilic balance of water-colloidal systems[7].

Hydrophobic interactions are known to play an important role in stabilization of protein and nucleic acids structure and in the regulation of their conformation changes, as well as in stabilization of biological membranes via regulating phase transitions in bilipid layers. The hydrophobic interactions are also important for

* Crimean Scientific Center of National Academy of Science and Vladimir Vernadsky Taurida National University, Vernadsky ave, 2, Simferopol, 95007, Ukraine, E-mail: mavis@science-center.net.
† Taras Shevchenko Kyiv National University, Vladimirskaya str., 64, 01033б Kyiv, Ukraine, E-mail: yuc@univ.kiev.ua

binding specific and non-specific ligands (hormones, vitamins, intracellular messengers, pharmacological substances, etc.) by enzymes, receptors and signaling proteins, as well as for extracellular, membrane, and intracellular transportation of hydrophobic substances. It is possible to expect that slight changes in hydrophobic-hydrophilic balance will result in appreciable biological effects on the molecular, cellular, and systemic level. However, this assumption must be verified on simple experimental models of hydrophobic interactions in biological structures. Therefore, the investigation of the influence of ecologically significant low frequency magnetic field upon hydrophobic interaction with proteins was carried out.

2. THE EXPOSURE OF EXPERIMENTAL SAMPLES TO EXTREMELY LOW FREQUENCY MAGNETIC FIELD (ELFMF)

In our experiments, we have used rectangular and different polarity magnetic field impulses with 8 Hz frequency and 5 or 25 μT induction generated by Helmholtz coils. Such a frequency was selected because of its ecological and geophysical importance and high biological activity[6,8]. Induction vector of magnetic field was parallel to the direction of geomagnetic field vector. Exposure time varied from 1 to 24 hours.

Experiments with sham exposure were carried out to estimate the possible influence of background magnetic fields in the samples' locations. In these cases, experimental samples were located in Helmholtz coils but magnetic field was not generated. A usual noise ELFMF intensity was in 500-1000 time lower than that of magnetic fields in Helmholtz coils.

3. INFLUENCE OF ELFMF UPON THE BINDING OF LOW MOLECULAR NON-POLAR SUBSTANCES WITH HYDROPHOBIC CAVITIES IN PROTEIN MOLECULES

3.1. Basic Experimental Model

The phenomenon of saturation of protein solutions by low-molecular non-polar substances described in Ref. 9 was used as the basis for our experiments. Non-polar substances are known to have an extremely low water solubility because the dissolution of such compounds in water is exothermal. Their dissolution proceeds with increase of temperature and dissipation of heat energy indicating the decrease of enthalpy of the system after mixing up non-polar substances with water ($\Delta H<0$). But the enthalpy decrease is compensated by a decrease of entropy ($\Delta S<0$) due to formation of ordered crystal-hydrates in which non-polar substances are located. Therefore, a hydrophobic dissolution of substances is accompanied by a small increase of free energy of the system ($\Delta G>0$).

Thus, the absolute value of enthalpy decrease is smaller than the entropy decrease, and as a consequence the dissolution of non-polar compounds in water is thermodynamically disadvantageous: $\Delta G = \Delta H - T\Delta S > 0$. The consequence of the big contribution of entropy component into change of free energy of the system is the repulsion of non-polar molecules by water molecules and its thermodynamically advantageous interaction with each other. Such effect of repulsion of non-polar substances by water molecules is named "hydrophobic interactions". The increase of concentration of non-polar substances in the water phase results to increase of van der Waals interaction between dissolved molecules and also to formation of associates "covered" with ordered layer of water on its hydrophobic surface. This phenomenon is the cause of division of the system to two phases - the phase that consist of molecules of solvent (water) and the phase of hydrophobic substance (Fig. 1). Thus, the primary factor in dissolution of non-polar substances on the hydrophobic mechanism is the ability of molecules of water to form ordered crystal-hydride structure around molecules of non-polar substance that result to decrease entropy of the system. Any factors that destroy or change the ordered structure of water change to some extent the solubility of hydrophobic compounds in water phase.

If the water solution contains large colloidal particles with hydrophobic cavities, for example protein molecules, then part of non-polar molecules bind with these cavities on the hydrophobic mechanism (Fig. 2). Such binding of the hydrophobic substances with proteins can be studied by means of different quantitative or semi-quantitative methods.

In these studies the chloroform and benzol (Fig. 2) were used as the non-polar substances that non-specifically bind with proteins on hydrophobic mechanism. The saturation of water and water protein solutions by non-polar substance made by means of layering of 3 ml protein solution on 1.5 ml organic phase of non-polar substance and incubation of samples under room temperature during 1-24 hours in dependence of the chosen experimental protocol.

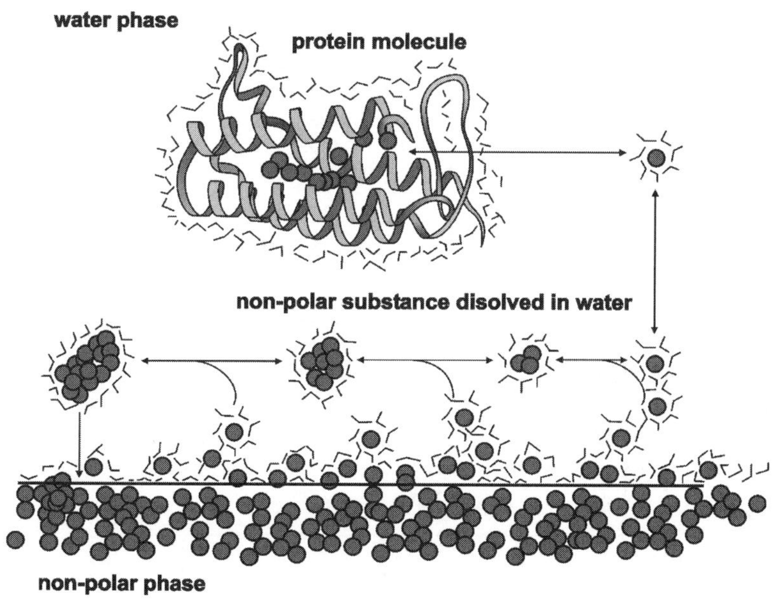

Figure 1. Experimental model of non-specific binding of non-polar ligands with protein molecules.

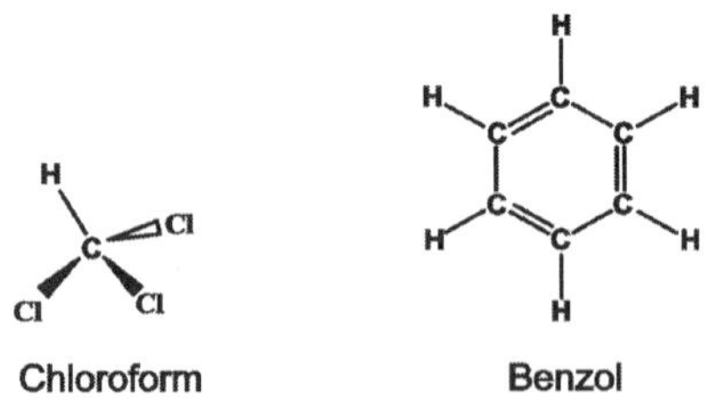

Figure 2. Non-polar ligands used in the experiments.

3.2. The Refractometric Study of Influence of ELFMF on Binding of Benzol with Serum Albumin

3.2.1. Method

The refractometric method of quantitative estimation of hydrocarbons, bound with protein, is based on dissolution of hydrocarbons on hydrophobic mechanism in water-protein solutions that results to the increase of refraction of experimental solutions up to its saturation (Fig. 3). The usage of aromatic hydrocarbons in such model systems is most convenient because their refraction strongly differ from the refraction of water systems that allows to increase the precision and sensitivity of this method.

The volume of the dissolved hydrocarbon in water and water solution of proteins is determined on the base of the rule of additivity of specific refraction[9]. Quantity of the hydrocarbon bound with molecules of protein is calculated as $\mathbf{V}_{\mathbf{prot}} = \mathbf{V}_{\mathbf{prot.\ solution}} - \mathbf{V}_{\mathbf{water}}$ (ml).

The serum albumin with concentration 0.5% was used in this experimental series.

The rectangular and different polarity magnetic impulses with frequency 8 Hz and induction 5 μT were used in these studies. The ELFMF influenced during saturation of experimental solutions for 5 hours.

The statistical significance of the differences between experimental series was assessed by 5% level Student T test.

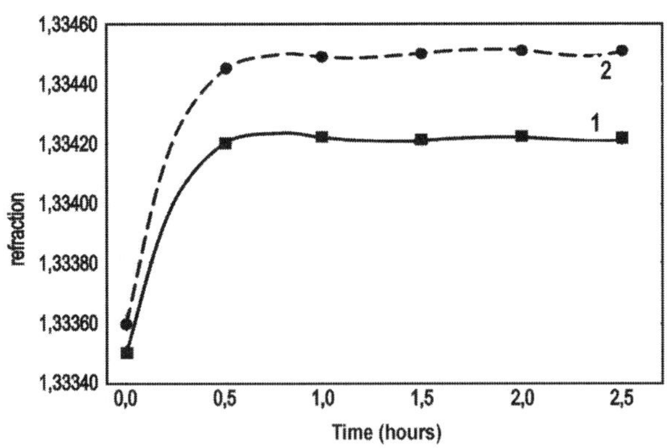

Figure 3. Time dynamics of refraction coefficient of water (1) and protein solution (2) under its saturation by benzol.

3.2.2. Results

Results of refractometric study on binding of benzol with serum albumin testified on significant increase of binding index (**V_{prot}**) under the influence ELFMF (Fig. 4). This statistically significant increase has been revealed already after 1 hour of MF exposure. Simultaneously with the increase of binding of benzol with protein the solubility of used hydrocarbon in the water statistically significant decreased (Fig. 5).

So, these experimental results testify on the following features of influence of ELFMF on hydrophobic interactions. In the used model conditions, the influence of impulse 8 Hz magnetic field changes the dynamical balance in the system *benzol : water : protein* that resulted in pushing out of benzol from water to hydrocarbon phase, on one hand, and on the other, pushing out of benzol to non-polar phase of hydrophobic cavities of protein molecules.

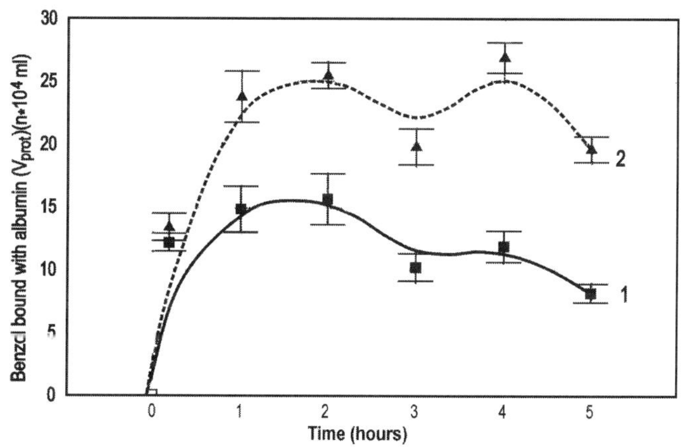

Figure 4. Dynamics of binding of benzol by serum albumin ($V_{prot} = V_{prot.\ solution} - V_{water}$ (ml)) in control samples (1) and under the influence of 8 Hz 5 μT magnetic field (2).

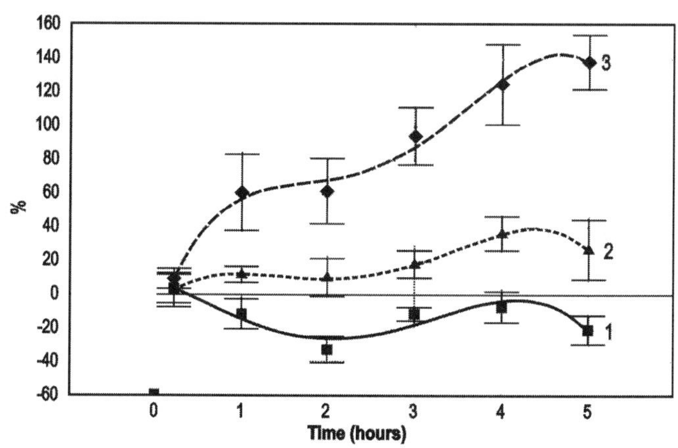

Figure 5. Influence of 8 Hz 5 μT magnetic field on benzol solubility in water (1) and in 0.5% solution of serum albumin (2) and also on binding of benzol by albumin (3) in ±% relative to control samples.

3.3. The Spectrophotometric Study of Influence of ELFMF on Binding of Chloroform with Serum Albumin

3.3.1. Method

The spectrophotometric method of studying of non-specific liganding of non-polar substances with biopolymers is based on registration of changes of light absorption spectra in biopolymers caused owing to changes of polarity of chromophore surrounding. The spectral changes can be caused both by the influence of hydrocarbon substances on chromophores and influence of solvent (water molecules), accessibility to chromophores of which can be raised, or reduced owing to liganding of non-polar substances or/and owing to ligand-induced conformation changes of macromolecule.

Thus, this method does not allow to obtain information about absolute amount of non-polar molecules that are bound with biopolymers, but allows to investigate qualitatively and semi-quantitatively the process of liganding and also the features of influence of non-polar and non-specific ligands on conformation state of macromolecules. With the purpose to increase the informational capacity of this method, the integral absorption spectera and also differential spectera were used for analysis. Differential spectera were obtained

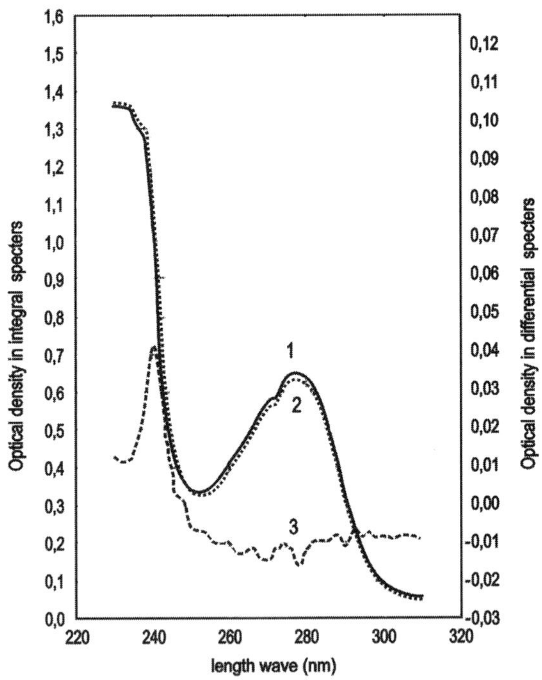

Figure 6. Integral (1, 2) and differential (3) spectera of 0.1% solutions of serum albumin under its saturation by chloroform. 1 - integral specterum of native *serum albumin*; 2 - integral specterum of *serum albumin* saturated by chloroform; 3 - differential spectera as difference between spectra of saturated and native proteins.

by subtraction of the absorption spectera of native protein of the integral spectrum of ligand-loaded protein (Fig. 6) or by means of direct instrumental registration using differential scheme.

The concentration of serum albumin in this experiment was 0.1%.

The rectangular and different polarity impulses with frequency 8 Hz and induction 25 μT used in these studies. The ELFMF influenced during saturation of experimental solutions for 24 hours.

The statistical significance of the differences between experimental series was assessed by 5% level Student T test.

3.3.2. Results

The differential spectera of 0.1% solutions of serum albumin after their incubation with chloroform are shown in Figure 7. The saturation of protein by chloroform realized in two spectral shifts: the long-wave shift (red-shift) in the range of 235-250 nm and also weak short-wave shift (blue-shift) in range of 260-300 nm. The formation of red-shift under the saturation protein molecules of proteins by chloroform testifies about the decrease of polarity of surrounding around the peptide bounds and aminoacid radicals located on protein surface and contacted with molecules of water.

At the same time the cause of the formation of blue-shift is the increase polarity in the internal hydrophobic cavities of molecules of albumin where majority on non-polar aromatic aminoacid radicals is located. The chloroform is considered as the non-polar substance, but dipole moment (μ) of this molecules is about $\mu = 1.06$ D (for comparison $\mu_{water} = 1.84$ D; $\mu_{benzol} \approx \mu_{heptan} \approx 0$). Therefore the presence of this substances in hydrophobic cavities induces the spectral shift in blue range.

The influence of ELFMF has statistically significantly ($p<0.05$) increased the amplitude of red shift but not changed the characteristics of weak blue shift (Fig. 7). There were no changes of differential spectera in the experiments with sham exposure. So, the increase of amplitude of red shift in range of 235-250 nm testifies on the MF-induced increase of binding of chloroform on protein surface.

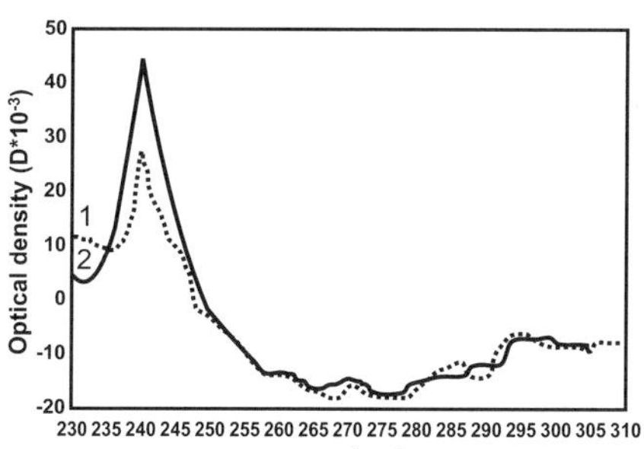

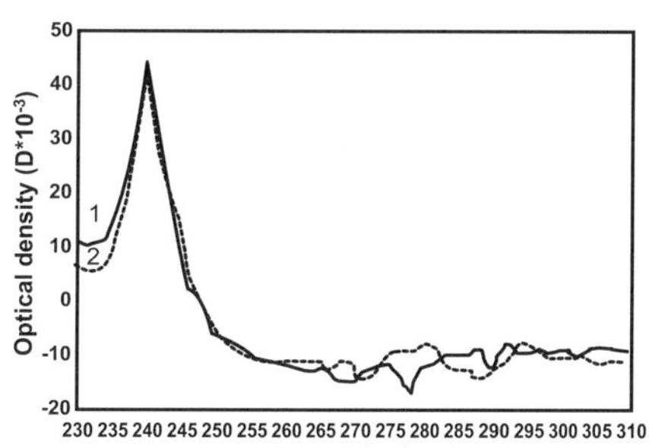

Figure 7. Differential spectera of 0,1% solutions of serum albumin after their saturation by chloroform in experiments with influence of 8 Hz 25 μT magnetic field (left figure) and sham exposure (right figure). The presented spectral lines are the average for 2, 4, and 24 hour exposure to ELFMF. Specters were obtained relative to protein solution without chloroform saturation. 1-control samples; 2 - magnetic field of sham exposure.

Being based on the literature data about the acceleration of coagulation colloidal particles under the magnetic field treatment and also about the increase of adsorption of colloidal particles on surface of solid bodies and air cavities on water[10] it can be supposed that this increase of binding of chloroform with protein molecules can stimulate the reversible formation of the protein associates by means of hydrophobic mechanisms that additionally amplifies the red shift in absorption spectera of studied protein.

3.4. The Spectrophotometric Study of Influence of ELFMF on Binding of Chloroform with Cytochrome *c*

3.4.1. Methods

The binding of chloroform with cytochrome *c* was studied on changes of optical absorption spectera of this protein in Soret-peak wavelength range 350-440 nm. The saturation of 0.05% solutions of cytochrome ***c*** has been resulted in formation of blue shift of Soret-peak (Fig. 8). The value of this shift is depended on time of incubation of protein with chloroform. The differential spectera were obtained by measurement of optical density of the chloroform-saturated protein solutions relativity to the protein solutions that do not contain chloroform.

The enzymatic activity of cytochrome c was studied by measurement of speed of rise of optical density on λ=550 during reduction of enzyme by ascorbic acid.

The concentration of cytochrome *c* in this experiment was 0.05%.

The rectangular and different polarity magnetic impulses with frequency 8 Hz and induction 25 μT were used in these studies. The ELFMF influenced during saturation of experimental solutions for 24 hours.

The statistical significance of the differences between experimental series was assessed by 5% level Student T test.

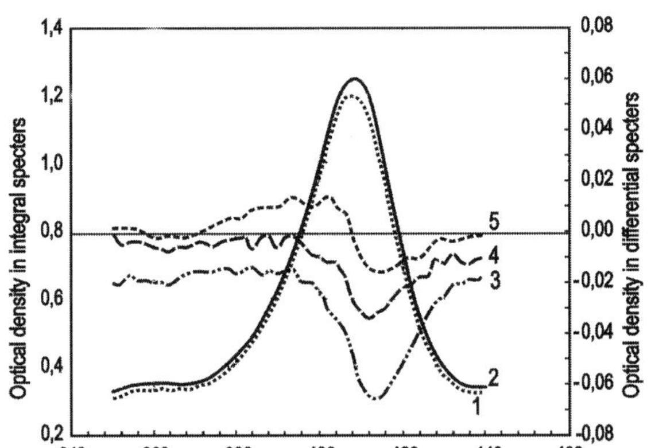

Figure 8. Integral (1, 2) and differential (3-5) spectera of 0.05% solutions of cytochrome ***c*** (oxidized form) under its saturation by chloroform. (1 - integral specterum of native cytochrome ***c***; 2 - integral specter of *cytochrome* ***c*** saturated by chloroform; 3-5 – differential spectera as difference between spectra of saturated and native proteins after 2, 4, and 24 hours of saturation.

3.4.2. Results

The analysis of differential spectera has shown that the influence of ELFMF was more significant during the first hours of saturation of cytochrome c by chloroform. ELFMF has increased the formation of blue shift (Fig. 9). The maximum of differential spectera and the difference λ_{min}-λ_{max} were spectral parameters that more sensitive to influence of studied physical factor (table 1). The statistical increase of D_{max}-D_{min} was revealed after daily exposure of chloroform-saturated solution of cytochrome ***c*** to magnetic field. This parameter characterizes the quantity of bound non-polar substance too[11] that also allows us to conclude about the increase of binding of chloroform with cytochrome c under the influence of ELFMF.

The study of enzymatic activity of cytochrome ***c*** has shown the small but statistically significant ($p<0.05$) inhibitory effect of the slow saturation of protein solution by chloroform on 10% after 2 hours. This fact of presence of the relatively high enzymatic activity testifies on insignificant conformational changes of studied protein under such soft and slow procedure of saturation.

The experiments with influence of ELFMF showed the accelerated inhibition of enzymatic activity due to MF-induced increase of binding of chloroform with protein. The statistically significant decrease ($p<0.05$) of enzymatic activity on 15-20% was shown already after 1 hour of incubation of protein solution with chloroform.

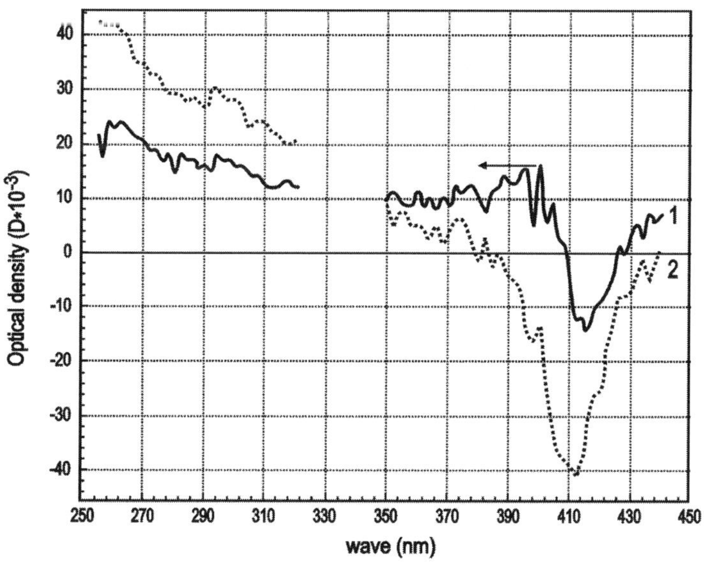

Figure 9. Differential spectera of 0.05% solutions of cytochrome ***c*** after their saturation by choloroform during 2 hours in control samples (1) and under the 8 Hz 25 μT magnetic field influence (2). Spectera were obtained relatively to protein solution without chloroform saturation. 1 – control samples; 2 - ELF MF of sham exposure.

Table 1. The basic characteristics of differential spectera of cytochrome *c*

Expo-sure time (hours)	Parameters	Control samples	Sham exposure	ELF MF influence	Changes relative to control samples (%)	Changes relative to sham exposed samples (%)
2	λ_{max}, nm	397.3 ± 1.9	394.0 ± 3,8	387.8 ± 5.0*	-2.9	-1.6
4	λ_{max},, nm	391.5 ± 3.0	3988 ± 1.6	390.3 ± 43**	-0.6	-2.1
24	λ_{max}, nm	393.2 ± 18	395.8 ± 2.2	391.0 ± 5.1	-1.1	-1.2
2	λ_{min}, nm	413.9 ± 0,5	415.0 ± 1.3	413.0 ± 0.9	-0.2	-0.5
4	λ_{min}, nm	412.7 ± 0.7	414.3 ± 2.6	413.9 ± 0.7	0.3	-0.1
24	λ_{min}, nm	414.4 ± 0.4	415.2 ± 0.4	415.0 ± 1.0	0.0	0.0
2	λ_{min}-λ_{max}, nm	16.7 ± 1.7	21.0 ± 3.4	25.2 ± 4.3*	76.2	20.0
4	λ_{min}-λ_{max}, nm	21.3 ± 2.9	15.5 ± 2.9	23.6 ± 3.7**	24.0	52.3
24	λ_{min}-λ_{max}, nm	21.2 ± 1.9	19.4 ± 2.4	24.0 ± 4.1	32.7	23.7
2	D_{max}-D_{min}, o.d.u.	0.034 ± 0.002	0.038 ± 0.007	0.038 ± 0.006	12.6	0.0
4	D_{max}-D_{min}, o.d.u.	0.032 ± 0.002	0.034 ± 0,004	0.033 ± 0.003	2.8	-4.4
24	D_{max}-D_{min}, o.d.u.	0.0463 ± 0.003	0.046 ± 0.004	0.055 ± 0.002*,**	18.3	20.1

* Statistically significant changes relative to control samples; ** statistically significant changes relatively to sham exposed samples.

3.5. The Spectrophotometric Study of Influence of ELFMF on Binding of Chloroform and Benzol with Methemoglobin

3.5.1. Method

The binding of chloroform and benzol with methemoglobin was studied on changes of optical absorption spectera of this protein in Soret-peak wavelength range 350-450 nm. The saturation of 0.02% solutions of methemoglobin by chloroform or benzol realized to formation of red shift of Soret-peak (Fig. 10). The value of this shift depended on time of incubation of protein with non-polar substances. The integral absorption spectera were registered in this experimental series. The differential spectera were calculated as the difference between integral spectera of saturated protein solution and non-saturated one.

The concentration of methemoglobin in this experiment was 0.02%.

The rectangular and different polarity impulses with frequency 8 Hz and induction 25 μT used in these studies. The ELFMF influenced during saturation of experimental solutions for 24 hours.

The statistical significance of the differences between experimental series was assessed by 5% level Student T test.

3.5.2. Results

The analysis of spectera of methemoglobin in the Soret's peak sowed the maximum absorption in 404 nm. It is necessary to note that the specter of native hemoglobin significantly differs from specter of methemoglobin. The maximum of Soret's peak of native hemoglobin is in the range 412-415 nm. The cause of such differences is the different conformation state of these protein molecules where hem of methemoglobin is more accessible to polar molecules of water, thereof maximum of Soret's peak of methemoglobin significantly shifted in blue range.

The spectral data testify on red shift to 405 nm during saturation of protein solution by chloroform (Table 2.3). In our opinion this fact testifies on binding of chloroform with cavities containing hem and also on decrease of accessibility of water molecules to this chromophore.

The analysis of the absorption specter of methemoglobin in Soret's peak has showed the statistically significant influence of ELFMF on absorption spectera of this protein (Table 2). The influence of ELFMF realized to increase of red shift caused by chloroform binding up to 5-6 nm. The maximal changes were revealed after two hour of exposure to magnetic field but mean value of effect of ELFMF for all exposure time was 1.5 nm (Table 2). The cause of these effects of ELFMF, to our mind, is the increased binding of chloroform with methemoglobin. Such changes were absent in the experiments with sham exposure (Table 3).

The analysis of differential spectera additionally testifies on the statistically significant influence of ELFMF on binding of chloroform with methemoglobin. In particular, the influence of ELFMF has changed also the form of Soret's peak that has realized in dynamical changes of value of maximum and minimum in the differential spectera (Table 4). At the same time the amplitude of differential spectera (D_{max}-D_{min}) was insensitive to the influence of studied physical factors. The statistically significant changes were not revealed in experiments with sham exposure (Table 4).

The saturation of methemoglobin solution by benzole realize to more weak spectral changes than in the experiments with chloroform (Tables 5 and 6). The

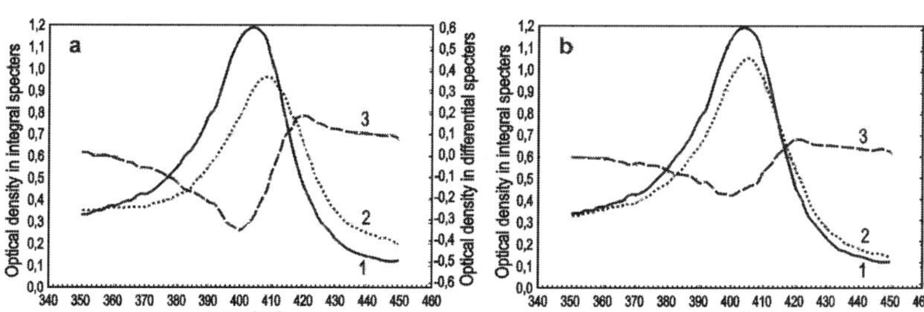

Figure 10. Integral (1, 2) and differential (3) spectera of 0.02% solutions of methemoglobin under its saturation by chloroform (a) and benzol (b) during 1 hour. 1 - integral specter of methemoglobin; 2 - integral specter of *met-hemoglobin* saturated by chloroform; 3-5 – differential spectera as difference between spectra of saturated and native proteins after 2, 4, and 24 hours of saturation.

Table 2. The basic characteristics of integral spectera in range 350-450 nm of met-hemoglobin after its saturation by chloroform and ELF MF influence

Exposure time (hours)	Parameters	Control samples	MF influence	Saturation by chloroform	Saturation by chloroform and MF influence	MF-induced spectral shift relative to samples saturated by chloroform (nm)
1	λ_{max}, nm	404.0±0.3	403.8±0.2	408.0±0.5*	409.3±05*	+1.3
2	λ_{max}, nm	403.8±0.2	404.8±0,3	406,8±0.3*	409.3±0,6*,**	+2.5
4	λ_{max},, nm	404.0±0.3	404.2±0.2	407.3±0.9*	408.2±0.4*	+0.9
24	λ_{max}, nm	405.0±0.5	404.6±0.3	407.6±0.6*	409.1±0.7*	+1.2
1 - 24	$\Delta\lambda_{mean}$, nm					+1,47±0,35 p<0.05
1	D_{max}, o.d.u.	1.196±0.070	1.219±0.065	0.978±0.072*	0.967±0.071*	
2	D_{max}, o.d.u.	1.254±0.062	1.230±0.055	0.957±0.057*	0.934±0.060*	
4	D_{max}, o.d.u.	1.193±0,058	1.229±0.057	0.947±0.053*	0.985±0.054*	
24	D_{max}, o.d.u.	1.206±0.062	1.185±0.085	0.849±0.090*	0.853±0058*	

* Statistically significant changes relative to control samples; ** statistically significant changes relatively to samples saturated by chloroform.

Table 3. The basic characteristics of integral spectera in range 350-450 nm of methemoglobin after its saturation by chloroform and sham influence

Exposure time (hours)	Parameters	Control samples	Sham influence	Saturation by chloroform	Saturation by chloroform and sham influence	Spectral ship relative to samples saturated by chloroform (nm)
1	λ_{max}, nm	404.4±0.3	404.2±0.2	408.5±04*	409.1±03*	+0.6
2	λ_{max}, nm	404.2±0.2	404.0±0.0	407.8±0.5*	408.4±04*	+0.6
4	λ_{max},, nm	404.4±0,4	404.0±0.0	407.6±0.4*	407.7±0.4*	+0.1
24	λ_{max}, nm	404.4±0.3	404.0±0.0	408.6±0.3*	408.2±0.2*	-0.4
1 - 24	$\Delta\lambda_{mean}$, nm					+0.22±0.23
1	D_{max}, o.d.u.	1.207±0,020	1.197±0,020	0.991±0.057*	1.006±0.046*	
2	D_{max}, o.d.u.	1.208±0.022	1.150±0.051	0.989±0.044*	0.992±0.043*	
4	D_{max}, o.d.u.	1.239±0.017	1.218±0.085	0.909±0.045*	0.958±0.056*	
24	D_{max}, o.d.u.	1.280±0.062	1.258±0.086	0.919±0.110*	0.946±0.097*	

* Statistically significant changes relative to control samples; ** statistically significant changes relatively to samples saturated by chloroform.

possible cause of smaller values of red shift is the lager size of molecules of benzol that prevent their penetration to hydrophobic caviars containing hem. Therefore the using of optical method for estimation of ELFMF influence on interaction of benzol with methemoglobin is not so effective and statistically significant changes were not revealed after separated exposure. But there was a stabile tendency to increase the red shift in integral spectera on 0,6 nm during all time of exposure to magnetic fields (Table 5) and, in general, this fact does not contradict the results obtained in experiments with chloroform (Table 2.3). On the other hand, there were no any changes or tendencies in experiments with sham exposure (Table 6).

The analysis of differential spectera has confirmed the statistically significant influence of ELFMF on spectral characteristics of methemoglobin loaded by benzol (Table 7). The influence of ELFMF has changed the form of

Table 4. The basic characteristics of differential spectera of methemoglobin after its saturation by chloroform

Exposure time (hours)	Parameters	Control samples	Sham exposure	ELF MF influence	Changes relatively to control samples (%)	Changes relatively to sham exposed samples (%)
1	λ_{max}, nm	419.8±0.7	419.4±1.1	419.4±0.7	-0.1	0.0
2	λ_{max}, nm	419.2±0.8	419.0±1.5	418.7±1.1	-0.1	-0.1
4	λ_{max}, nm	421.2±1.3	421.8±1.4	419.8±1.1	-0.3	-0.5
24	λ_{max}, nm	422.7±1.6	424.0±2.0	**417.9±1.0 *,****	**-1.1**	**-1.4**
1	λ_{min}, nm	399.1±0.7	400.0±0.8	**395.7±1.4 *,****	**-0.9**	**-1.1**
2	λ_{min}, nm	398.7±0.5	398.4±0.9	398.8±0.6	0.0	+0.1
4	λ_{min}, nm	397.9±1.23	399.2±0.7	398.6±0.4	+0.2	-0.2
24	λ_{min}, nm	398.9±0.9	397.8±1.7	398.0±0.7	-0.2	+0.1
1	λ_{min}-λ_{max}, nm	20.7±0.6	19.1±0.9	**23.7±1.8 *,****	**+14.5**	**+22.2**
2	λ_{min}-λ_{max}, nm	20.5±0.9	20.6±1.6	19.9±1.2	-2.9	-3.4
4	λ_{min}-λ_{max}, nm	23.3±1.1	22.6±0.9	21.2±1.1	-9.0	-6.2
24	λ_{min}-λ_{max}, nm	23.8±1.4	26.2±2.0	19.9±1.4 **	-15.3	-24.0
1	D_{max}-D_{min}, o.d.u.	0.589±0.036	0.670±0.049	0.639±0.055	+8.5	-4.6
2	D_{max}-D_{min}, o.d.u.	0.543±0.032	0.552±0.055	0.592±0.038	+9.0	+7.2
4	D_{max}-D_{min}, o.d.u.	0.590±0.029	0.591±0.034	0.546±0.036	-7.5	-7.6
24	D_{max}-D_{min}, o.d.u.	0.602±0.028	0.558±0.039	0.594±0.058	-1.3	+6.5

* Statistically significant changes relative to control samples; ** statistically significant changes relative to sham exposed samples.

Soret's peak more significantly that resulted in statistically significant decrease of value of minimums (λ_{min}) and increase amplitudes (D_{max}-D_{min}) in the differential spectera in the first hours of exposure to magnetic fields (Table 7). The statistically significant changes were not revealed in the experiments with sham exposure (Table 7).

Table 5. The basic characteristics of integral spectera in range 350-450 nm of methemoglobin after its saturation by benzol and ELF MF influence

Exposure time (hours)	Parameters	Control samples	MF influence	Saturation by benzol	Satutation by benzol and MF influence	MF-induced wave shift (nm)
1	λ_{max}, nm	404.0±0.3	403.8±0.2	404.6±0.7	405.6±0.7	+1.0
2	λ_{max}, nm	403.8±0.2	404.8±0.3	404.8±0.3	405.6±0.6	+0.8
4	λ_{max},, nm	404±0.3	404.2±0.2	405.0±0.3	405.3±0.3	+0.3
24	λ_{max}, nm	405.0±0.5	404.6±0.3	405.4±0.5	405.6±0.4	+0.2
1 - 24	$\Delta\lambda_{mean}$, nm					**+0.57±0.19** p<0.05
1	D_{max}, o.d.u.	1.196±0.070	1.219±0.065	1.175±0.044	1.183±0.039	
2	D_{max}, o.d.u.	1.254±0.062	1.230±0.055	1.132±0.037	1.173±0.019	
4	D_{max}, o.d.u.	1.193±0.058	1.229±0.057	1.150±0.036	1.174±0.024	
24	D_{max}, o.d.u.	1.206±0.062	1.185±0.085	1.106±0.041	1.009±0.063	

* Statistically significant changes relative to control samples; ** statistically significant changes relative to samples saturated by chloroform.

Table 6. The basic characteristics of integral spectera in range 350-450 nm of *met-hemoglobin* after its saturation by benzol and sham influence

Exposure time (hours)	Parameters	Control samples	Sham influence	Saturation by benzol	Satutation by benzol and sham influence	Wave shift (nm)
1	λ_{max}, nm	404.4±0.3	404.2±0.2	405.6±0.6	405.4±0.4	-0.2
2	λ_{max}, nm	404.2±0.2	404.0±0.0	405.2±0,3*	405.6±0.4*	+0.4
4	λ_{max},, nm	404.4±0.4	404.0±0.0	405.8±0,4*	405.4±0.3	-0.4
24	λ_{max}, nm	404.4±0.3	404.0±0.0	407.0±0.6*	406.0±0.3*	-1.0
1 - 24	$\Delta\lambda_{mean}$, nm					-0.30±0.28
1	D_{max}, o.d.u.	1.207±0.020	1.197±0.020	1.077±0.058	1.090±0.051	
2	D_{max}, o.d.u.	1.208±0.022	1.150±0.051	1.116±0.033	1.118±0.038	
4	D_{max}, o.d.u.	1.239±0.017	1.218±0.085	1.105±0.025	1.107±0.036	
24	D_{max}, o.d.u.	1.280±0.062	1.258±0.086	1.110±0.112	1.054±0.054	

* Statistically significant changes relative to control samples; ** statistically significant changes relatively to samples saturated by chloroform.

Table 7. The basic characteristics of differential spectera of *met-hemoglobin* after its saturation by benzol

Expo-sure time (hours)	Parameters	Control samples	Sham exposure	ELF MF influence	Changes relative to control samples (in %)	Changes relative to sham exposed samples (in %)
1	λ_{max}, nm	422.3±1.8	420.2±1.6	419.4±1.4	-0.7	-0.2
2	λ_{max}, nm	421.5±1.5	423.0±1.7	421.9±1.4	+0.1	-0.3
4	λ_{max},, nm	419.9±1.3	422.0±1.2	419.1±1.3	-0.2	-0.7
24	λ_{max}, nm	422.1±0.9	424.0±1.6	417.9±1.0	-1.0	-1.4
1	λ_{min}, nm	398.1±0.8	400.2±1.3	**394.0±1.6 *,****	**-1.0**	**-1.5**
2	λ_{min}, nm	399.4±1.0	397.3±0.7	**394.5±1.7 ***	**-1.2**	**-0.7**
4	λ_{min}, nm	399.3±1.6	399.8±0.6	397.4±1.0	-0.5	-0.6
24	λ_{min}, nm	397.9±1.2	398.5±2.0	398.9±1.1	+0.3	+0.1
1	λ_{min}-λ_{max}, nm	24.2±1.8	20.0±1.8	**25.4±1.7****	+5.0	**+27.0**
2	λ_{min}-λ_{max}, nm	22.1±1.5	25.7±1.7	**27.4±2.3***	**+24.0**	+6.6
4	λ_{min}-λ_{max}, nm	20.6±1.7	22.2±1.9	21.7±1.7	+5.3	-0.17
24	λ_{min}-λ_{max}, nm	24.2±1.6	25.5±1.9	**19.0±0.9*,****	**-21.5**	**-25.5**
1	D_{max}-D_{min}, o.d.u.	0.344±0.032	0.344±0.042	0.338±0.051	-1.7	-1.7
2	D_{max}-D_{min}, o.d.u.	0.283±0.023	0.275±0.028	**0.365±0.029*,****	**+29.0**	**+32.7**
4	D_{max}-D_{min}, o.d.u.	0.341±0.020	0.278±0.037	**0.382±0.025 ****	+12.0	**+37.4**
24	D_{max}-D_{min}, o.d.u.	0.461±0.026	0.474±0.038	0.430±0.036	-6.7	-9.2

* Statistically significant changes relative to control samples; ** statistically significant changes relative to sham exposed samples.

4. DISCUSSION

The analysis of results that were obtained in the different experimental models and by means of different methods testifies to the binding of low-molecular hydrophobic substances with protein molecules. Is this binding specific or not? On our opinion, the binding of substances on hydrophobic mechanism in general is non-specific because it takes place in different colloidal systems[9]. At the same time, the sorption capacity of biomacromolecules can strongly depend on specific structural-functional properties. For example, the basic function of serum albumin is the binding and transportation of hydrophobic compounds such as fatty acids, bilirubin and others.

The next important question is about conformational changes induced by filling of hydrophobic cavities in protein molecules. According to[9], the saturation of proteins by hydrocarbons not results in significant changes of secondary structure of protein. Researchers speculatively concluded that tertiary conformation does not change too. But, in our opinion these conclusions were hasty. Our experiments showed that the saturation of protein solutions by non-polar substances is the dynamical and fluctuative process (fig. 4 & 5; table 1) accompanied by spontaneous cooperative and synchronous conformational transition in protein molecules. Such dynamical behavior of proteins in solution is known for a long time[12, 13]. At the same time it is known for a long time that interaction of ligands with proteins induces specific and non-specific conformational changes in protein molecules[14]. On our opinion the non-specific binding of non-polar substances by proteins induces conformational changes too. Therefore the "blue" shift in the absorption spectera of oxidized cytochrome ***c***, caused by binding of chloroform, can be caused by two reasons - by introduction of molecules of chloroform with nonzero dipole momentum (μ=1,06 D) in hydrophobic cavities of protein or/and by conformational changes macromolecule that increase availability of chromophore (gem) to strongly polar molecules of water (μ=1,84 D). It is necessary to note that similar effect was revealed in our experiments with native oxygenated hemoglobin[15]. But the features of conformation of methemoglobin are those that initially gem more strongly interacts with molecules of water. This realizes to shift of Soret peak in the absorption spectrum to short-wave range ("blue" shift) on 8-10 nm (λ_{max} = 404 nm) in comparison to absorption spectrum of native hemoglobin (λ_{max}=412-414 nm). Therefore the saturation of molecules of methemoglobin by chloroform results in the decrease of polarity of around gem and shift of maximum in absorption spectrum to the red range (fig.10). It is important to note that slow saturation of protein solution by non-polar low-molecular substances without active hashing not results in strong denaturation of proteins. The experimental results on high enzymatic activity of cytochrome *c*[16] and on absence of significant changes of electrophoretic mobility of serum albumin[17] after saturation by chloroform testify to it.

The carried out researches do not show statistically significant influence of ELF MF on properties of native proteins. At the same time, all statistically significant effects of ELF MF were revealed in the experiments with saturation of proteins by non-polar substances. So, it can be assumed that the "conformation stressing" of proteins, i.e. its structural modification, caused by loading by low-molecular non-specific ligands, is the factor which increase the sensitivity of protein molecules to the influence of ELF MF. The acceleration and increase of binding of hydrophobic substances by protein molecules is one of the features of the influence of ELF MF.

What are possible mechanisms of influence of ELF MF that result in increase of sorption capacity of protein molecules? Probably, the main cause of it is the decrease of solubility of hydrophobic substances in water. The obtained by refractometric methods experimental data testify to it (see part 2.2). The decrease of solubility, as a consequence, resulted in to moving of non-polar from water to basic volume of the same substance and also to the hydrophobic cavities of protein molecules dissolved in water. So, ELFMF changes the dynamical balance in the system *non-polar substance : water : protein*. The cause of this shift is connected with the changes of hydrophobic interactions in the protein solutions that directed to increase of hydrophobicity of low-molecular substances. As it is noted above, the properties of water is the basic factor that determines the dissolution of substances on hydrophobic mechanism. The dissolution of non-polar substances results in structurization of water and formation of crystal-hydrates covering hydrophobic molecules. Thus, it is possible to assume, that ELFMF influence on stability of such crystal-hydrates that results in sticking of non-polar molecules to each other and their raised adsorption in hydrophobic cavities of protein molecules. Analysis of literature data[10] confirms the revealed in our experiments facts of changes of solubility of substances under the treatment by magnetic field. So, all these facts confirm the hypothesis that water is primary acceptor of influence of ELFMF, but according to our data the MF-induced changes of properties of water first of all results in changes of hydrophobic interactions in the water-colloidal systems.

It is necessary to note one important circumstance. The mechanism of dissolution of gases in water media is similar to dissolution of hydrophobic compounds. The dissolution of gases is accompanied by structurization of water, too, and also by formation of clatrate cavities inside which molecules of gases are located[18,19]. Therefore it is possible to expect the influence of ELFMF on solubility of biologically important gases. On the other hand the MF-induced changes of structural-dynamical properties of water and its ability to form clatrate structures can be one of the causes of changes of binding of Ca^{2+} with proteins that are revealed in numerous studies[20].

The changes in the hydrophobic interactions caused by the influence of ELFMF can result in far-reaching biological consequences. It is known that the binding of hormones and other biologically active non-polar substances with receptors, enzymes and transport proteins and also reversible interactions of proteins with each other and with biological membranes is realized on hydrophobic mechanisms. The hydrophobic interactions play an important role in structural-functional organization and regulation of biological macromolecules and membranes[21]. Therefore, we can assume that the any small changes in hydrophobic-hydrophilic balance finally should result in shifts of activity of various molecular, cellular and system processes and changes of time organization of biological processes. These system changes should strongly depend on the functional specialization of cells and biological tissues.

Thus, the results of the researches that have been carried out by us at different times on base of the different protein models and the methods of investigations, allows us to make the following conclusions:

1. The saturation of protein solutions by low molecular hydrophobic substances results in non-specific binding of these substances with proteins on hydrophobic mechanisms accompanied by changes of conformation in molecules of protein.
2. ELFMF does not significantly influence on the structure of native proteins but changes the structure of proteins with modified conformation.
3. The influence of ELFMF increases the non-specific binding of non-polar substances with proteins and amplifies the conformation changes induced by such changes.

4. The influence of ELFMF decreases the solubility of low-molecular non-polar substances in water as through raising their non-polarity. More probably, the cause of this phenomenon is the changes of structural-dynamical properties of water that result in shifts of hydrophobic-hydrophilic balance in water-colloidal systems.

REFERENCES

1. A. Gurwitsch, Ueber den Begriff des Embryonalen Feldes, *Arch. Entw.-Mech der Organismen.* **51**(a), 383-415. (1922).
2. A. Gurwitsch, Ueber Ursachen der Zellteilung, *Arch. Entw.-Mech der Organismen.* **52**(a). 167-181, (1922).
3. G. Frank, Das mitogetische Reizminimum und-maximum und die Wellenlange mitogenetscher Strahlen, *Biol. Zbl.* **49**(H.3), 129-141 (1929).
4. O. V. Betsky, Millimeter waves in medicine and biology, *Raditechnika I radioelectronika (Radiotechnics and radioelectronics).* **38**(12)**,** 1760-1782 (1993).
5. A. S. Presman, *Electromagnetic fields and living nature* ("Nauka" Publisher, Moscow, 1968).
6. N. A. Temuryants, B. M. Vladimirsky, O. G. Tishkin, *Extremely low frequency signals in biological world* ("Naukova dumka" Publisher, Kiev, 1992).
7. V. S. Martynyuk, O. G. Shadrina, The influence of extremely low frequency magnetic field on solubility of benzol in water and protein solutions, *Biomeditsinskaya radioelectronika (Biomedical radioelectronics),* **2**, 56-60 (1999).
8. B. M. Vladimirsky, N. A. Temuryants, *Influence of solar activity on biosphere* (MNEPU Publisher, Moscow, 2000).
9. V. N. Izmajlova, P. A. Rebinder, *Structural formation in protein systems* ("Nauka" Publisher, Moscow, 1974).
10. V. I. Klassen, *Water and magnet* ("Nauka" Publisher, Moscow, 1973).
11. A. A. Akhrem, E. I. Tishenko, P. A. Kiselev, D. I. Metelitsa, Spectral characteristics of interaction of cytochrom c and hemoglobin with methanol and aniline, *Biochemistry.* **43**(11), 2033-2037 (1978).
12. F. R. Chernikov, Influence of some physical factors on oscillation on light dispersion in water and water solutions of biopolymers, *Biophysics.* **35**(5), 711-715 (1990).
13. L. I. Kyajveryajnen, *Dynamical behavior of proteins in water media and its functions.* ("Nauka" Publisher, Leningrad, 1980).
14. D. A. Sorkina, Conformational changes of blood serum proteins during performance transport function by them, *Voprosy meditsinskoj khimii (The Questions of Medical Chemistry).* **13**(3), 263-270 (1967).
15. V. S. Martynyuk, Yu. V. Tseyslyer, The influence of some hydrophobic ligands on spectral characteristics of hemoglobin, *Uchenye zapiski Tavricheskogo natsional'nogo universiteta (Scientific Works of Vernadsky Taurida Nationla University).* **17**(1), 150-155 (2004).
16. V. S. Martynyuk, P. S. Kalinovsky, Yu. V. Tseyslyer, The influence of weak extremely low frequency magnetic field on spectral characteristics of cytochrome c under its loading by chloroform, *Uchenye zapiski Tavricheskogo natsional'nogo universiteta (Scientific Works of Vernadsky Taurida Nationla University).* **14**(3), 121-128 (2001).
17. Yu. V. Tseyslyer, P. S. Kalinovsky, V. S. Martynyuk, Influence of extremely low frequency magnetic field on spectral characteristics of serum albumin under its interaction with hydrophobic ligands, *Uchenye zapiski Tavricheskogo natsional'nogo universiteta (Scientific Works of Vernadsky Taurida Nationla University).* **16**(3), 8-12 (2003).
18. S. I. Aksenov, *The water and its a role in regulation of biological processes.* ("Nauka" Publisher, Moscow, 1990).
19. S. P. Gabuda, *The connected water. The facts and hypothesises.*("Nauka" Publisher, Novosibirsk, 1982).
20. V. N. Bingi, *Magnetobiology. Experiments and Models.*("MILTA" Publisher, Moscow, 2002).
21. A. B. Rubin, *Biophysics* ("Knizhny dom" Publisher, Moscow, 1999).

9

FEATURES OF REACTIONS OF BIOLOGICAL AND PHYSICO-CHEMICAL SYSTEMS TO EXTERNAL FACTORS

Ivan A. Stepanyuk*

1. THEORETICAL PRESENTATIONS

Today the interest to ecological role of Cosmogeophysical factors (CGF) associated mainly with solar activity (SA) is highly increased. Investigations are usually run by finding statistical connections between reactions of biological systems (BS), or their physicochemical analogs (physicochemical systems - PCS), with cosmogeophysical factors (CGF) or with SA directly). Commonly the proof of the influence effect is thought to be sufficient if the coincidence of the characteristic of BS or PCS marked periods with corresponding periods of CGF ("cosmic rhythms" - by convention) is revealed.

Commonly the appearance of any foreign periods is thought to been caused by some more unaccounted physical factors.

Indeed, the scheme of typical spectrum analysis necessarily includes detection of typical periods in autospectrum (with corresponding filtrations "at the left" and "at the right", with the averaging for distortion registration "at the left" and "at the right " etc) in the beginning. After this the cross-spectral analysis and calculation of coherence functions meanings of which are serious only for coincident autospectral periods is possible to carry out.

But there are several specific peculiarities that may beget artifacts. Especially, that concerns false periods.

During examination of biological and phase characteristic (PC) systems, such as "black box" with input and output signals, the transfer functions are necessary to take into account. In the general case, the transfer function is complex. Amplitude-frequency characteristic (AFC) and phase-frequency characteristic (PFC) are marked out of transfer function. Signal transmission function in static mode - static connection function (SFC), and in dynamic mode - dynamic connection function (DFC) can be deduced from AFC.

In the experiments SFC is usually considered as linear *a priori*. However data series obtained during last years demonstrate that SFC is nonlinear, and and even non-monotone (the presence of "amplitude window").

---

* Saint-Petersburg, Russian State Hydrometeorological University

Dynamic non-linear features of DFC in experiments should be also taken into account.

A special problem is an infringement of experimental conditions. Naturally, we can agree with[1], that all the conducted experiments are partially "bad". But, except outspoken arguments of these authors, there is another methodical feature of experiments that has not been yet reviewed. It is in neglecting rules of choosing the discretization of observation of BS or PCS.

Due to this, the appearance of false ("illusory") periods in spectra and cross-spectra, as well as significant errors in determination of connections between reactions and influencing factors may take place.

2. MATERIAL AND METHODS

First of all the risk errors and artifact forming due to nonlinearity is considered in this work. The theoretical consideration is carried out using of simple mathematical apparatus, at the same time the references to experimental data are also used.

3. RESULTS AND DISCUSSION

3.1. Estimation of the Influence of SFC Nonlinearity

In case of summing the constant and quasiperiodic factors at the input in the system that is represented as "black box" (fig.1) with nonlinear SFC in the form of

$$R = R(0)\left[1 + \alpha x + \beta x^2\right], \quad (1)$$

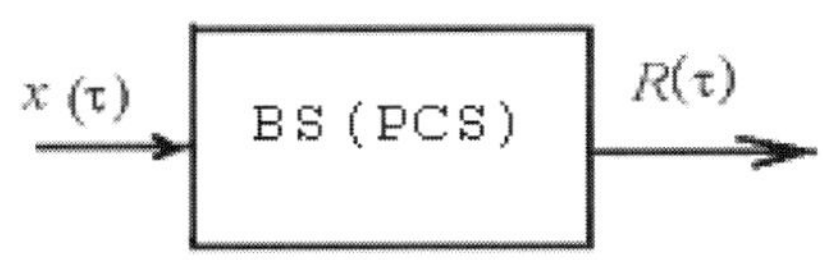

Figure 1. Conditional scheme of the system.

where α and β – coefficients, in the reaction of system R (output signal) the specific distortions will appear in additions to the typical reaction for constant signal. These additions will be proportional to the square of the alternating part of amplitude in the input signal (statistic error Δ_{0c}, see fig. 2). Moreover, the typical reaction on the alternating signal becomes dependent on the intensity of the constant part of input signal. In addition the false quasi-periodical reaction ("clear" artifact) appears in the second input signal harmonic. In mathematical interpretation of this is following:

$$R = R(0)[1 + (\alpha x_1 + \beta x_1^2 + 0{,}5\beta x_m^2) + (\alpha x_m \sin\omega\tau + 2\beta x_1 x_m \sin\omega\tau) - (0{,}5\beta x_m^2 \cos 2\omega\tau)], \quad (2)$$

where x_1 – constant component of input factor, x_m – amplitude of periodical component of input factor

The example of distortion is seen on fig. 2.

If the structure of the alternating input signal part is more complicated (two harmonics), then "clear" artifacts appear on both second harmonics and also in summary and differential frequencies. The situation when the affecting signal is amplitude-modulated have a special interest. In this case the nonlinearity of SFC leads to partial detection and, as a consequence, to the appearance of amplitude modulation frequencies and their harmonics in complex spectrum of system reaction (this is also a "clear" artifact).

The considered combinations are not speculative. The estimations of influence of geomagnetic disturbances (alternating part of signal) on biological and

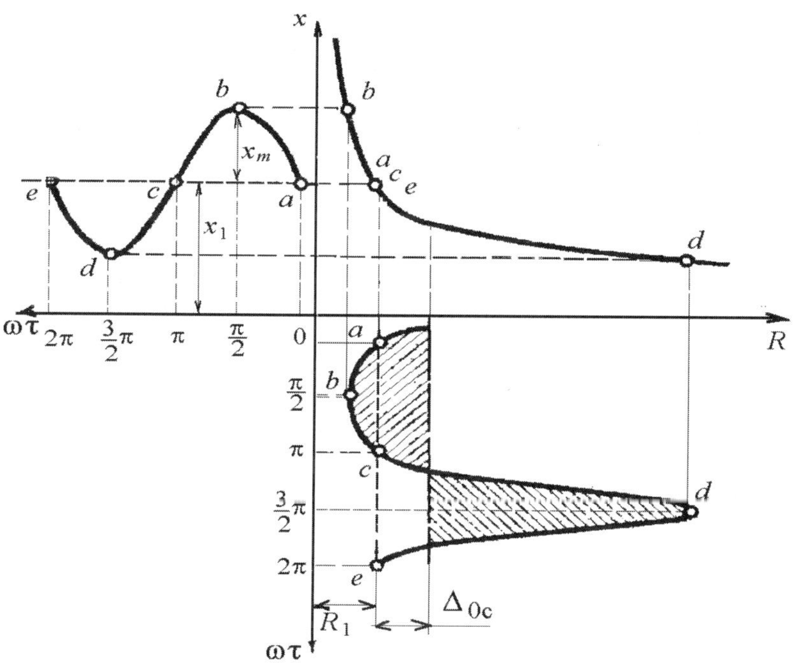

Figure 2. The distortion of reaction **within the** complex of constant and periodical factors at nonlinear SFC condition

physical-chemical systems are held under the Earth magnetic field (constant part of signal). But, in some experiments there were full or partial compensation of the Earth magnetic field (see collected articles[2]).

Usual (monotone) nonlinearity of SFC seems to be very widespread. Nonmonotone nonlinearities are more hypothetical. However, the existing of "amplitude windows" is a circumstantial evidence of such non-monotone nonlinearities. At the same time non-monotony of SFC and "amplitude windows" conditioned by it cannot be the artifacts but in the case of non-monotonous SFC the usual nonlinearity is indispensable on some curve segments, and neglect of it causes effects similar to those indicated above.

3.2. The Estimation of Influence of the Dynamical Features of BS (PCS) on the Results

If the dynamic features of BS (PCS) are linear (i.e. when they are described by differential equations of 1-st, 2-nd order or higher), unexpectedness are possible including conditions of quasiresonance. In particular, if the (DFC) of the system is described by the equation of the 2-nd order:

$$\tau_{e2}^2 \frac{d^2R(\tau)}{d\tau^2} + \tau_{e1}\frac{dR(\tau)}{d\tau} + R(\tau) = X(\tau), \tag{3}$$

A sufficient condition of quasiresonance is a significant exceeding of the time constant in the second derivative over that in the first one. Such situations are not speculative too. In particular, in our investigations of the influence of GMF on infectious diseases of gastrointestinal tract[3] we have came to the necessity of estimation of an organism as a second order system containing, first, the sluggishness of variability of enterobacteriums, and, second, the immune system sluggishnes.

The nonlinearity of dynamic properties of BS (PCS) is the cause of peculiarities of its reactions on external factors. The results of experiments often represent this. The usual, i.e. the most simple case of nonlinearity of DFC appears when the reaction time constant under the increase of intensity of the affecting factor differs from the reaction time constant under the decrease of this factor. There are no general solutions here. Some simple examples show how the "additions" to a quasiconstant signal, depended on the alternating period and time constant, appear under combining the alternating and the quasiconstant input signal. This is presented in our work[4].

If the alternating part of the signal is frequency-modulated then due to nonlinearity of DFC the detection of this part is provided and the "clear" artifact – the appearance of frequency modulation in the spectra of the reaction is formed. This artifact shows the difference of the role of DFC nonlinearity from the role of SFC noted above where the amplitude modulation is produced. The example of the effect of frequency-modulated detecting factors is shown on fig. 3.

So, there is no necessity to suppose the presence of neglect extra external factors in experiment after determination of discrepancy of specters of the reaction spectra to the spectra of the affecting physical factor. This extra factor can simply be absent and distortions can be caused by the effects concerned here.

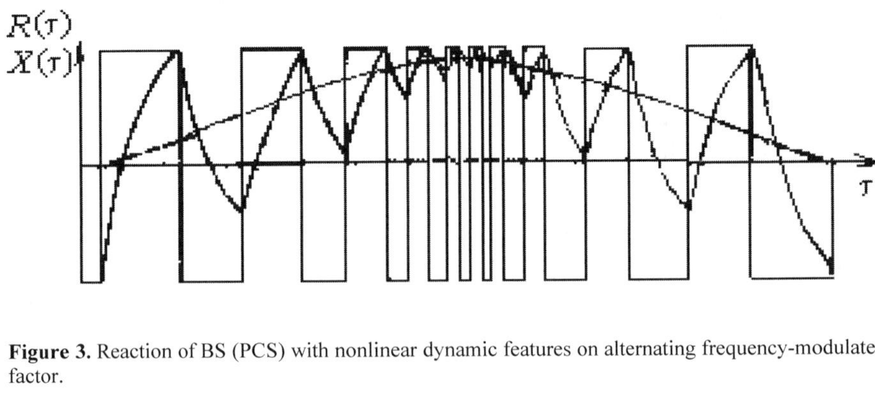

Figure 3. Reaction of BS (PCS) with nonlinear dynamic features on alternating frequency-modulated factor.

3.3. Influence of Experiment Conditions

The separated problem is a neglect of experimental conditions. There is another methodical feature of experiments that, seemingly, has not reviewed earlier.

During the observation of natural processes with non-limited spectrum the effect of illusion of discretization (effect of "misplaced frequencies") often becomes apparent. Usually the measure discretization during the experiment is determined a priory according to necessity or experimental possibilities. In this case the conventional limiting frequency for spectral results analyses is $(2\Delta\tau)^{-1}$, where $\Delta\tau$ - discretization. However, oscillate phenomena outside this frequency area do not disappear. Their energy is transferred to the low frequencies area, either increasing whole noise, **or** forming false "illusory" oscillating modes in the spectrum.

An example of the illusory periods is shown on fig. 4.

In this example the period of affecting factor X(τ) is accepted as 103 min (one of the periods of solar oscillations), while the observation of BS (or PCS) reactions are held every 2 hours. Smoothed curve of this observations (discontinuous) shows periodical R(τ) reaction with 12,4 hours period. Of cause, this result can be wrongly interpreted as the reaction of the system to the lunar influence (12.4h is the semi-diurnal lunar period). By examining the general regularities of this effect like a vibration energy transfer on the frequencies $(2kf_k \pm f)$ where $f_k = (2\Delta\tau)^{-1}$ and k=1,2,3... etc to a certain frequency f we, unfortunately, do not see the possibilities to take the distortions into account correctly. But we have no primary information about spectrum nature at frequencies

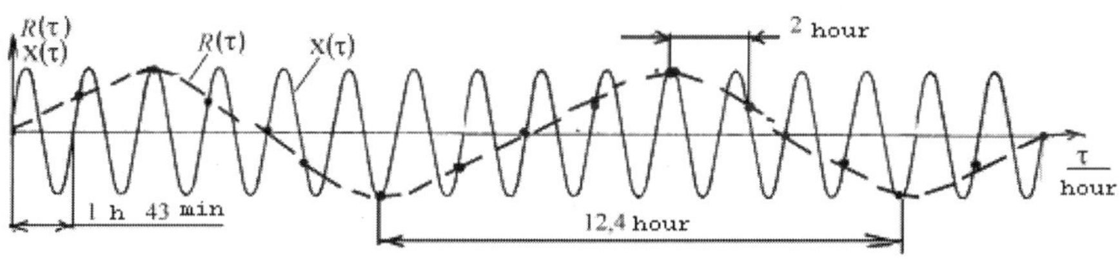

Figure 4. Formation of the effect of of illusion of discretization caused by discrepancy of period of experimental observation and period of affecting factor.

above $(2\Delta\tau)^{-1}$. This leads to a conclusion that the main problem in the studies of influences of cosmogeophysical factos upon the earth processes is a necessity to make more frequent observations for estimating spectrum features in high frequency areas.

An example of such a transfer during spectral data analysis is shown in fig. 5. Here in the real spectrum $S(f)$ maximum on frequency f_R, the outside frequency $f_k = (2\Delta\tau)^{-1}$ is presented. But due to considered effect the maximum appears in the calculated spectrum at the ($2f_k - f_R$) frequency.

The considered peculiarity, similarly to a preceding one, is not speculative. A clear example, where the illusion of discretization is possible, is the long-term monitoring (once per 24 hours) of any physiological characteristics of human and animal organism with the well known diurnal rhythms.

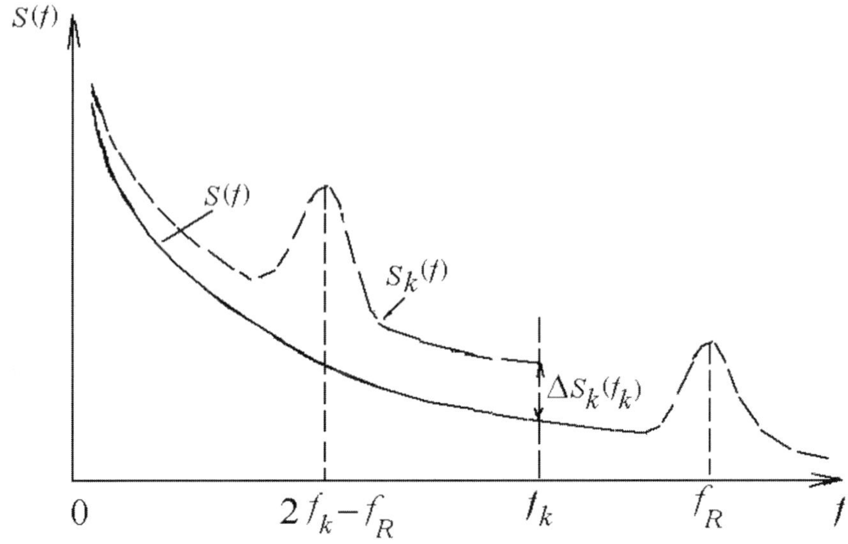

Figure 5. Example of energy transfer from high-frequency area to low-frequency at the expense of discretization.

4. RESUME

The considered peculiarities are not principally new. In electric engineering (for example, dealing with linear and nonlinear electric circuits[5, 6]) in informational-measuring technique («metrological features of measurements ») and so on, this peculiarities are estimated[7, 8]. However, with the reference to biophysical problems the similar observation had never been carried out.

In more details all of these problems were discussed in[9], including the dynamic properties of complex biological systems under the influence of external factors.

5. REFERENCES

1. B. M. Vladimirskiy, N. A. Temuryants, *Influense of solar activity on biosphere-noosphere* (Moscow, Pub. MNEPU, 2000).
2. Themes of reports of Int. Crimea Conference "Cosmos and biosphere" Partenit. Crimea, Ukraine, 28.09-4.10.2003, (Crimea Scientific Center NAS UA and IOS UA, 2003).
3. I. A. Stepanyuk and others, The influence of cosmogeophysical factors on infectious diseases of gastrointestinal tract. *Materials of the final scientific council session of RSHU. Part 2. Sections: oceanology, ecology and physics of the environment.* (RSHU, 2004), p. 128-129.
4. I. A. Stepanyuk, Dynamic features of bioobjects while affecting them by outer factors and methods of analogy parameter creating. *Cosmos and biosphere. Themes of reports Int. Crimea Conference* Partenit. Crimea, Ukraine 28.09-4.10.2003. (Crimea Scientific Center NAS UA and IOS UA. 2003), p. 28-29.
5. L. A. Bessonov, *Nonlinear electric circuits* (Petrozavodsk, Higher school, 1964).
6. L. A. Bessonov, *Linear electric circuits* (M.: Higher school, 1974).
7. I. A. Stepanyuk, *Oceanologic measuring transferes* (L.: Hydrometeoizdat, 1986).
8. I. A. Stepanyuk, *Informational measuring systems in oceanology* (St-Petersburg, Pub. RSHU, 1998).
9. I. A. Stepanyuk, *Features of reactions of biological and physico-chemical systems to external factors*, (St-Petersburg, Pub. RSHU, 2004).

10
NEWS AND VIEWS IN UVA-LASER– INDUCED ULTRAWEAK DELAYED LUMINESCENCE OF CULTURED MAMMALIAN CELLS

Hugo J. Niggli, Salvatore Tudisco, Giuseppe Privitera, Lee Ann Applegate, Agata Scordino, and Franco Musumeci [1]

1. HISTORY IN QUANTUM OPTICS

Light is basic in our physical universe. In studying the radiance of black bodies, Max Planck applied the principles of statistical mechanics and hypothesized that energy is not continous but given off in packets or quanta (photons), the smallest part of light. Based on this postulation, Einstein used in 1905 this idea in order to publish his insight on the photoelectric effect in order to show how much energy you need to get an electron out of a metal[1]. He received the Nobel prize in the year 1921 for this breakthrough in modern physics. These important findings are the bases of modern quantum physics, which shows that elementary particles no longer represent separate, but complementary terms. In 1954, Richard H. Dicke pointed out that groups of molecules can act as antenna systems with coherent emission when their distances are smaller than the wavelength of the radiation they emit[1]. This coherence concept has been introduced by Herbert Fröhlich into biological fields in the late 1960s[1]. Most recently it has been shown that correlations in the quantum world have their own causes, which cannot be reduced to those of the events, and are insensitive to space and time[2].

2. BIOPHOTONIC EMISSION

Photons can be regarded as the information carriers of matter because they can be exchanged within atomic and molecular interactions. In the early 1920s,

[1] Hugo J. Niggli, BioFoton AG, rte. d'Essert 27, CH-1733 Treyvaux, Switzerland E-mail: biofoton@swissonline.ch (corresponding author). Lee Ann Applegate, Department of Orthopedics, Laboratory of Oxidative Stress and Ageing, University Hospital, CHUV, CH-1011 Lausanne, Switzerland. Salvatore Tudisco, Giuseppe Privitera, Agata Scordino, Franco Musumeci, Laboratori Nazionali del Sud, I.N.F.N. and Dipartimento di Metologie Fisiche e Chimiche per l'Ingegneria, University of Catania, I-95129 Catania, Italy.

the Russian biologist Alexander Gurwitsch suggested that ultraweak photon emission (biophotons) transmit information in cells[1], which has been refuted by Hollaender and Klaus[3] as reviewed before[1]. A breakthrough showing the presence of biological radiation by physical instrumentation was obtained by the development of photomultiplier tubes in the mid-1950s, which has been introduced into modern biology by Facchini and co-workers[4]. Highly sensitive photon counting systems have been developed for biophotonic measurements in a variety of different cells by Quickenden in Australia[5], Popp in Germany[1,6] and Inaba in Japan[7] in the 1970s. This modern biophotonic research showed that plant, animal, and human cells emit very weak or ultraweak photons[1,4-14].

The sources of biophotons has been discussed intensively. Since ultraweak photon emission has been detected in both the visible and ultraviolet region[10], radiation in these regions has been connected with excited carbonyl groups and/or excited singlet oxygen dimers arising from lipid peroxidation, which in turn are associated with an increase in various reactive oxygen species such as the superoxide anion, hydrogen peroxide, hydroxyl radical and singlet oxygen. There also has been shown substantial evidence for DNA playing a key role in these emissions[12,13].

3. SPONTANEOUS AND LIGHT-INDUCED ULTRAWEAK PHOTONS IN CULTURED CELLS

Experiments with cultured human cells were reported[14] in which normal and DNA excision repair deficient Xeoderma pigmentosum (XP) cells were UV-irradiated in medium and balanced salt solution (EBSS) and assessed for ultraweak photon emission. These investigations showed that an important difference between normal and XP cells was present and that XP cells are unable to store ultraweak photons which are efficiently absorbed by normal cells.

Using the fibroblastic differentiation system of Bayreuther and co-workers[15], which resemble, in their design, the hemopoietic stem-cell differentiation system, we were able to demonstrate that no significant difference exists in the rate and the extent of the excision-repair response to thymine-containing pyrimidine dimers following UV-irradiation shortly after mitomycin C treatment of distinct strains of human skin fibroblasts and in the mitomycin C-induced PMF stage of these cells[16]. In addition, aphidicolin inhibits excision repair of UV-induced pyrimidine photodimers in low serum cultures of mitotic and mitomycin C-induced postmitotic fibroblasts of human skin[17]. We have also shown in distinct stages of fibroblastic differentiation that bone growth factors induce proliferation in this system and change the emission of ultraweak photons[6]. Based on our finding that the most important induction range for these very weak photon emission is in the UVA range[6], we developed a highly sensitive technique for UVA-laser-induced ultraweak photon emission[18,19]. This new biological model system will open new dimensions on the importance of light in cell biology.

4. SPECTRAL ANALYSIS OF LASER-INDUCED ULTRAWEAK DELAYED LUMINESCENCE IN CULTURED MAMMALIAN CELLS: TEMPERATURE DEPENDENCE

Yield of ultraweak photon emission in the fibroblastic differentiation cell culture model depended on the temperature of photonic measurements[20]. It was also found that after several UVB-irradiations normal cells begin to absorb the ultraviolet light, while cells from patients with the disease Xeroderma Pigmentosum (these cells are unable to repair UV-induced DNA changes), loose this capacity of light storing[20].

Several laboratories showed that light-induced ultraweak photon emission relaxation dynamics are closely connected with the functional state of the investigated biological system[8,9,13]. Figure 1 represents such a DL dynamic in normal and tumor cells at two different temperatures (10°C vs. 32°C).

As depicted in Figure 1, there is a strong temperature dependence of DL in both cell types. From the data of DL dynamics mentioned above no clear picture has been evolved concerning the chromophore involved for DL induction. It can not be excluded that temperature dependent changes in DNA- and chromatin-conformation will influence the DL dynamics. More than 20 years ago, we have

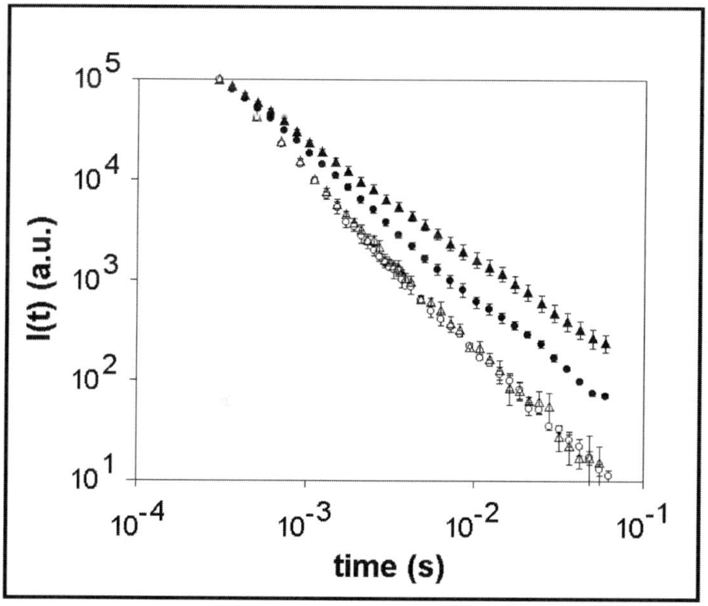

Figure 1. DL dynamic of human cells samples at several temperatures: (●) human fibroblasts at 32°C, (▲) human melanoma at 32°C, (○) human fibroblasts at 10 °C, (Δ) human melanoma a 10°C (average of three independent measurements and for each sample 100 repetitions have been performed).

shown that more DNA-photoproducts during ultraviolet B irradiation are produced by raising the temperature[21]. In view of different photochemical effects in chromatin and proteins during ultraviolet irradiation at various temperatures as basis for DL processes, the observed DL-changes may involve photosensitization[20]. The differences in structure between tumor and normal cells are well known and therefore our finding of a difference at higher temperature of DL dynamics between the two cell types tend to identify structural proteins as primary chromophores involved. In this respect, Scordino et al.[22], studying DL from unicellular algae, discovered that changes in DL on increasing temperature could be connected to changes in the structure's order during the polymerization process, showing a sharp transition analogous to the behavior observed in DL from liquid crystal samples when they undergo, on changing the temperature, a phase transition from the ordered solid state to the less ordered nematic liquid crystal state. At a higher temperature we observed a considerable change of the slope between normal and tumor cells which may be due to probability changes of the decay from the excited states produced. This difference between the various time trends can be better described by means of an analysis of decay times probabilities. In fact, relaxation of complex systems from non-equilibrium state towards equilibrium cannot be characterised by single rate coefficient, but more appropriately a distribution of relaxation kinetics exists, which in turn can be related to the hierarchically organised structure of the energy landscape in proteins.

Figure 2 depicts the result of the spectral measurements on the total number of photons emitted by normal fibroblastic and melanoma cells following UVA-laser induction upon temperature. There is a difference between normal and melanoma cells which is depending on temperature and wavelength. It has to be noted that these histograms does not represent the effective spectrum of the samples, because it does not take in account the sensitivity factor which is the product between the optical transmittance of the filtering system and the quantum efficiency of the photomultipier used (S-20 cathode).

The corrected emission spectrum shows that most of the emitted photons are detected in the visible red region (data not shown). It has to be noted however, that due to the low quantum efficiency of our S-20 cathode in this region, there exist a remarkable systematical error. Experiments are in progress using photomultipliers with higher efficiencies in the red and infrared region.

5. SPECTRAL ANALYSIS OF LASER-INDUCED ULTRAWEAK DELAYED LUMINESCENCE FOR DISCRIMINATION OF NORMAL AND CANCER CELLS

A recent paper illustrates a remarkable difference between the intensity of the DL of normal fibroblasts and of tumor cells[19]. Nevertheless the intensity is an extrinsic parameter depending on various factors as serum or passage number of culture which can lead to remarkable differences in the determination of ultraweak photon yields. Therefore it is important to include in the analysis other

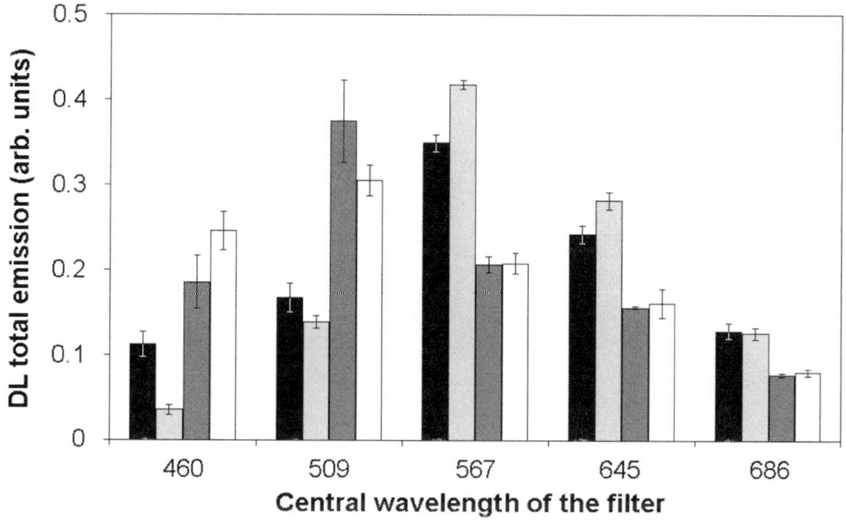

Figure 2. DL emission spectrum of different human cells at different temperatures, obtained normalizing the single spectral intensities to the value of their sum: (black bars) human fibroblasts at 32°C; (gray bars) human melanoma at 32°C, (dark gray bars) human fibroblasts at 10°C, (white bars) human melanoma at 10°C. Average values and standard deviations on independent measurements performed on four samples.

parameters as the time trend and the spectral distribution of DL, in order to obtain further information.

Based on this aim, we have determined in the present studies the components of the time trend the emission spectrum of DL in fibroblastic and human melanoma cells. These measurements have been repeated several times with distinct cell culture-samples in order to increase the statistical significance. The results of these determinations are shown in figure 3 and figure 4.

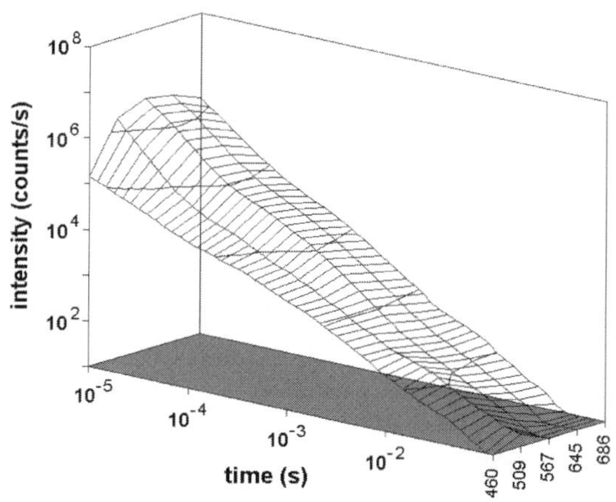

Figure 3. Intensity of DL as a function of time and wavelength for human fibroblasts.

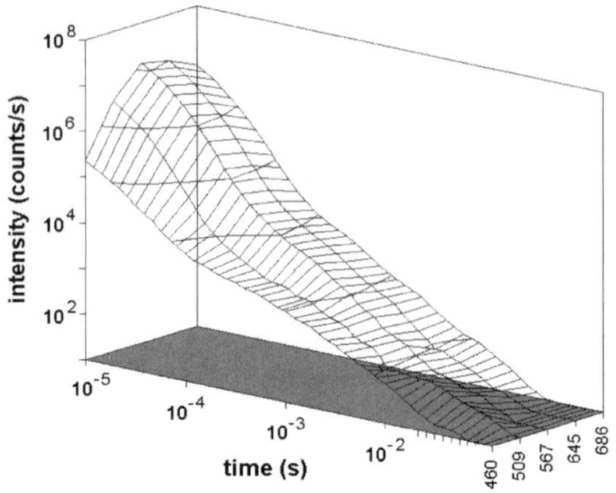

Figure 4. Intensity of DL as a function of time and wavelength for melanoma cells.

In contrast to investigations performed in algae, the various components of the emission spectrum are shown which exhibit a different time trend. It can be concluded that the measured emission spectrum depends on the time window of our measurement. Furthermore, due to the increase of the time interval of our measurements, we confirm for t <100 μs our previously published finding of higher biophotonic emission of fibroblasts compared to that of melanoma cells[19]. In the range t >100 μs the ultraweak photon emission of melanoma cells is higher as published more than a decade ago[10]. As pointed out before, this information on the DL intensity requires a quite precise evaluation of the quantity of cells as well as precise culture conditions in order to be used as a discrimination parameter. This evaluation, which is possible in "in vitro" experiments become quite difficult in "in vivo" investigations. Therefore it is necessary to find parameters that are able to give information on that state of the cells independently from their cell number. These parameters can be extracted by analyzing the time trend of the various spectral components.

Due to the fact that the emission spectrum is measured on the same sample and then the determination is repeated, using another sample to increase the statistic reliability, the spectral information does not depend on the calculation of the cell number. It is clearly shown that the emission spectrum depends strongly on the time window in which the measurements are performed. In particular, the two spectra seem to be quite different in the shorter time region, while, increasing the time range, it seems they will converge to similar values.

We have previously found that UV-induced ornithine decarboxylase response decreases with age and may therefore be used as a marker of aging[23]. As shown in Fig. 5, initial rates of DL for t <100 μs may be used in a similar manner as an aging parameter. The same Fig. 5 indicates that UVA-induced DL for t <100 can serve as a marker of carcinogenesis as published elsewhere[19]. It has to be noted that for this testing, well defined fibroblastic cells as well as melanomic cells have been used.

As reported by Grasso et al.[10], normal cells emit in the range t >100 μs significantly less ultraweak photons than cells obtained from cancer tissue. Similar results have been obtained by Van Wijk and Van Aken[9] in the early nineties in cultured murine cells.

In conclusion, in present studies, it is suggested to DL as a parameter for cell identification. In particular, our results show that DL is a new powerful non-invasive tool to determine biophysical changes within normal and tumor cells. It is also found that the differences between the DL of several biological systems strongly depended from the time and energy intervals in which measurements are performed. For this reason, our foreseen future developments will concern also the improvement of instrumentation with the aim to collect simultaneously the luminescence in several spectral bands, extended up to 1 μm, and to reduce the time delay of acquisition up to 100 ns. In this respect, future research is needed using a monolithic micro-device, such as Single Photon Avalanche Diode, in order to be able to determine at the same time the temporal decay trends in the region of the emission spectrum of interest.

6. FUTURE STRATEGIES USING SINGLE PHOTON AVALANCHE PHOTODIODES AS PHOTODETECTORS IN CELL RESEARCH

Avalanche photodiodes (ADP's) (see Scheme 1) are basically photodiodes with a high electric field region in which photoproduced carriers can gain enough energy to themselves produce electron-hole pairs as recently reviewed by Swain[24]. This provides a mechanism whereby, depending on the electric field, an initial photon can produce an electron hole pair which then initiates an avalanche of thousand of pairs. In this sense, the device provides gain in manner

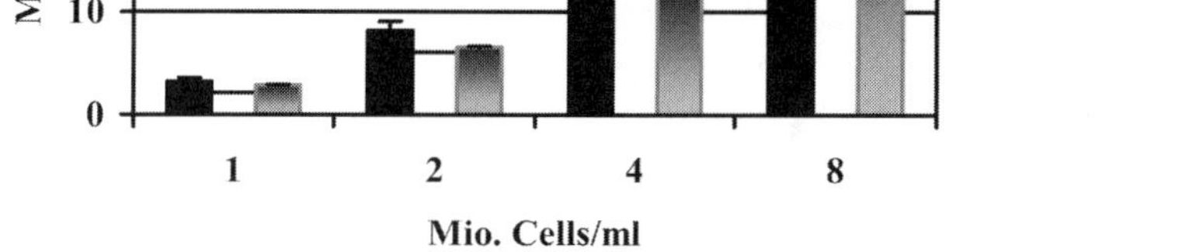

Figure 5. UVA laser light-induced ultraweak photon emission dynamics of emission spectra of normal fibroblast and melanoma cells. Initial rates DL of normal young skin fibroblasts (3229 p 13 and 117), adult fibroblasts (GM 1717; p 9) and melanoma cells (CCL53.1 p 40; CRL 1585 p 5) after pre-illumination with UVA-laser light in nanoseconds at suspension densities ranging from 0 to 8x10^{6} cells. One experimental point included in the graphic represents the average of 100 determinations with standard deviation less than 5%.

analogous to that afforded by photomultipliers, but with a significant added advantage: the amplification takes place in the same place as the light is absorbed. This means that the quantum efficiency (the probability that an incident photon will be detected) can be extremely high, and values above 90% are readily attainable. This is an enormous improvement over what can be achieved with devices such as photomultipliers which use photocathodes, where 20%, even over a rather narrow spectral range, is already quite good. Since light comes in in single quantam there is no way to compensate for missed photons, regardless of amplification, other than to extend the duration of measurements proportionately. For dynamical systems, such as living cells this is not always possible, so quantum efficiency can well be a limiting factor in making real-time studies. As an added bonus, the bandgap in silicon is much lower than the typical work function for any photocathode meaning that ADP's can retain this high quantum efficency over an extremely large spectral range, from infrared to ultraviolet. The main barrier to ADP's replacing photomultipliers has been the dark current generated by thermally produced carriers. This will be amplified in the same way that photoproduced carriers are, but recent research has shown that the effect can be largely circumvented by new techniques even at room temperature. This is ongoing research for applications in high energy physics, but now the time seems to be very appropriate to begin investigating how best to use these novel devices in biology and medicine in order to open new perspectives in scientific biophotonic research.

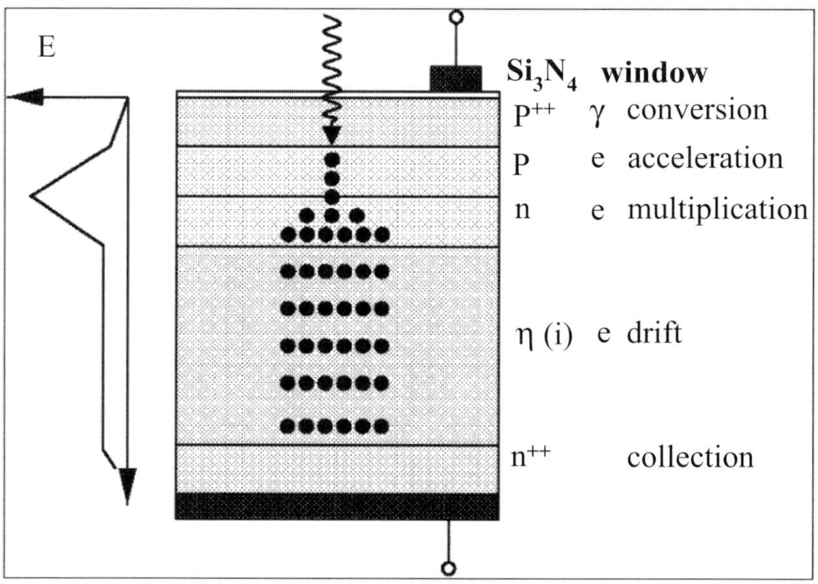

Scheme 1. Schematic of a basic avalanche photodiode structure and the corresponding electric field profile. The silicon nitiride window is for radiation hardness and can be replaced with silicon dioxide.

ACKNOWLEDGEMENTS

This study was funded in part by grants from the Swiss League Against Cancer (KFS 695-7-1998) and the Erwin Braun Foundation. HN was generously supported by his parents Alfred and Emmy Wehrli (Aarau, Switzerland). We thank also John Swain (Northeastern University,Boston) for helpful discussion on ADP-technique.

REFERENCES

1. F. A. Popp and L. Beloussov, *Integrative Biophysics: Biophotons* (Kluwer Academic Publishers, Dordrecht, 2003).
2. N. Gisin, A. Stefanov, A. Suarez and H. Zbinden, Quantum correlations that are not sensitive to space and time, Press-communiqué, Geneva 31 October 2001.
3. A. Hollaender and W. Klaus, An experimental study of the problem of mitogenetic radiation, *Bull. Nat. Res. Council* **100**, 3-96 (1937).
4. L. Colli, U. Facchini, G. Guidotti, R. Dugnani Lonati, M. Arsenigo and O. Sommariva, Further measurements on the bioluminesense of the seedlings, *Experientia* **11,** 479-481 (1955).
5. T. I. Quickenden and S. S. Que-Hee, The spectral distribution of the luminescence emitted during growth of the yeast Saccharomyces cerevisiae and its relationship to mitogenetic radiation, *Photochem. Photobiol.* **23**, 201-204 (1976).
6. H. J. Niggli, C. Scaletta, Y. Yan, F. A. Popp and L. A. Applegate, Ultraweak photon emission in assessing bone growth factor efficiency using fibroblastic differentiation, *J. Photochem. Photobiol. B: Biol.* **64**, 62-68 (2001).
7. H. Inaba, Y. Shimizu, Y. Tsuji and A. Yamagishi, Photon counting spectral analyzing system of extra-weak chemi- and bioluminescence for biochemical applications, *Photochem. Photobiol.* **30**, 169-175 (1979).
8. A. Scordino, A. Triglia, F. Musumeci, F. Grasso and Z. Raifur, Influence of the prescence of Atrazine in water on in-vivo delayed luminescence of acetabularium acetabulum, *J. Photochem. Photobiol. B: Biol.* **32**, 11-17 (1996).
9. R. Van Wijk and H. Van Aken, Spontaneous and light-induced photon emission by rat and by hepatoma cells, *Cell Biophys.* **18**, 15-29 (1991).
10. F. Grasso, C. Grillo, F. Musumeci, A. Triglia, G. Rodolico, F. Cammisuli, C. Rinzivillo, G. Fragati, A. Santuccio and M. Rodolic, Photon emission from normal and tumor human tissues, *Experientia* **48**, 10-13 (1992).
11. W. B. Chwirot, G. Cilento, A. A. Gurwitsch, H. Inaba, W. Nagl, F. A. Popp, K. H. Li, W. P. Mei, M. Galle, R. Neurohr, J. Slawinski, R. V. Van Wijk and D. H. J. Schamhart, Multi-author review on Biophoton emission, *Experientia* **44**, 543-600 (1988).
12. B. Devaraj, R. Q. Scott, P. Roschger and H. Inaba, Ultraweak light emission from rat liver nuclei, *Phochem. Photobiol.* **54**, 289-293 (1991).
13. H. J. Niggli, The cell nucleus of cultured melanoma cells as a source of ultraweak photon emission, *Naturwissenschaften* **83**, 41-44 (1996).
14. H. J. Niggli, Artificial sunlight irradiation induces ultraweak photon emission in human skin fibroblasts, *J. Photochem. Photobiol. B: Biol.* **18**, 281-285 (1993).
15. K. Bayreuther, H. P. Rodemann, R. Hommel, K. Dittman, M. Albiez and P. I. Francz, Human skin fibroblasts in vitro differentiate along a terminal cell lineage, *Proc. Natl. Acad. Sci., USA* **85**, 5112-1516 (1988).
16. H. J. Niggli, K. Bayreuther, H. P. Rodemann, R. Röthlisberger and P. I. Francz, Mitomycin C-induced postmitotic fibroblasts retain the capacity to repair pyrimidine photodimers formed after UV-irradiation. *Mutation Res.* **219**, 231-240 (1989).
17. H. J. Niggli, Aphidicolin inhibits excision repair of UV-induced pyrimidine photodimers in low serum cultures of mitotic and mitomycin C-induced postmitotic human skin fibroblasts. *Mutation Res.* **295**, 125-133 (1993).
18. S. Tudisco, F. Musumeci, A. Scordino and G Privitera, Advanced research equipment for fast ultraweak luminescence analysis, *Rev Sci Inst* **74**, 4485-4490 (2003).
19. H. J. Niggli, S. Tudisco, G. Privitera, L. A. Applegate , A. Scordino and F. A Musumeci, Laser-ultraviolet A-induced ultraweak photon emission in mammalian cells, *Journal of Biomedical Optics* **10**, in press (March/April issue 2005).
20. H. J. Niggli, Temperature dependence of ultraweak photon emission in fibroblastic differentiation after irradiation with artificial sunlight, *Indian Journal of Experimental Biology* **41**, 419-423 (2003).
21. H. J. Niggli and P. A. Cerutti, Temperature dependence of induction of cyclobutane-type pyrimidine photodimers in human fibroblasts by 313 nm light, *Photochem. Photobiol.* **37**, 467-469 (1983).
22. A. Scordino, A. Trigla and F. Musumeci, Analougus features of delayed luminescence from Acetabularia acetabulum and some solid state systems, *J. Photochem. Photobiol. B: Biol.* **56**, 181-186 (2000).
23. H.J. Niggli and P.I. Francz, "May ultraviolet light-induced ornithine ecarboxylase esponse in mitotic and postmitotic human skin fibroblasts serve as a marker of aging and differentiation?," *Age* **15**, 55-60 (1992).
24. J. Swain, Detectors for the quantizied electromagnetic field, in: *Integrative Biophysics: Biophotons*, edited by F. A. Popp and L. Beloussov (Kluwer Academic Publishers, Dordrecht, 2003), pp. 261-285.

11 ULTRAWEAK PHOTON EMISSION AS A TOOL FOR ANALYSING COLLECTIVE PROCESSES IN CELLS AND DEVELOPING EMBRYOS

L.V. Beloussov*

Abstract An ultraweak photon emission (UWPE) of the living systems, while being registered in certain experimental situations and/or demonstrating some specific dynamics, should indicate the collective properties of entire organisms and their populations. To these situations and dynamic properties belong: a degradational UWPE, largely synchronized (rhythmic) UWPE patterns (especially if accompanied by a set of harmonics), and non-additive relations between UWPE of the whole sample and its components. We describe the experiments performed on fish, amphibians, and hen's eggs and embryos and on the monolayer cell cultures fitting the above categories. A special emphasis is made upon the autocorrelation analysis of rhythmic patterns which demonstrate the ordered and functionally-dependent features of the UWPE frequency spectra within seconds - dozens of seconds scale. An adequate UWPE analysis opens new ways for non-invasive and non-inertial exploration of collective properties of the living organisms including macroscopic objects and populations of cells and embryos.

1. INTRODUCTION

Modern science uses three main epistemological categories for classifying the worlds' objects. Namely, the latter can be treated as: (1) collections of individual non-interacting particles with specific inside located properties; (2) sets of particles whose behavior is to a large extent determined by their interactions; (3) singularities of a common field. In this latter case, the concept of an individual material particle is ultimately rejected and replaced by a notion of a field, namely an overall space-temporal continuum obeying a certain common law. It is well-known that the fundamental physics of today takes the third viewpoint: "Modern physics can no longer base its description of physical phenomena on the classical concepts of particles as basic building blocks; rather, its "construction elements" are "elementary processors", complex-valued field "operators" that depend on time and location " (Dürr, 2002). Meanwhile, the

* L.V. Beloussov, Department of Embryology, Faculty of Biology, Moscow State University Moscow 119899 Russia.

conventional biology is still far from this paradigm. Up to now, an overwhelming amount of empirical studies in cell and developmental biology (not to say about a molecular one) takes the notion of a specific particle (let it be a gene, or a protein, or a cell organelle, or a specifically differentiated entire cell, etc.) as a fundamental one. True, the interactions between particles are also largely accepted, but the field approaches, in spite of some remarkable contributions (e.g., Gilbert et al., 1996) are still in a rudimentary state. Such a situation cannot last for long, because a very logic of an "interaction approach" shifts the researchers toward a more holistic viewpoint: it is a mere increase of a number of interacting particles which erases each particle's individuality, replacing it by certain global laws. There are certainly also many more direct arguments (some of them belonging to classical science) arguing for a similar conceptual shift. As concerning developing embryos, those are so-called embryonic regulations, that is, a capacity to develop in the normal way after removal, addition or a substantial rearrangement of embryonic parts. Such events directly indicate a subordination of the individual properties of constituent parts to a whole (which is probably at the given time moment only in a *status nascendi*). Coming to individual cells, we can also observe some cases of a behavior which is holistic in a classical sense. To these belongs, for example, an apico-basal electrical polarization of the epithelial and egg cells (Stern and MakKenzie, 1983), or a capacity of microtubules to establish a new assembly center after the removal of a centriole. Even more important is however a vast category of the processes to which a term "holistic" is rarely applied, and which are usually treated as collective, that is, irreducible to short-range interactions and uncorrelated activities of individual particles (molecules). To these belong:

- Spreading of excitation waves through individual cells and cell populations as through a homogeneous medium. Worth mentioning, those are not the diffusion waves of a certain substance, but a relay transmission of some excitation (for example, induction of a Ca^{2++} release – Jaffe, 1993). Usually these waves are accompanied by periodic oscillations (Rensing, 1993). In several remarkable cases the correlated oscillations cover extremely large frequency ranges (up to 74 octaves), namely, from days (circadian rhythms) to seconds (Ho, 2004). They consist of a series of harmonics (integer frequencies).
- Capacity of many types of cells for rapid and extensive amplification of weak and ultraweak signals. For example, in the photoreceptor cells 1 photon-excited rhodopsine molecule can, in its turn, excite no less than 1600 molecules of a transducine per 1 s (Alberts et al., 2003). From a conventional point of view the excitation of each next transducine molecule requires its lateral movements (those oriented along the plane of a photoreceptor cell membrane) towards and away of a rhodopsine molecule. Assuming that the rate of the lateral movements of protein molecules are of 1 μm/s magnitude order, the corresponding shifts of the individual transducine molecules cannot exceed 1/1600 μm < 1 nm, which is too small for producing the observed effects. It is difficult to avoid a conclusion, that such a rapid and extensive amplification is associated with some generalized transformations of a cell's electromagnetic field (mostly expressed in its submembraneous regions) providing an almost simultaneous excitation throughout large areas, rather than with lateral movements and short-range

interactions of individual molecules. Same way of reasoning can be applied to the events of an "explosive exocytosis": for example, in neuro-muscular synaptic junction no less than 100 acetylcholin (AC)-containing vesicles (each one containing about 10^4 AC molecules) per millisecond are extruded from a so called motor plate (a small part of a neuro-muscular synaptic membrane) (Slawinski, 1988). Direct electromagnetic influences of a subthermal ($< kT$) energy are also among those considerably affecting cell behavior. In these cases the effects are, as a rule, of a resonance nature (Webb, 1983).

- Cells' capacity to react in a specific way (up to switching on and off some definite genetic programs) to non-specific and smoothed mechanical forces (Huang and Ingber, 2000), which may be extremely small: thus, the acoustic receptor cells are able to percept a mechanical force not exceeding 2×10^{-13} N and caused by a 0.04 nm stretching of microfibrils (which is less than a diameter of a hydrogen atom) (Alberts et al., 2003). Again, these data invite us to consider a cell as a continuum, extremely sensitive to mechanical deformations, rather than a mere collection of specific molecules allowed to diffuse in a random way.

Within last decades several important theoretical and experimental studies of collective processes on supramolecular and cell levels have been performed (Fröhlich, 1968; Webb, 1983; Welch and Berry, 1983; Bistolfi, 1991; Wu, 1994; Ho, 2004) unfortunately almost unnoticed by a conventional science. Among the mentioned authors' conclusions is that about $10^{11} - 10^{12}$ Hz common oscillations of DNA, proteins and the bound intercellular water are taking place (Webb, 1983). By Fröhlich, a thermal energy is transformed into this frequency range coherent electromechanical oscillations of highly polarized medium, located within and below cell membrane. Emphasized is also a possibility of a directed spread of energy via so called electrets, the chains of dipoles. To these belong microtubules, composed of tubuline dimers. They are able to transmit coherent excitations with 8 m/s rate (Bistolfi, 1991). In such a way, "macroscopic vectorial flows" (Welch and Berry, 1983) can be established. To conclude: "the living cell is a unique ensemble of macromolecules which act as a single unit. Also, its ability to perform each of its many functions in a set of time sequence, at rapid rates and at what must be considered as low temperatures suggests that it employs a form of electrodynamic property analogous to that of solid-state systems" (Webb, 1983).

Being of a high heuristic potency, these views however do not have up to now a strong enough experimental support. To a great extent this is due to a deficiency of non-invasive and non-inertial techniques permitting to reveal the collective processes in vivo conditions. Already *a priori*, the analysis of UWPE signals should be regarded in this respect as a very promising one. In no way occasionally, a discoverer of the biological photon emission, Alexander Gurwitsch (1922), started this line of investigations from assuming that a "cell surface" percepts the signals for cell division in a cooperative way, rather than by excitation of individual receptors. On the other hand, it is obvious that a mere detection of UWPE is not enough for making any conclusions about the existence and properties of collective processes: taken *per se*, photon emission can be easily attributed to individual uncorrelated chemical reactions producing free radicals. Only if UWPE is detected under certain experimental conditions and/or shows some specific properties it can really point to collective processes.

2. CRITERIA FOR REGARDING UWPE AS INDICATORS OF COLLECTIVE PROCESSES

Here we formulate several experimentally verifiable criteria which fulfilling permits to consider UWPE as an indicator of collective processes.

2.1. UWPE Flashes After Damaging Influences (Degradational Radiation, DR)

This kind of UWPE was discovered long ago in Gurwitsch's lab (Gurwitsch and Gurwitsch, 1945, Gurvich, 1968). The authors observed brief impulses of mitogenetic radiation (stimulating yeast budding) after abrupt cooling, mechanical agitation, or an exposure to alternate electrical current in all the living samples studied in this respect with an important exception of cancer tissues. (Meanwhile, a list of samples demonstrating a spontaneous UWPE emission was much more brief). Also, contrary to a spontaneous UWPE, this kind of photon emission (called by the authors degradational) was not inhibited by a so-called cancer extinguisher, a protein body known to bound free radicals. The authors concluded, that DR is associated with a destruction of certain associations of molecules (which they called unequilibrial molecular constellations, UMC), non-connected by any chemical bonds but maintained due to a permanent energy flow. Until a given UMC exists, relatively small portions of metabolic energy (for example, 0.3 – 0.5 eV relieved by a hydrolysis of ATP molecule) can migrate within UMC and being concentrated up to the level of visible and even UV quanta. UMC hypothesis was a first indication of a presence of essentially collective processes in the living cells. A modern concept of exiplexes (excited molecular complexes, see Li, 1992) is a direct development of the UMC concept. From a more general view, UMC can be regarded as a preview of dissipative structures.

Since Gurwitsch times several other examples of DR after damaging and necrotic influences have been presented (Popp et al., 1994; Slawinsky, 2003). In parallel with that, a remarkable feature of a slow and ultra-slow relaxation of energy (from 10^{-6} to 10^{3} s) combined with a high intermolecular energy density (about 0.5 eV per 5 nm length of a protein chain) have been discovered (Blumenfeld, 1983; Chernavskii and Chernavskaya, 1999). This largely increases the probability of energy accumulation up to the levels largely exceeding those of macroergic phosphate bonds. Here are some examples of collective processes where such an accumulation is mostly probable:

As mentioned before, about 10^6 acetylcholine (AC) molecules are released per 1 ms from a small part (15 – 20 nm diameter) of a presynaptic membrane. Hydrolysis of a single acetylcholine (AC) molecule by cholinesterase takes < 0.4 ms and produces ≈ 1.2 eV (what corresponds to 1000 nm wave length). About 10^6 AC molecules are hydrolyzed almost synchronously. "Therefore the probability of an energy accumulation of 3-10 single hydrolysis reactions in close proximity to achieve the threshold of electronic excitation is high" (Slawinsky, 1988).

Under relaxation of a supertwisted DNA, 9 supertwists are relieved within 10^{-2} liberating 2 eV (corresponding to photon emission in the red spectral range (Slawinsky, 1988).

It is known (Alberts et al., 2003), that immediately after polymerization of actin monomeres into F-actin fibers a hydrolysis of the monomeres-bound ATP molecules is taking place. Similar process of GTP hydrolysis is taking place after polymerization of tubulin subunits into microtubules. Rate of actin polymerization is from 10 to 1500 monomeres/s. Correspondingly, the average time of hydrolysis of 10 ATP molecules (which is enough for photon emission in the visible range at least) will be from 1 to 6×10^{-3} s. If taking into consideration that F-actin fibers are usually collected into dense bundles, such a time can be diminished at least in an order. Hence, a probability of accumulation of the energy of several split macroergic bonds into one large quanta is high enough.

2.2. Rhythms, Harmonics, and Interactions of Oscillators, as Revealed by Fourier Analysis of UWPE Records

Modern computer techniques for recording and analyzing prolonged time series opened new perspectives for using UWPE records as indications of collective processes. First, a very existence of rhythmic patterns (as revealed by Fourier analysis) either of spontaneous or degradational UWPE from any large enough samples or entire populations of embryos points to the existence of collective interactions involving many individual photon emitters. Next, the presence of well expressed harmonics (doubling frequencies) is very instructive, pointing to the ordered nature of the frequency spectra and making it probable the coherent properties of UWPE (Mae Wan Ho, 2004). Moreover, by tracing autocorrelation patterns of Fourier spectra, one can detect, whether a given spectrum is a mere collection of independent lines (frequency maxims) or the individual oscillators of similar frequencies interact with each other, obeying to few dominating centers. A "saw-like" spectral structure (presence of many narrow lines) points to the first possibility, while the existence of few broad lines to the second one.

2.3. "Non-Additivity" of UWPE Intensity

Suggest that a given part of a sample emits *A* photons, while the rest of it *B* photons per time unit. The question is whether the emission of the whole system (*AB*) will be a mere sum of its components emission. As shown below, contrary to reductionist expectations, in many biologically important cases a considerable inequality $(AB) \neq A + B$ is taking place. Moreover, in many cases a part of a system (even if it looks as a "passive" one, like for example an egg's envelope) emits more photons than the entire system. That points to a so called "subradiance" (Dicke, 1954), which indicates that the system under study consists of closely arranged coherent photon emitters. This is a very important collective property beforehand unknown for biological systems.

2.4. Analysis of a Delayed Luminescence

A so-called delayed luminescence (DL) of a sample after its illumination by an external source of light (Popp, 1992) can be considered as a particular case of a more general phenomenon described in Gurwitsch's lab as a secondary photon emission (Gurwitsch and Gurwitsch, 1945; Gurvich, 1968). Already at that time, it was noticed that the secondary emission has some obvious properties of organized collective processes. Namely, it can be induced only by periodic impulses of the "primary" irradiation, possesses itself an oscillatory pattern, is spread throughout a sample as a collective excitation and is not damped out by inhibitors of free radical processes. Further evidences on the collective DL properties came from Popp and Li (1992) approval that by determining DL slope one may conclude whether a given sample emits coherent or non-coherent light. Namely, a similarity of DL slope to a hyperbolic curve is sufficient for concluding that the emitted light is coherent.

3. EXPERIMENTAL DATA CORRESPONDING TO ABOVE FORMULATED UWPE PROPERTIES

3.1. Specific Frequencies and Harmonics, as Revealed by Fourier and Autocorrelation Analysis

3.1.1. Fish eggs and embryos

As seen from Fig. 1, Fourier spectra of the different stages fish embryos (taking a loach, *Misgurnus fossilis* L. as example) significantly differ from each other and from the control samples (see Beloussov et al., 2003, Beloussov, 2003 for more details). Most clearly these differences can be revealed in the autocorrelograms of the spectra (middle column): an almost non-correlated pattern typical for non-fertilized eggs (upper row) is exchanged in several dozens minutes after fertilization by a highly correlated one consisting of a series of extended and equidistant narrow peaks (row second from above). At the later stages the correlation decreases again (rows 3rd and 4th from above), but after hatching a peculiar wave-like patterns appear (lower row). These data can be interpreted as following. Immediately after fertilization a single powerful oscillator with many harmonics (like a string with many obertones) comes into action. After the end of cleavage a single oscillator becomes exchanged by a large set of smaller ones which are at first non-correlated with each other. Later on meanwhile these oscillators become again coupled together around few dominating frequencies that entangle the neighboring ones in a more or less loose but holistic way. These latter events are associated with the start of active behavior and hence the functioning of a central nervous system. We can see that a proper analysis of UWPE rhythmical patterns can tell us much about the holistic state of an entire organism and even of a large embryonic population (note that all the measurements were performed on the populations consisting of about 50 samples).

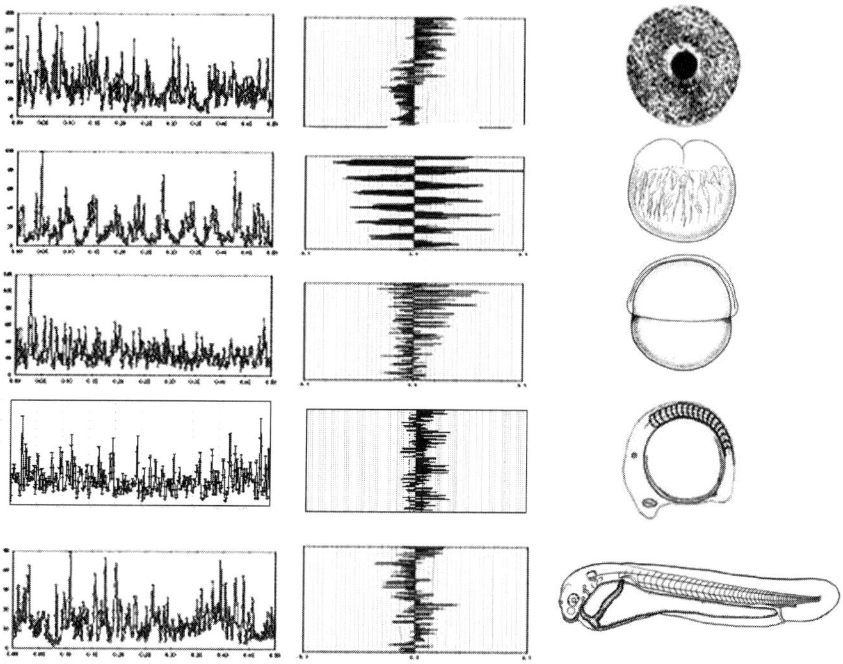

Figure 1. Fourier spectra (in terms of spectral densities) of loach eggs and embryos at successive developmental stages (left column), these spectra autocorellograms (middle column), and the pictures of the corresponding stages (right column). From top to bottom: non-fertilized eggs, start of cleavage divisions (1 h after fertilization), end of epiboly, formation of somites, and freely swimming larvae. Full horizontal scale of Fourier spectra is 1 Hz.

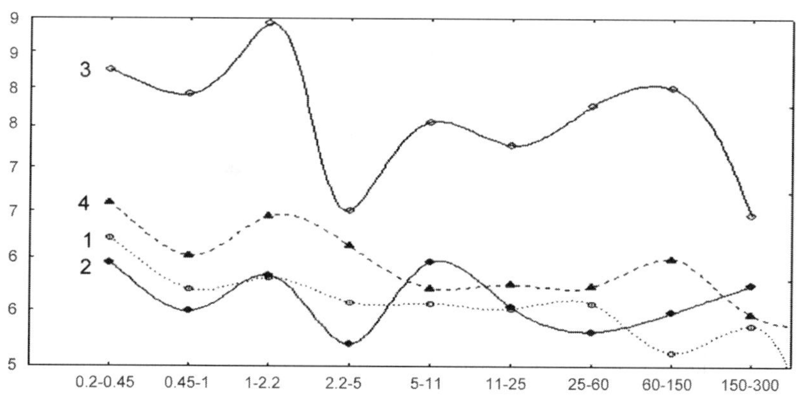

Figure 2. Extensively aggregated (up to 9 periods) Fourier spectra of different samples. 1: Dulbecco medium without cells; 2: non-treated fibroblast cell cultures; 3: similar culture immediately after addition of FGF; 4: non-treated cardyomyocytes. All the cultures were grown as monolayers on the vertical walls of quartz cuvettes filled by the same medium. Spectra 1, 2, and 4 are the averages of those recorded from 5 independent samples (8000 measurements per sample, dwell time 0.1 s). Spectrum 3 is taken from a single sample and normalized according to other spectra. Horizontal axis: limits of the periods (in seconds) for each measurement point. Vertical axis: UWPE intensity, logarithmic scale.

3.1.2. Monolayer cell cultures

We compared also Fourier patterns of UWPE from cell cultures of mouse fibroblasts and cardyomyocytes seeded as monolayers onto the internal surface of quartz cuvette walls, (48 x 22 mm square) filled with Dulbecco cultural medium, with those of the same cell-free medium. (The total amount of cells seeded onto a cuvette wall ranged from 2 to 6 $x10^5$). In the case of fibroblasts, we took those non-treated and treated with 0.05 mg/ml FGF-1 (fibroblast growth factor). 5 independently measured samples of each cell type and of a cell-free medium have been recorded during 12 min periods (8000 measurements with dwell time 0.1 s). We constructed Fourier periodograms for each of these periods, took the averages from periodograms values for each period and aggregated the obtained graphs in about 250 times. In such a way we got very much averaged but representative frequency spectra (Fig. 2).

Although the average UWPE rates of cell-free medium and non-treated fibroblasts were almost the same, the spectra of cell cultures significantly differed from each other and from a cell-free medium (Fig. 2, cf 1 and 2). At the

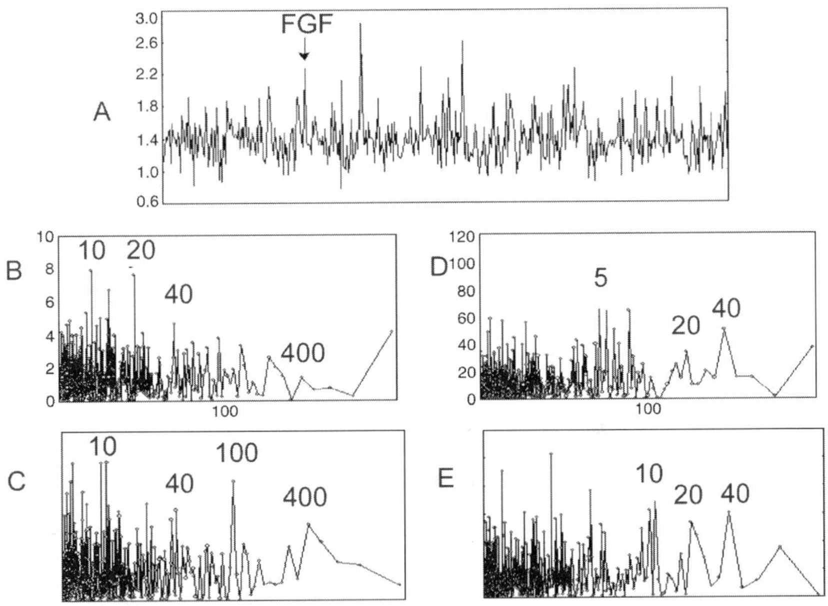

Figure 3. Effects of FGF-1 addition upon UWPE dynamics in fibroblast cultures. A: an experimental record showing the appearance of several UWPE peaks soon after FGF addition (full horizontal scale is 66 min). B, C: periodograms covering 30 min periods before and immediately after FGF addition to 1 h age fibroblast culture, correspondingly. Note the increase of low frequencies peaks (100 and 400 s^{-1}) after FGF addition while all the peaks still belong to the same "family" of harmonics. D, E: similar periodograms covering about 3 min periods of the same cultures recorded 7 h later. D: intact 7 h culture, E: that incubated for 7 h in FGF solution. Note the increase of peaks corresponding to 10, 20, and 40 s^{-1} frequencies correspondingly. In this and other pictures, the numbers inside spectrograms indicate the periods of mostly expressed spectral peaks in seconds.

same time, the spectra of the both non-treated and FGF-treated fibroblasts, in spite of considerable intensity differences, were similar to each other by having the peaks under 1-2,2 s and 5-11 s periods; their only differences belonged to greater periods (see below for more details). Cardyomyocytes shared with fibroblast cultures a spectral peak at 1-2,2 s but did not possess that under 5-11 s. Their overall UWPE intensity was also very close to that of a cell-free medium. One may conclude that the untreated cells themselves do not emit photons, but are able to concentrate ("canalyze") onto definite frequencies a photonic level energy produced by a culture medium (due, most obviously, to oxidative reactions).

As shown by more detailed analysis, the cells spectra differed from each other also by a set of harmonics and by a width of spectral lines. These differences were correlated with the functional state of cells. For example, just after FGF addition new low frequency spectral peaks appeared, being the harmonics of pre-existed ones (Fig. 3, cf B and C). Such a tendency was retained at 7 h of incubation in FGF solution (Fig. 3, cf D and F). As to the autocorrelation values of the intact fibroblasts, most of them did not go beyond the limits of significance (Fig. 4A), although if revealing sometimes a peculiar long-range pattern indicating a far-going coupling of individual oscillators (Fig. 4B). It is of interest, that these ordered patterns correlate with a decreased intensity of the individual spectral peaks (compare periodograms B and A). However, they never exhibited patterns discernible in 7 min after FGF addition (Fig. 4C), immediately after transferring cells into a hunger medium (Fig. 4D) or in 3 min after addition of trypsin (Fig. 4E). These specific autocorrelation patterns show that the coupling of UWPE oscillators is a very sensitive and rapid index of the collective cell responses to the stressful factors (see Discussion).

3.2. UWPE of "Degradational" Types

3.2.1. Mechanical perturbations of developing fish eggs

In each experiment, several dozens of loach eggs at blastula-gastrula stages maintained in a quartz cuvette have been slightly compressed by a vertically arranged coverglass. UWPE has been recorded before, during and after the end of compression. In 15 experiments, either a brief impulse or a prolonged increase of photon emission during pressure application has been observed (Fig. 5 A, B). The reasons of the differences between A and B types responses remained unclear. Meanwhile, as shown by Fourier analysis, the prolonged pressure-induced UWPE increases have been accompanied by extensive oscillations of emission intensity shifted towards lower frequencies and being multiple to pre-existed ones (Fig. 5, cf C, D, and E). Neither non-fertilized eggs, nor those treated by cytochalasine D, showed any reaction to pressure. We suggest therefore that the reactions observed are associated with a (reversible?) damage of a cortical layer of the microfilaments.

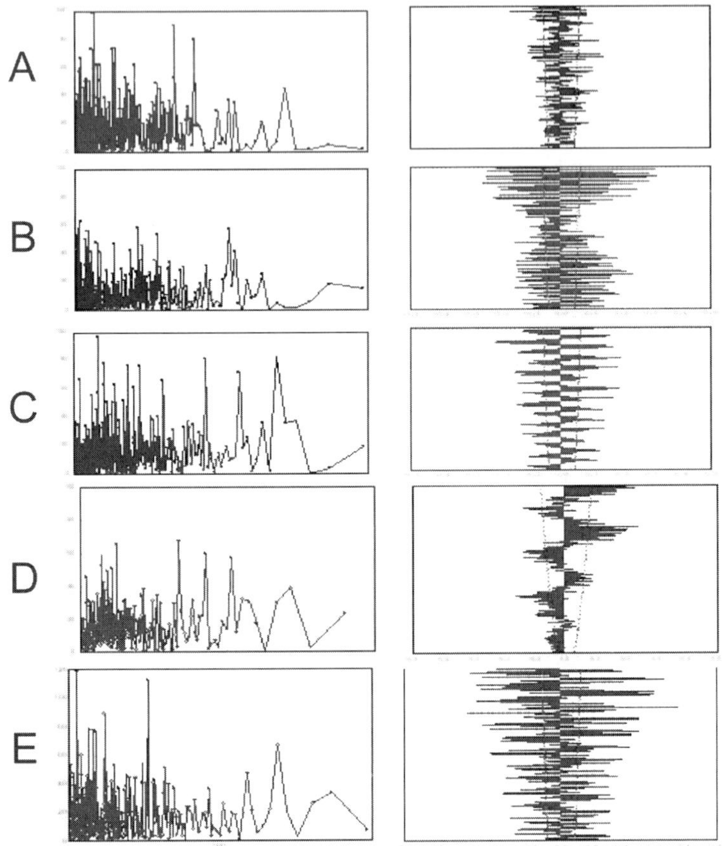

Figure 4. Periodograms and the corresponding autocorrelation patterns of fibroblast cultures maintained under different conditions. A, B: examples of intact fibroblast culture recorded in successive time periods; C: same culture, 7 min after FGF addition; D: a similar culture immediately after transfer into a hunger medium (phosphate buffer solution); E: 3 min after addition of trypsine (E). A, C, and E frames correspond to about 3 min periods while D frame to 2 min period.

3.2.2. Inhibitory influences upon cell culture

Fibroblasts and cardyomyocytes cultures were affected either by a sharp cooling (a transfer from 37^0C to 4^0C solution), or by addition of cytochalasine D (10 mkg/ml), colchicine (10^{-4}M) and 0.025% trypsin (which under the concentration employed did not detach the cells from the substrate; the cells only lost most of their mutual contacts and approached a spherical shape). Changes in UWPE intensity and in the oscillatory regimes have been traced.

As to the cooling experiments, in 5 cases out of 8 a significant increase of UWPE level has been observed (Fig. 6 A, E, I). Meanwhile in all the experiments including those where an integral UWPE increase was not detected (Fig. 6 B, F), the cooling led to increased oscillations with the periods either coinciding or multiple to pre-existed ones (Fig. 6 cf. C and D; G and H; J and K). Interestingly, (in accordance with Gurwitsch and Gurwitsch (1945)

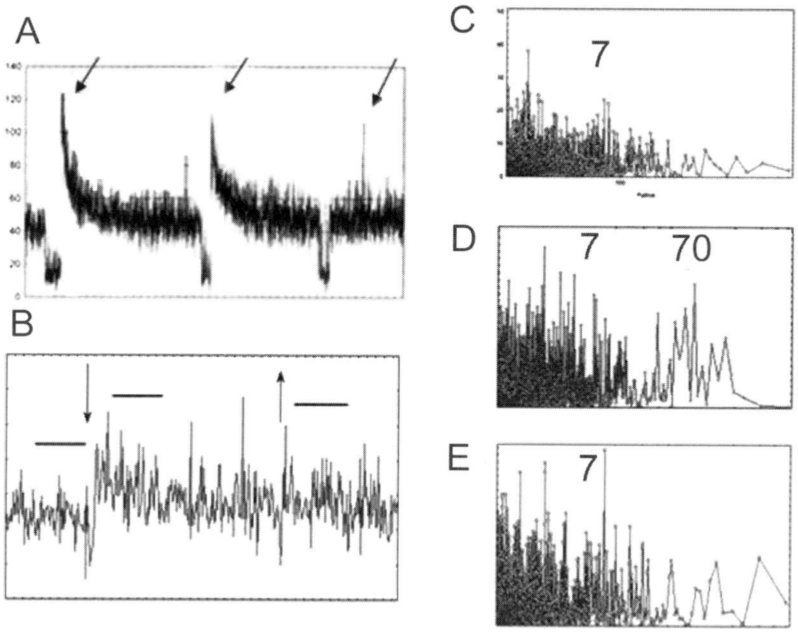

Figure 5. Mechanoemission of loach eggs. A: brief emission bursts after two first pressure impulses and almost no reaction to the third impulse. B: an example of a more prolonged response to the pressure. The moment of its application is shown by a downward directed arrow and the moment of a force release by the upward directed one. C-E: Fourier spectra (in terms of periodograms) before (C), immediately after the force application (D) and after its release (E) (marked by horizontal bars in B). Note an extensive increase of a low frequency harmonics (70 s = 7 s x 10) in D and its gradual damping in E, while a 7 s period is retained.

data) a second round of cooling greatly reduced the oscillations amplitudes (Fig. 6 cf. K and L). Noteworthy, neither UWPE intensity, nor the oscillatory pattern of a cell-free medium was affected by cooling.

A considerable increase of UWPE level was observed immediately after the addition of cytochalasine D and trypsine, but not after the addition of a colchicine (Fig. 7, upper row). As to the changes in Fourier spectra, they consisted, as before, in considerable enhancement of some spectral maxims either coinciding with or being multiple to the preexisting ones (Fig. 7, from top to bottom).

3.3. Non-Additivity and Subradiance

We compared UWPE intensities of the whole hen eggs of different incubation times, their freshly isolated shells and the yolks together with embryonic discs (see Beloussov et al., 1997 for details). First of all, it was noticed that UWPE intensity of both whole eggs and their freshly isolated shells (immediately after been taken from room light) in 2 – 2.5 orders exceeded that of their yolks. The remaining egg component, an egg white, did not emit at all. From this one might assume that UWPE intensity of the whole eggs should be equal to that of their shells. However, this turned out to be true only for non-fertilized eggs (Fig. 8A, white boxes).

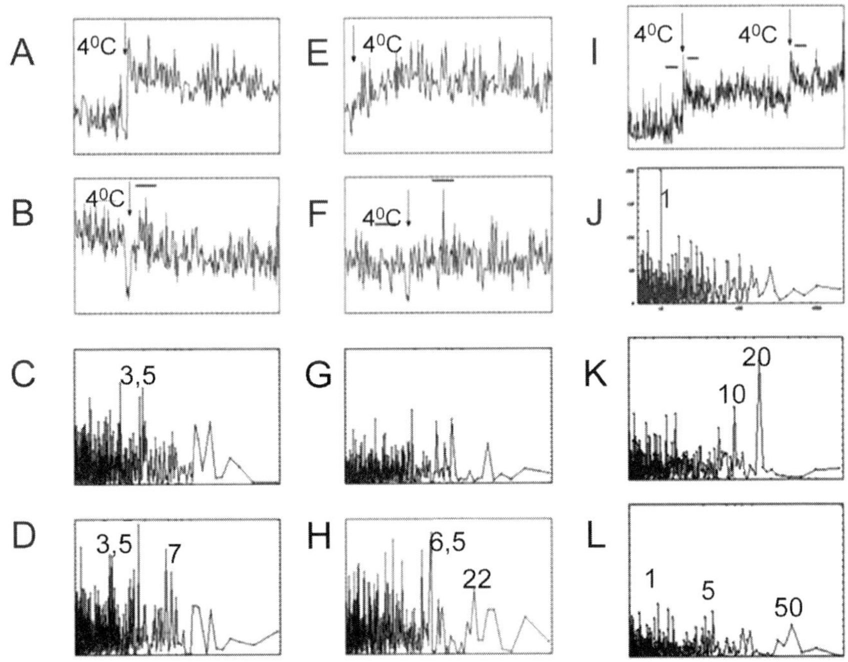

Figure 6. Features of a degradational UWPE after cooling fibroblast (A, B, F) and cardyomyocytes (E, I) cultures. A, B, E, F, and I are the records showing the changes in UWPE intensity after transferring cell cultures from 37⁰C Dulbecco medium to that at 4⁰C (downward arrows). In the case I, such a transfer was made twice. (Note that after each transfer to a cold medium the samples were returned to the measuring chamber under 37⁰C. Full horizontal axes equal 60 min. C, D are the periodograms corresponding to the time intervals shown as bars on B. G, H are the same for F and J-L the same for I. Note an extensive increase of some spectral peaks (mostly harmonics) in D, H, and K even if an overall UWPE intensity did not increase after cooling (as it took place in B and C). On the contrary, after second cooling (I) spectral peaks have been decreased (L).

As to developing eggs, in most of 2 incubation days samples the whole eggs UWPE was significantly higher than UWPE of the same eggs shells, while for 9 incubation days eggs, the reverse was true (Fig. 8A, hatched and black boxes, respectively). In other words, the photon emission intensity of an entire macroscopic system "egg + shell" differed significantly from a mere sum of its components emissions. This conclusion is confirmed by another experiment where the isolated yolk from 1 incubation day egg was put into quartz cuvette and either covered or non covered by a piece of a same egg shell. As one can see (Fig. 8B), the UWPE intensity of a system "yolk+shell" was significantly greater than the sum of the emissions of its components. As a result, one may conclude that at the beginning of incubation (1-2 days) the developing embryo in a highly non-linear way stimulates its shell's UWPE while at 9 days stage, on the opposite, a subradiance is taking place, as if a developing embryo "sucks out" from a shell some amount of the latter's emission.

Similar results have been obtained on amphibian (*Rana temporaria*) embryos (Beloussov and Louchinskaia, 1998). While at the gastrula stage the

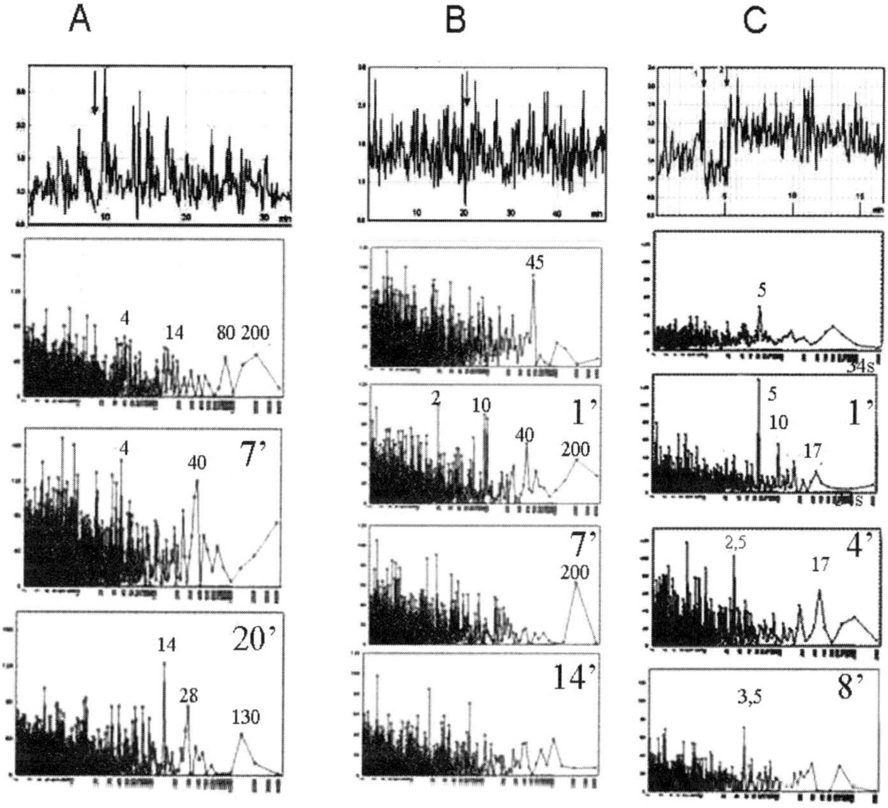

Figure 7. Changes in UWPE patterns of fibroblast cultures after addition of a cytochalasine (A), colchicine (B), and trypsine (C). Upper row: records, single vertical arrows in A and B and a right vertical arrow in C indicate the time of drugs application. Second row from the top depicts periodograms of intact samples and the next rows to below those measured in successive times after drugs application (starting times are shown in minutes). Each periodogram corresponds to a 3.5 min time period. Note an extensive increase of several spectral peaks, which are mostly the harmonics of each other and the preexisted peaks of the intact samples.

UWPE intensity of the isolated egg membranes was lower than that of the membranes + embryos, for more advanced stage embryos the reverse was true (Fig. 9A). One may conclude that the embryos stimulate their envelopes emission at the early stages and are "sucking off" the photonic level energy from the envelopes at the later stages.

Remarkably, the developing samples (in this case just hatched larvae of a loach, *M. fossilis)* can "suck up" photons even from a nylon net, into which they were put: an empty net (if taken from room light) emits much more photons than the same net (also taken from room light) containing several dozens larvae (Fig. 9B).

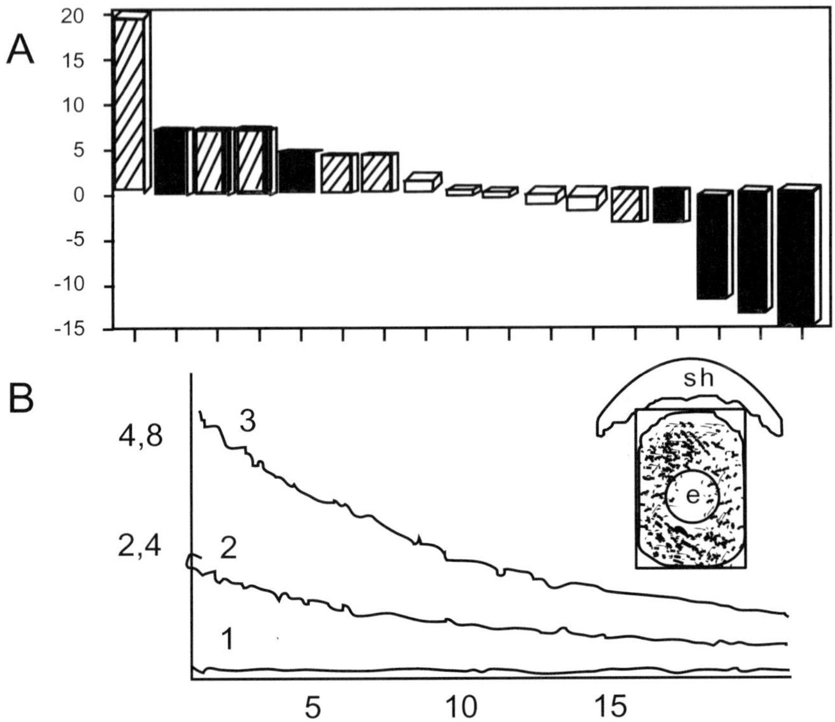

Figure 8. UWPE non-additivity in developing hen eggs. A: differences in UWPE intensities between whole eggs and their freshly isolated shells in non-fertilized eggs (empty boxes), 2 days incubation eggs (hatched boxes) and 9 days incubation eggs (filled boxes) The differences (number of impulses per 3 min x 1000) are plotted along vertical axis. Only for non-fertilized eggs they are close to zero. For most of 2 days samples they are positive while for most of 9 days samples negative. B: UWPE records after room lightning: (1) from a yolk + embryonic disc isolated from 1 incubation day egg poured into quartz cuvette and covered by a piece of shell from the same egg; (2) from a piece of a shell alone; (3) from yolk alone. To the right is a sketch of experimental mount Sh: shell, y: yolk, e: embryonic disc. One can see that (1) >> (2) + (3) (Beloussov et al., 1997).

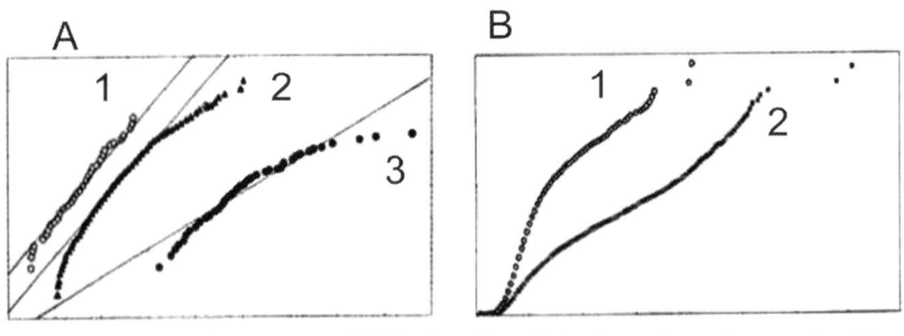

Figure 9. Examples of UWPE non-additivity in embryonic batches of neurula stage frog (*Rana temporaria*) embryos (A) and just hatched loach larvae (B). Given are cumulative distribution curves of UWPE intensities. Horizontal axis: photon counts per 5 s. A:1 is UWPE values for control samples (cuvettes with water), 2 is for embryos in vitelline membranes, 3 is for freshly isolated same eggs vitelline membranes. These latter emit much more than the embryos + membranes. B: 1 displays UWPE signals from 62 just hatched loach larvae put into a nylon net and 2 shows UWPE signals from an empty net, which is much higher.

Same effects could be also observed during optical interactions of two populations of the different age *M. fossilis* embryos. Embryos were put into two quartz cuvettes arranged one behind another in respect to a photomultiplier cathode. Firstly, UWPE of the frontal cuvette alone has been measured (Fig. 10, section 1) and then the hind one was placed. Immediately after putting the hind cuvette (section 2) the common UWPE almost doubled (as it should be if the emissions from the both populations are simply summed up). Soon however it went down in a gradual fashion, approaching the initial UWPE level of the frontal cuvette alone (section 4). If measuring now again UWPE of this latter (section 5) it turned out to be considerably lower than the same cuvette initial one. Important in this experiment is a graduality of UWPE decrease after the beginning of optical communication between the both cuvettes. This permits to exclude purely physical artifacts (like an abrupt change of optical conditions)

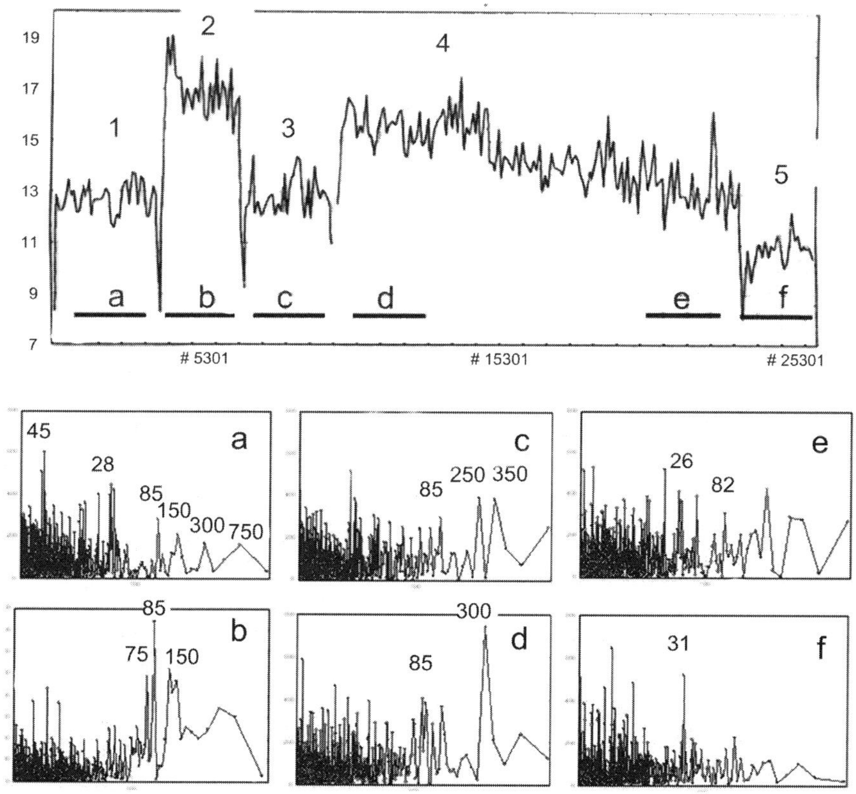

Figure 10. Subradiance under optical communication of two populations of loach embryos. Above shown is 200 min duration experimental record (dwell time 0.5 s).1: UWPE from a single embryonic batch at the start of cleavage divisions (30 min after fertilization); 2: cuvette with embryos at postgastrula stage is put behind; 3: hind cuvette removed; 4: hind cuvette put again; 5: hind cuvette removed again. Note a gradual decrease of UWPE intensity during period 4. While after a brief optical interaction with embryonic population from the hind cuvette the frontal one still keeps its initial UWPE level (cf. 1 and 3), this is not so after more prolonged interaction (cf. 1 and 5). Below are periodograms corresponding to 25 min duration periods a-e (shown by bars on the experimental record). Vertical scale for periodogram b is twice as large as for periodograms a, c-e.

making thus obvious that the subradiance is mediated by relatively slow biological interactions.

These conclusions were confirmed by experiments on the optical communications between the embryonic batches put into two symmetrically arranged cuvettes. No subradiance was observed if preventing their optical communications by screening the cuvettes (Beloussov et al., 2003).

4. DISCUSSION

The above presented data show that UWPE signals, while properly analyzed, can be used as unique non-invasive and non-inertial tools for exploring the collective properties of the living systems. At least two classes of such properties can be derived from our experiments. First, such are the specific, stage- and functionally-dependent rhythmic patterns taking place within at least seconds – dozens of seconds range and accompanied by a set of harmonics (multiple frequencies). Next, there are the phenomena of non-additivity (subradiance). We will start our discussion from the first class of events.

A possibility to register specific rhythmic patterns (frequency, or Fourier spectra) in the large embryonic populations and in cell monolayers consisting of billions of cells indicates, first of all, that these samples behave as integrated continuums, effectively synchronizing their activities. On the other hand, such interventions as a mechanical pressure, rapid cooling, and administration of cytoskeletal inhibitors make the rhythmic patterns more pronounced. This feature (rather than a mere increase of the average UWPE intensity) looks as a most reproducible property of the degradational radiation. Worth mentioning, in Gurwitsch's labs it was also noticed that the degradational radiation consists, most probable, of few number of flashes which affect yeast budding to a greater extent than a constant radiation (Gurwitsch and Gurwitsch, 1945).

In this respect it is of a special interest that the addition of FGF to a fibroblast culture already in few minutes is followed by several UWPE flashes which show, in comparison with non-treated fibroblast samples, an enhanced rhytmicity, mostly in the low frequency range (Fig. 2, cf. 2 and 3; Fig. 3, cf. B and C). Consequently, already the first steps of the cells' reaction to growth factors addition (associated, most probably with clusterization of membranes receptors) lead not only to the appearance of new electronically excited states, but also to the changes of the cells' collective properties.

The following two properties of the rhythmic UWPE patterns seem to be mostly important. First, as mentioned above, the patterns consist, as a rule, of several harmonics (multiple frequencies), covering sometimes very large frequency intervals (up to 1:100 ratio). Second, as it was most clearly demonstrated by autocorrelation diagrams, the patterns differ from each other by a width and a smoothness of the spectral maxims. Sometimes (Fig. 1, line 2 from above; Fig. 4D) the maxims are narrow, while in other cases (Fig. 1, lowest line; Fig. 4B) they are not so pronounced, but more extended. One may conclude, that in the latter cases the spectral maxims are acting like dominating centers entangling the oscillators with similar frequencies. As a result of such an entanglement, each individual oscillator's amplitude becomes reversely proportional to its spectral distance from a dominating center. In a rough approximation, this may be illustrated by a simple physical example. Suggest that we have a set of strings of the different flexibility and to a different degree bound (by some kind of friction) with each other. If a given string has a great flexibility and one can neglect its bounding with other strings, it will produce, after being excited, a series of pure obertones (harmonics). This corresponds to a series of narrow peaks, which are best of all seen in the autocorrelogramms. On the other hand, a string which is less flexible but is bound more firmly with other oscillators, will produce more smooth spectral patterns. As a development of fish embryos proceeds, the first kind of "strings" seems to be exchanged by the second one.

In any case, the presence of a large set of harmonics makes it probable, that the basic oscillators are similar, in their essence, to the real strings, rather than, for example, to the chemical reactors, non-linked with any spatially extended structures. We make such a conclusion because just in the string-like entities the harmonics are generated automatically while in chemo-diffusion systems they have to be each time settled *de novo* by adjusting the parameters. Now, it would be just naturally to identify the strings with the cytoskeletal structures, regarding the enhancement of the harmonics after destruction of a cytoskeleton simply as the sounds of the breaking strings. However, by suggesting this, we become confronted with the following problem: the seconds - dozens of seconds periods detected in our experiments are 10-11 orders greater than those of the assumed self-oscillations of, say, microtubules (the dipole oscillations in the tubuline subunits are estimated as 10^{-9} s periods: Bistolfi, 1991). Is it possible to imagine that the oscillators of so different frequencies are nevertheless correlated with each other?

This problem has been already discussed by Ho (2004). By her suggestion, a coupling of very much removed frequencies is possible within highly coherent systems which transmit phase relations without a considerable distortion to many magnitudes orders upwards. A striking example is the correlation between the frequency of wing beatings and circadian rhythms in insects which are removed from each other by 7 magnitude orders. In light of this, it is not so improbable to assume a transmission of the phase relations of some supramolecular "strings" up to a seconds – dozens of seconds frequency level.

The phenomena of non-additivity and subradiance are also compatible only with a property of coherency. At the first glance, it looks quite improbable to ascribe a coherence to such large bodies as whole hen's eggs or two optically communicating embryonic batches, each of them consisting of several dozens samples. However, the notion of a macroscopic coherency becomes increasingly accepted in the modern science (Del Guidice, 2000; Preparata, 2000) and is largely associated with the properties of a "bound water" (see Voeikov, this volume). With these concepts in mind, a border between the living and the surrounding non-living substance (for example, between a hen's embryo and its egg shell) becomes largely smoothed. "Alles lebendigen verbreitet sich die Atmosphaere um sich selbst" (J.-W. Goethe).

It is also worth mentioning, that one of the first manifestations of the optical communications is the enhancement of a certain frequency, beforehand pre-existing in one of the communicating populations (see Fig. 10 and the corresponding comments). Therefore, the introduction of an optical partner is appreciated by another one as a kind of a stress, to some extent similar to that induced by the above mentioned damaging factors or by growth factor.

While suggesting that the living bodies even of macroscopic dimensions can reveal coherent properties, we do not think that such a coherence is maintained perpetually. On the contrary, quite a rare presence of well correlated spectra in the fish embryos of advanced stages or in the intact (non-affected) cells indicates that the vibrational modes are as a rule dissipated towards lower energy levels, rather than come into coherent regimes associated with periodic production of electronically excited states. These latter events are most typical for the stressful conditions, although not necessarily damaging (remind FGF effect).

In any case, a proper analysis of UWPE signals opens us a large and sometimes strange world of the beforehand unknown collective processes in the living systems, most of which are at the moment quite far from being understood.

ACKNOWLEDGEMENTS

I express my thanks to Prof. R. van Wijk for the discussions and for his help in performing experiments on cell cultures and to Mrs N.N. Loochinskaia for her skillful technical assistance.

REFERENCES

Alberts, B., Johnson, A., Lewis, J., Raff, M., Roberts, K., Walter, P., 2002, *Molecular Biology of the Cell*, Garland Science, Taylor & Francis Group.

Beloussov, L.V., 2003, Exploring the dynamic background of the developmental processes and cell reactions with the use of an ultraweak photon emission. *BioSystems* **68**: 199.

Beloussov, L.V., Burlakov, A.B. and Loouchinskaia, N.N., 2003, Biophotonic patterns of optical interactions between fish eggs and embryos. *Indian J. Exp. Biol.* **41**: 424-430.

Beloussov, L.V., and Loouchinskaia, N.N., 1998, Biophoton emission from developing eggs and embryos: non-linearity, holistic properties and indications of energy transfer, in *Biophotons*, Jiin-Ju Chang, J. Fish and F.-A. Popp ed., Kluwer Acad Publ. Dordrecht, Boston, London, pp. 121-142.

Beloussov, L.V., F.-A. Popp and N.I. Kazakova, 1997, Ultraweak photon emission from hen eggs and embryos: non-additive interaction of two emitters and stable non-equilibricity. *Ontogenez (Russ. J. Devel. Biol.)* **28**: 377-388.

Bistolfi, F., 1991, *Biostructures and Radiation Order-Disorder*, Edizioni Minerva Medica. Torino.

Blumenfeld, L.A., 1983, *Physics of Bioenergetic Processes*, Springer, Berlin.

Chernavskii, D.S. and Chernavskaya, N.M., 1999, *Protein-machine. Biological macromolecular constructions*. Moskva, Moscow University Press (in Russian).

Del Giudice, E., 2000, What an electromagnetic biology could teach us. *Rivista di Biologia/Biology Forum* **93**: 492-495.

Dicke, R.H., 1954, Coherence in spontaneous radiation processes. *Phys. Rev.* **93**: 99.

Dürr, H.-P., 2002, Inanimate and animate matter: orderings of immaterial connectedness – the physical basis of life in: *What is Life? Scientific Approaches and Philosophical Positions*, H.-P. Dürr, F.-A. Popp and W.Schommers ed., World Scientific, New Jersey, pp. 145-166.

Fröhlich, H. ,1968, Long-range coherence and energy storage in biological systems. *Int. J. Quant. Chem.* **2**: 641.

Gilbert, S.F., Opitz, J.M. and Raff, R.A., 1996, Resynthesizing evolutionary and developmental biology. *Dev. Biol.* **173**: 357.

Gurvich, A.A. (1968). A problem of mitogenetic radiation as an aspect of molecular biology. Leningrad, Izdatelstvo "Medicina" (in Russian).

Gurwitsch, A., 1922b, Ueber Ursachen der Zellteilung. *Arch. Entw.-Mech der Organismen,* **52**: 167.

Gurwitsch, A.G., Gurwitsch, L.D. ,1945, *Mitogenetic radiation, its physico-chemical basis and applications in biology and medicine.* Moskva, Medgiz, 283 p. (in Russian).

Huang, S., Ingber, D.E., 2000, Shape-dependent control of cell growth, differentiation and apoptosis: switching between attractors in cell regulatory networks, *Experimental Cell Research* **261**: 91.

Ho, Mae Wan , 2004, *Rainbow and the Worm.* World Scientific, Singapore.

Jaffe, L.F., 1993, Classes and mechanisms of calcium waves, *Cell Calcium*, **14**: 736-745.

Li, Ke-hsueh, 1992, Coherent radiation from DNA molecules, in: *Recent Advances in Biophoton Research and its Applications*, F.-A. Popp, K.H. Li and Q. Gu ed. World Scientific, Singapore, pp. 157-196.

Popp, F.-A., 1992, Some essential questions of biophoton research and probable answers, in: *Recent Advances in Biophoton Research and its Applications*, F.-A. Popp, K.H. Li and Q. Gu ed. World Scientific, Singapore etc. pp. 1-46.

Popp, F.-A. and Ke-hsuch Li, 1992, Hyperbolic relaxation as a sufficient condition of a fully coherent ergodic field, in: *Recent Advances in Biophoton Research and its Applications*, F.-A. Popp, K.H. Li and Q. Gu ed. World Scientific, Singapore, pp. 47-58.

Preparata, J., 2000. Quantum electrodynamics and medicine. *Rivista di Biologia/Biology Forum* **93**: 470-481.

Rensing L. (ed) *Oscillations and Morphogenesis*, Marcel Dekker, New York, Basel, Hong Kong.

Slawinski, J., 1988, Luminescence research and its relation to ultraweak cell radiation. *Experientia* **44**: 559.

Slawinski, J., 2003, Photon emission from perturbed and dying organisms – the concept of photon cycling in biological systems, in: *Integrative Biophysics*, F.-A. Popp and L. Beloussov ed., Kluwer Acad Publ. Dordrecht, Boston, London, pp. 307-330.

Stern, C.D. and MacKenzie, D.O.,1983. Sodium transport and the control of epiblast polarity in the early chick embryo. *J. Embryol. Exp. Morphol.* **77**: 73-98.

Van Wijk, R.,2003, Cellular and molecular aspects of integrative biophysics, in: *Integrative Biophysics*, F.-A. Popp and L. Beloussov ed., Kluwer Acad Publ. Dordrecht/Boston/London., pp. 179-202.

Webb, S.J. (1983). Nonlinear phenomena in bioenergetics and oncology as seen in 25 years of research with millimeter microwaves and Raman spectroscopy, in: *Coherent Excitations in Biological Systems* , H. Fröhlich and F. Kramer ed., Springer-Verlag, Berlin, pp. 549-566.

Welch, G.R. and M.N. Berry, 1983. Long-range energy continua in the living cell: protochemical considerations, in: *Coherent Excitations in Biological Systems,* H. Fröhlich and F. Kramer ed., Springer-Verlag, Berlin, pp. 95-116.

Wu, T.M., 1994. Fröhlich's theory of coherent excitation – a retrospective, in: *Bioelectrodynamics and Biocommunication*, Mae-Wan Ho, F.-A. Popp and U. Warnke, ed. World Scientific. Singapore, pp. 387-410.

12
DISTANT INTERACTION DURING GERMINATION OF *BACILLUS SUBTILIS* SPORES

Yury A. Nikolaev, Galina I. El'-Registan, and Seshu B. Desu*

1. SUMMARY

The effect of vegetative bacterial cultures (*Pseudomonas putida* and *Bacillus subtilis*) on *B. subtilis* spores' germination mediated by their physical fields (distant interaction – DI) was investigated. We used two devices, of "flask-in-flask" and "sandwich" types. Construction of experimental setups excluded air exchange and chemical communication between signal emitting and signal receiving cultures. We found that actively growing bacterial cells stimulated germination of spores under unfavorable conditions (mineral medium M9 with glucose or rich LB medium supplemented by 6% NaCl). Stimulating effect was as high as few hundred percent under certain conditions. DI effect depended on experimental conditions: the worse the growth conditions, the higher the DI effect. There was some specific (competent) stage in detector life cycle, which was susceptible to DI signal, i.e., after short time of contact, emitter culture was removed from detector culture and stimulating effect of DI was still pronounced. Because the growth-stimulating physical signal was transmittable through acrylic plastic, therefore, it could not be of UV nature. It might be of electro-magnetic waves of non-UV regions or of sonic nature. About its mode of action, we can propose that DI signal caused conformational changes in the cell membranes and/or proteins, not in DNA, because DNA replication and transcription process do not take place during first stages of spores germination.

2. INTRODUCTION

Live organisms possess their own physical fields, of electromagnetic and sonic nature[1-3]. These fields are under thorough investigation. They can carry important information about an organism. The question is whether living

* Yury A. Nikolaev, Galina I. El'-Registan, Winogradsky Institute of Microbiology Russian Academy of Science, Pr. 60-letya Oktyabrya, 7, kor. 2, Moscow 117312, Russia, e-mail: nikolaevya@mail.ru. Seshu B. Desu, Department of Electrical and Computer Engineering, University of Massachusetts, Amherst. 201 Marcus Hall, Amherst, MA. 01003, USA, e-mail: sdesu@ecs.umass.edu.

organisms for their own profit can retrieve this information. Only biotests can answer this question surely. Therefore, progress in investigation of living organisms communication via their physical fields (distant interaction, DI) is restricted by availability of reliable experimental systems (biotests). In early works, it was demonstrated that DI influences many sides of microbial life – their growth rates, adaptation to new environment, metabolism, morphology etc.[4]. Phenomenon of DI was quite capricious and data sometimes were not reproducible. Therefore, many researchers are not convinced in the ability of bacteria to communicate with each other via physical fields. Recently, few publications appeared firmly demonstrating DI effect in bacteria, yeast, and plants. Stimulating effect of DI was observed for after-thermal-shock recovery of yeast[5], for *Pseudomonas* adhesion[6], for plant seeds germination[7], and for bacilli spores' germination and growth[8]. In the last case, the nature of signal was revealed and signal happened to be of sonic nature[9]. Other authors believe that DI is of electromagnetic nature[1].

In this work, we aimed to find a good test-system for bacterial DI investigation and to reveal some of its features.

3. MATERIALS AND METHODS

Strains: *Pseudomonas putida* KT2440 ATCC47054 and *Bacillus subtilis* ATCC 23857 were purchased from ATCC. Nutrient media: LB and M9 media were used[10]. M9 medium was supplied by 0.2% of glucose. Both media were supplied by NaCl when indicated and by 1.5-2% of agar for making solid media. Growth was monitored by counting of colonies on the surface of solid media. Cultures at solid surfaces were incubated at different temperatures stated below. Spores of *B. subtilis* were obtained in M9 medium supplied with 1% of glucose and 10% of LB medium. They were collected (Centra CL5R centrifuge, Thermo|EC, USA), washed with sterile M9 medium without carbon source, treated by lysozyme (Sigma) (50 mkg/ml) for 14 hrs at 4°C, washed by sterile distill water, collected, and kept at 4°C. Microscopy investigation was performed using Nikon eclipse E600 microscope with phase contrast and CCD camera Camamatsu C4742-95.

Experimental set-ups: Flask-in-flask set-up was described in principle earlier[11] and modified for this work (Fig. 1). Device consisted from small inner flask fixed by cotton-wool stopper to a tall glass beaker (250 ml). Flask was hand-made of quartz glass, with long neck and spherical bottom part, volume was 50 ml. Ten millilitres of spores' suspension was put inside and silicone rubber stopper was used to prevent air exchange with outer space. Sampling of spores was performed using syringes. Complete set-up was covered by black paper and Al-foil. Emitter culture was an external one, and a receiver culture was in a small internal flask.

Sandwich set-up (Fig. 2) consisted of flat tissue culture flask (Falcon), volume 250 ml with screw cap and containing 60 ml of solid LB medium with 12 hrs old bacterial lawn (emitter), and two Petri dishes (diameter 60 mm, Fisher) with 5 ml of appropriate solid medium for plating detector spores). Petri dishes

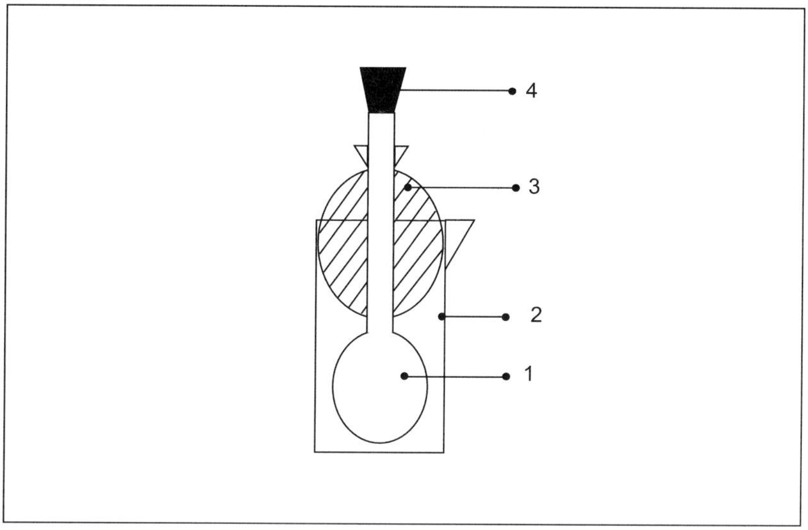

Figure 1. Scheme of flask-in-flask set-up, cross-section. Small inner quartz glass flask (1) is fixed to a tall glass beaker (2) by cotton-wool stopper (3). Silicone rubber stopper (4) was used to prevent air exchange.

without lids were fixed on a flat surface in up-side-down manner by tape. Pile of 3-4 bottles with attached Petri dishes was placed into a plastic bag to prevent drying of agar in Petri dishes. Caps of tissue culture flasks were tightly closed to prevent air exchange between emitter and detector cultures.

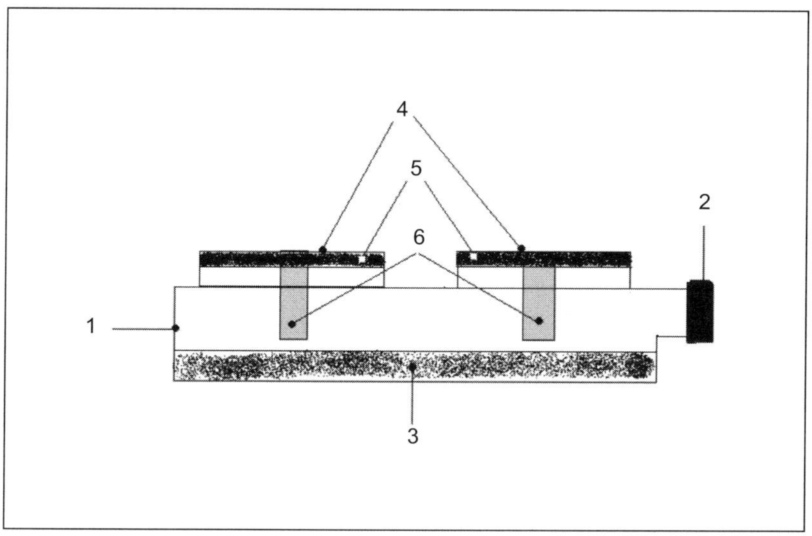

Figure 2. Scheme of sandwich set-up, cross-section. Flat tissue culture flask (1) with tight screw cap (2) contained solid LB medium (3) with bacterial emitter culture. Two Petri dishes (4) without lids with solid medium (5) for plating detector spores were fixed by tape (6).

Schemes of biotest systems. B. subtilis spores' germination process under unfavorable conditions was estimated both visually – using plate count method and microscopically. 20-2000 viable spores were spread on agar surface using glass spreader and then Petri dish was fixed to tissue bottle in face-to-face manner by tape. Grown colonies were counted after 1-7 days of incubation depending on conditions. When microscope was used for monitoring spores' development, 0.5 ml of stock spore's suspension was placed to 10 ml of nutrient medium and immediately pored to quartz flasks fixed in the beakers containing 35 ml of emitter culture or water (in control).

Statistics. Mean values presented in the figures were calculated using Excel 2003 package (Microsoft). For non-parametrical statistics calculations, Statistica 6.0 (StatSoft, Inc. USA) was used. It was necessary to use non-conventional statistical approaches, as data scattering (standard deviations) between control and experimental series was comparable with this parameter within control or experimental groups of data. Hence, classical parametrical statistics would lead us to wrong conclusions. Further, it is advised to use nonparametric statistics if we do not know for sure the type of data distribution[12].

4. RESULTS AND DISCUSSION

Under optimal conditions (LB medium, 37 or 44°C), 100% of seeded *B. subtilis* spores recover to vegetative cells as revealed by plate count and microscopy techniques. 40-70% of spores outgrow in LB medium with 6-12% NaCl at 30-44°C for 24 hrs. In M9 medium with or without 2.5% NaCl, only 4-10% of spores germinate at 37°C and grow to dwarf colonies for 6-8 days. In the presence of emitter culture (*P. putida* or *B. subtilis*) in a "sandwich" type set-up, the number of grown colonies (which reflects mainly spores' germination efficiency) was higher (Fig. 3), 60% for LB medium or 330% for M9 medium.

Statistical calculations have proven high reliability of control-experiment difference. t-tests for dependent samples, sign and Wilcoxon tests gave 0.02-0.04 values for p-level, which is enough to prove positive effect of emitter culture on spore's germination efficiency.

From data presented in Fig. 3, we see that the more stressful are the conditions, the higher is the DI effect (M9 medium versus LB medium, according parameters of growth time, colonies size, and percentage of outgrown spores). The same tendency was observed within LB series, i.e., DI effect was higher in the presence of 12% NaCl if compared with 6% NaCl (figures are not shown). This observation is in agreement with data of other researchers, who demonstrated great DI effect under very bad conditions (less than 10^{-5} of spores germinated)[8].

DI effect was independent on type of emitter culture, as we can conclude from the fact that lawns of *B. subtilis* and *P. putida* emitter cultures exhibited the same effect on spores' germination efficiency. Therefore, DI can be the mean of inter-species communication.

Another interesting observation is: it was possible to get stimulation of spores by their irradiating by emitter culture only for the first 7-12 hrs of incubation (while complete incubation time was up to 8 days). Hence, there is

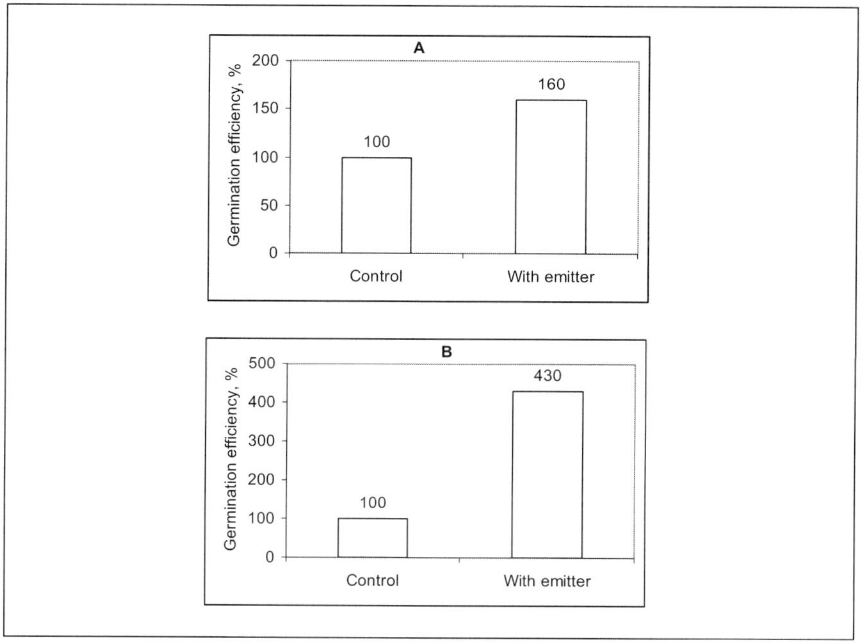

Figure 3. Effect of emitter cultures on *B. subtilis* spores germination efficiency according plate count method. Amount of grown colonies in the control experiment was taken for 100%. A – for spores seeded to LB medium with 6-12% of NaCl at 30-44°C. B - for spores seeded to M9 medium with or without 2.5% NaCl at 37°C.

some competent stage in spores' germination, which makes them susceptible to stimulating DI signal.

To find out this susceptible (competent) stage, we undertook one more series of experiments, where we investigated the very first stages of germination. *B. subtilis* spores' germination represents a very convenient bio-test system as during their germination they undergo a few very distinctive and easily observable stages. Originally dormant spores (bright white under phase contrast microscope because of dehydrated protoplasm) convert to activated spores (reversible state, they look gray due to partially re-hydrated protoplasm), then become initiated spores (look black due to fully hydrated protoplasm, irreversible state). Final stages are hatching of swollen spores, outgrowth of protoplasm to outside and active cells formation[13]. In the presence of emitter culture spores developed quicker, i.e., amount of dormant spores decreased quicker, amount of initiated and outgrown spores increased faster, if compared with control. The most profound DI influence took place soon after the beginning of experiment, i.e., difference between experimental and control curves according initiated spores content was maximal. Hence, for all the following experiments, we calculated the percentage content of initiated (black) spores after 20-30 min of spores' incubation in LB germination medium. In case of M9 medium, this time was 45-100 min. Amount of black spores was a very convenient and adequate parameter to use to detect control/experiment difference. Figure 4 represents one couple of control/experiment photographs.

It was found that rate of black spores formation was much higher in the presence of a neighboring culture (Fig. 5). In LB+6% NaCl medium for 20-30 minutes, 25-70% of spores become re-hydrated (black), and their amount was 1.5 times higher in the presence of a growing culture in "flask-in-flask" device. In M9 medium for 45-100 minutes, only 2-10% of spores became initiated (black), and their amount was 2 times higher in the presence of emitter culture.

Hence, this observation again shows stimulating effect of DI on spores' germination; and its effect is higher in the worst growth conditions (M9 vs. LB media).

The data obtained during this research confirm good positive effect of DI on *B. subtilis* spores' germination rate. Because the growth-stimulating physical signal was transmittable through acrylic plastic, it could not be of UV nature. It can be electromagnetic waves of other regions (VIS, IR, etc.) or sonic.

Based on our data, we can reasonably speculate about the primary target for DI signal. It is definitely not macromolecules synthesis, because their syntheses do not take place during early stages of spore germination[13]. Most probably, this signal causes conformational changes in the cell membranes and/or in proteins followed by the change of their function and spores' germination. This presumption is in agreement with facts of non-nutrient germination phenomenon. This phenomenon is manifested in spores' germination under the influence of such factors as high pressure, high salt concentration, and cationic surfactants[13]. General mechanism of this type of germination is activation (i.e., conformational change) of specific receptors. We believe that conformational change (CC) is the key point in DI. First, many stress factors cause CC in proteins and membranes. Second, resting forms formation and germination is followed by CC due to drastic changes of hydration level and ion content in protoplasm[14]. Then, all trustworthy examples of DI takes place specifically under stress conditions and when life strategy changes. Then, CC can be triggered by small amount of energy and from the other hand, can provoke eruption of substantial amount of excessive energy from structure under the change. Finally, CC in membranes theoretically can result coherent signal formation, and coherency is one of specific DI characteristics[2]. The latter possibility is in agreement with the fact, that DI in our experiments was not

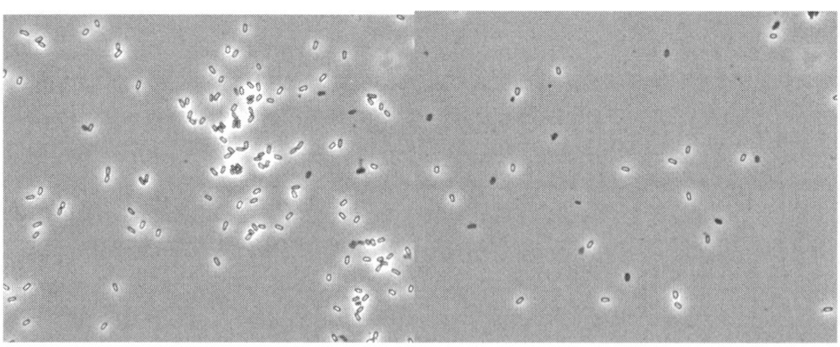

A B

Figure 4. *B. subtilis* spores germination in M9 medium, 45 min after inoculation to growth medium, as revealed by phase contrast microscopy. Percentage of black spores is 10% for control (A) and 25% for experiment (i.e. in the presence of emitter culture) (B).

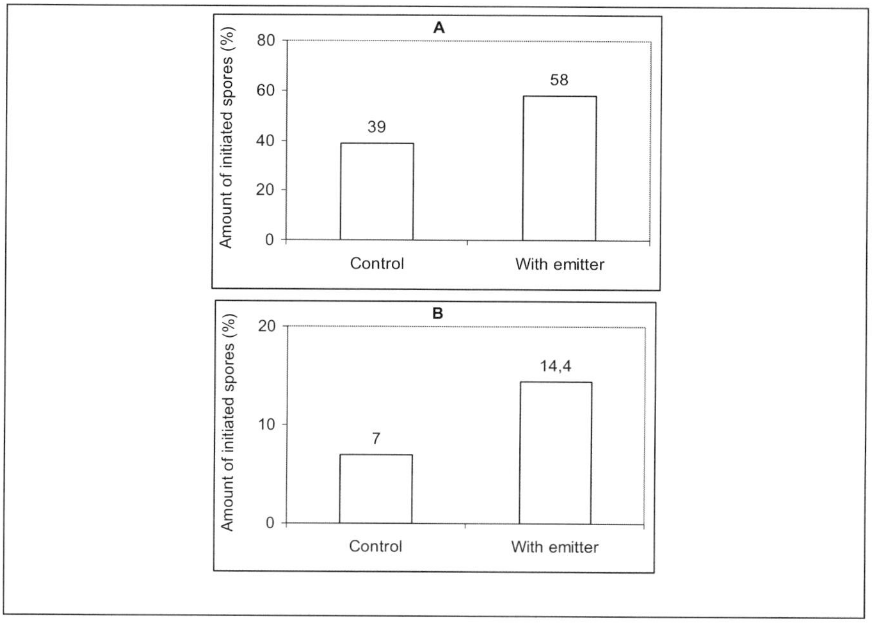

Figure 5. Effect of emitter cultures on *B. subtilis* spores germination efficiency according to microscopy observation. Given is the percentage content of initiated (black) spores. A - spores were put to LB medium with 6% of NaCl at 25-30°C after 20-30 min of incubation. B - spores were put to M9 medium at 30°C after 45-100 min of incubation. p-levels of these data were 0.004-0.04 depending on the test used (with the exception for t-test for independent samples).

sensitive to external light, noise, electromagnetic fields or other possible hindrances existing in laboratory. Such signal robustness is typical for coherent signals.

5. CONCLUSIONS

- Spores of bacterium *B. subtilis* are susceptible to physical signal producing by vegetative culture of the same and other species; i.e., distant interaction was demonstrated for this biotest;
- Growth-stimulating physical signal was transmittable through acrylic plastic; therefore, it could not be of UV nature. It can be electromagnetic waves of non-UV regions (VIS, IR, etc.) or sonic.
- Most probably, this signal caused conformational changes in the cell membranes and/or proteins, not in macromolecules synthesis, because their synthesis does not take place during early stages of spore germination.
- DI can be the means of inter-species communication.

ACKNOLEDGEMENTS

Authors are thankful to Professor Derek Lovley, University of Massachusetts, Amherst, USA, for the possibility to conduct experiments in his laboratory.

REFERENCES

1. *Biophotonics and Coherent systems,* Proceedings of the 2nd Alexander Gurwitch conference, edited by L. Beloussov, F.-A. Popp, V. Voeikov, R. van-Wijk, (Moscow Univ. Press, Moscow, 2000).
2. F.A. Popp, Properties of biophotons and their theoretical implications, *Indian J Exp Biol.* **41**(5), 391-402 (2003).
3. K. Zandonella, Dying cells dragged screaming under the microscope, *Nature,* **423**, 106-107 (2003).
4. O. Rahn, *Invisible radiations of organisms* (Gebruder Borntaeger, Berlin, 1936).
5. F. Musumeci, A. Scordino, A. Triglia., G. Blandino and I. Milazzo, Intercellular communication during yeast cell growth, *Europhys. Lett.*. **47**(6), 736-742 (1999).
6. Yu. A. Nikolaev, Role of distant interactions in the regulation of the adhesion of *Pseudomonas fluorescens* cells, *Microbiology (Transl. from Russian),* **69**(3), 291-295 (2000).
7. G.N. Surkenova and A.M. Kuzin, Some characteristics of secondary biogenic radiation (SBR) of living systems exposed to gamma-radiation at low doses, *Biophotonics and Coherent systems. Proceedings of the 2nd Alexander Gurwitch conference,* edited by L. Beloussov, F.-A. Popp, V. Voeikov, R. van-Wijk (Moscow Univ. Press. Moscow. 2000), pp. 249-451.
8. M. Matsuhashi, A.N. Pankrushina, K. Endoh, H. Watanabe, Y. Mano, M. Hyodo, T. Fujita, K. Kunugita, T. Kaneko, S. Otani, Studies on carbon material requirements for bacterial proliferation and spore germination under stress conditions: a new mechanism involving transmission of physical signal, *J. Bacteriol.*, **177**(3), 688-693 (1995).
9. M. Matsuhashi, A.N. Pankrushina, S. Takeuchi, H. Ohshima, H. Miyoi, K. Endoh, K. Murayama, H. Watanabe, S. Endo, M. Tobi, Y. Mano, M. Hyodo, T. Kobayashi, T. Kaneko, S. Otani, S. Yoshimura, A. Harata, Production of sound waves by bacterial cells and the response of bacterial cells to sound, *J. Gen. Appl. Microbiol.*, **44**, 49-55 (1998).
10. J. Sambrook, E.F. Fritsch and T. Maniatis, *Molecular cloning, 2nd edition,* (Cold Spring Harbor Laboratory Press. 1989), **3**. p. A3.
11. Yu. A. Nikolaev, Distant interactions between bacterial cells, *Microbiology (Transl. from Russian)*, **61**(6), 751-754 (1992).
12. M. Hollander, D.A. Wolfe, *Nonparamertic statistical methods* (John Wiley and Sons, New York, Sydney, Toronto,1973).
13. P. Setlow, Spore germination, *Current Opinion in Microbiology*, **6**, 550-556 (2003).
14. W.G. Murrell, Chemical composition of spores and spore structures, in: *The Bacterial Spore*, edited by G.W. Gould and A. Hurst (Academic Press, New York, 1969), pp. 215-273.

13
INFLUENCE OF RADIOFREQUENCY EMF ON THE YEAST *SACCHAROMYCES CEREVISIAE* AS MODEL EUKARYOTIC SYSTEM

E.N. Gromozova, and S.I. Voychuk*

1. INTRODUCTION

EMFs of natural and technogenic origin render significant influence on processes taking place in biosphere of the earth. The important role in a spectrum of electromagnetic radiation belongs to a radiofrequency range (RF EMF). Despite the long period of researches of RF EMF action on biological systems, the question of non-thermal mechanisms of influence remains open [1, 8]. For today the sharp necessity for development of modelling organisms for realization of researches in the given area has ripened [24].

It is well-known that many aspects of biology are similar in most or all organisms, and it is frequently much easier to study some aspects in particular organisms - for instance, genetics is easier in small organisms that breed quickly, and very difficult in humans. The most popular model organisms have strong advantages for experimental research and become even more useful when other scientists have already worked on them, discovering techniques, genes, and other useful information. A model organism is a species that has been widely studied, usually because it is easy to maintain and breed in a laboratory setting and has particular experimental advantages.

Saccharomyces cerevisiae is a unicellular eukaryote whose cellular structure and functional organization has much similarity with cells of higher-level organisms. It can be cultured easily, rapidly grows, its entire genome is known and it can be easily transformed with genes from other sources. (It is important to note that *S. cerevisiae* was the first eukaryote to have its genome sequenced. The genome is composed of about 13,000,000 base pairs and 6,275 genes.) [4, 12]. Unlike most other microorganisms, strains of *S. cerevisiae* have both a stable haploid and diploid state. Thus, recessive mutations can be conveniently isolated and manifested in haploid strains, and complementation tests can be carried out in diploid strains. Cells cycle phases are easily detected by bud-size.

Thus the purpose of our investigation was studying the influence of EMF radiofrequency on physiological-biochemical characteristics of *Saccharomyces cerevisiae* and further analysis of received data with known from literature

* E.N. Gromozova, & S.I.Voychuk, Institute of Microbiology and Virology, NAS Ukraine 03143 Kyiv, Ukraine, Zabolotny str., 154.

effects of EMF action of another frequency ranges on different level organization organisms.

2. MATERIALS AND METHODS

The yeasts *Saccharomyces cerevisiae* strain Y-517 from the Ukrainian collection of microorganisms were used. Yeasts were cultivated on a solid medium at 28°C during 24 hours.

As a source of electromagnetic radiation, we used the generator of EMF with frequency 40.68 MHz and capacity of radiation 15 and 30 W. Procedures of yeast cells preparation for exposure with EMF and further irradiation were controlled and carried out in strict conditions. The irradiation of cells was performed in a liquid medium (sterile distilled water or nutrient medium) in thermostatic conditions (28° C) during 5-60 minutes at shielding of seen light. The control tests were in the same conditions without irradiation.

At once after irradiation, the yeast *S. cerevisiae* placed in conditions distinguished from optimum: (i) the cells brought in acidic (pH 2.7) and in alkaline (pH 8.7) liquid medium (for *S. cerevisiae* an optimum pH within the limits of 4-5); (ii) carried out consecutive freezing (at -4°C) and thawing (at +25°C) of cells suspension; (iii) subjected to action of fungicide antibiotic nystatin in dozes causing lethal damages at half of cells of a population (1.0 μg/106 cells). Influence of EMF on yeast cells sensitivity to the adverse factors of external environment indicated by the quantity in the population of the viable undamaged cells that was obtained with a standard method of staining with methylene blue dye [9].

Biomass increase was controlled by optical density measuring (OD540) and speed of growth (μ, h-1) was calculated in the log-phase.

Effect of EMF action on sensitivity of yeast to fungicidal antibiotics investigated with agar diffusion method (nystatin (80 μg/ml), amphotericin B (40 μg/ml), clotrimazol (10 μg/ml), fluconazol (40 μg/ml), itraconazol (10 μg/ml) (SRCF, Russia)). Cultivation carried out at 28°C. Effect of EMF action indicated after 72 h by zones of inhibition of growth. Influence of EMF on sensitivity of yeast to antibiotics studied in three variants of experiments: (i) a simultaneous irradiation of yeast cells with disks of antibiotics; (ii) an irradiation of yeast suspension before adding of antibiotics; (iii) an irradiation of antibiotics disks.

Influence of solar and geomagnetic activities on the yeast growth, antibiotic sensitivity and stochastisity of effects of RF EMF action on microorganisms was studied by correlation analyses. Data about solar and geomagnetic activities during the period of experiment were taken from the Web site: http://www.dxlc.com /solar/indices.html.

Changes in biochemical processes of yeast were observed with measurements of dehydrogenase and catalase activity. Dehydrogenase activity (general, endogenous and substratum) determined with method described [10]. The source of carbon was 0.1 M glucose solution. The yeast cells were previously irradiated in 1/15 M phosphate buffer during 30 minutes at capacity of radiation 15 W. Catalase activity determined with Samner method in modification of Shestakov and Elchits [20].

Pull of external nucleotides determined with method of Spirin [21]. A preliminary irradiation of cells carried out in 0.9% NaCl solution.

Ability of yeast cells to acidificate cultivation medium studied with method described by Kulakovskaya et al. [11].

Influence of EMF on the cell cycle kinetics of yeast population was registered by calculating of cells in the G1, S, G2, and M phases of cell cycle during the time after EMF exposure. Samples of cell suspension were taken with 30 min intervals and were investigated with the dark-field microscopy using images of population with digital photocamera Nikon 950 (Japan) (Fig. 1). Analysis of images was done with morphometric program Image J 1.29. Cell cycle phases were detected by bud-size [16] (Fig. 2).

Statistical data processing carried out with the computer program Statistica 6.0.

3. RESULTS AND DISCUSSION

It was observed that previous exposure with EMF has protective effect on yeast cells to the influence of stress-factors. The action of acidities of cultivation medium (pH 2.7), freezing-thawing, and nystatin (1.0 μg/106 cells) on culture of the not exposed yeast of *S. cerevisiae* resulted in significant (to 80%) increase in number of damaged and dead cells (cells stained with methylene blue). It shows that some changes in cells structure occurring under influence of these factors. At the same time in suspensions of yeast previously exposed with EMF, the quantity of the damaged cells was increased insignificantly and depends on the type of the negative factor (Table 1). So at freezing-thawing the quantity of the damaged cells in control samples was increased up to 70%, while in the previously irradiated population this parameter changed in a range 5-10%. In a similar way, percentage of the stained cells in not irradiated and irradiated populations of yeast at presence of nystatin (LD50) and acidic medium (pH 2.7)

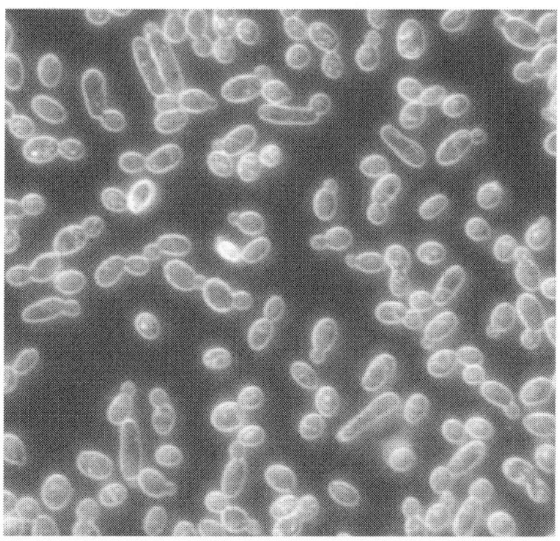

Figure 1. *Saccharomyces cerevisiae* population (x400).

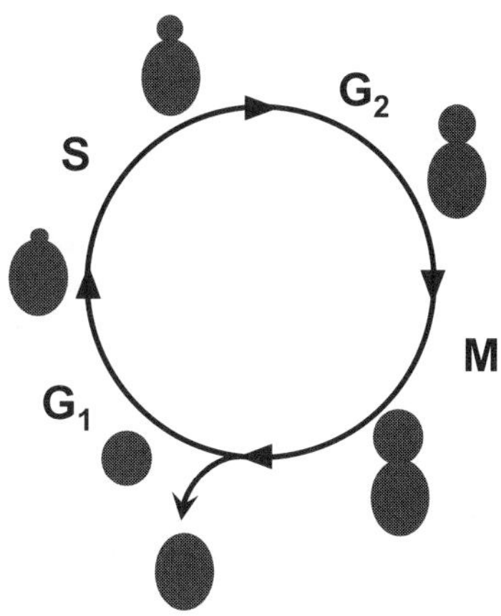

Figure 2. Cell cycle of *Saccharomyces cerevisiae.*

Table 1. Influence of a preliminary irradiation of *S. cerevisiae* strain Y-517 with EMF (40.68 MHz, 30 W, 5-30 minutes) on sensitivity to action of the stress-factors

Experiment conditions		Percent of the damaged cells ± σ_{n-1}	
		Not irradiated population	Irradiated population
Control (pH 4.5)		7.5±1.7	6.9±1.4
Stress factors	pH 2.7	47.5±10.6	15.5±6.5
	Freezing-thawing	60.0±14.1	7.5±3.5
	Nistatin (1.0 μg/10^6cells)	55.0±7.1	15.0±5.0

made, correspondingly, 55% : 15% and 50% : 17%. It was established, that the duration of an irradiation does not render essential influence on the effects of EMF, which were the same after 5, 15, 30, and 60 minutes of exposure.

Studying the influence of RF EMF on stability of the yeast population to the adverse stress-factors of the environment, it was revealed the protective character of irradiation action, which was seen in increase of a number of the alive undamaged cells [19]. The obtained effect of a protective action of EMF did not depend on the nature of the stress-factor (freezing-thawing, adverse pH, action of nistatin). It is necessary to note that another scientists working with various biological organisms established similar effects of EMFs action of different frequency ranges and power density. For example, Tambiev et al. [23] marked decrease of toxic action of zinc sulfate on *Spirulina platensis* cells as a result of their preliminary exposure to microwave EMF. Petin et al. [17] observed the delay of aging and significant increase of yeast and bacteria cells survive as a result of a unitary sharp irradiation and chronic action of ionizing radiation exceeding on intensity of a natural radiating background in 2 - 10000 times.

Research of possible reasons of RF EMF protective action on yeast was revealed, that the basic role in perception of EMF signal belongs to structures of cell wall and membranes. The role of membrane structures and in particular ion-channels in realization of effects of EMF influence on the live organisms was shown in many articles of other researchers [5, 13, 14]. However we mark some differences in effects caused by the action of RF EMF from earlier received by other researchers who studied permeability of membranes. We showed that as a result of yeast cells exposure to EMF the permeability of membranes is reduced (pull of external nucleotides become lower: 5.46 mg/ml against 6.49 mg/ml in control). This data was confirmed by the yeast cells acidification power measuring (Fig. 3) and the measurement of activity of dehydrogenase enzymes complex. We observed the increase of endogenous dehydrogenase activity in the yeast cells previously exposed to RF EMF (Fig. 4) while general activity of ferments was equaled to control.

Catalase activity of exposed yeast cells does not differ from control: so for *S. cerevisiae* in the control it was $(0.93 \pm 0.44) \times 10^{-7}$ units of fermentative activity on the cell (UFA/cell), and after an exposure – $(0.88\pm0.32) \times 10^{-7}$ UFA/cell. It specifies that this type of EMF exposure does not result in increase of a peroxide compounds level in the cells and probably has no destructive character.

Previously we showed the influence of the radiofrequency (RF) electromagnetic field (40.68 MHz) on physiological characteristics of growth of

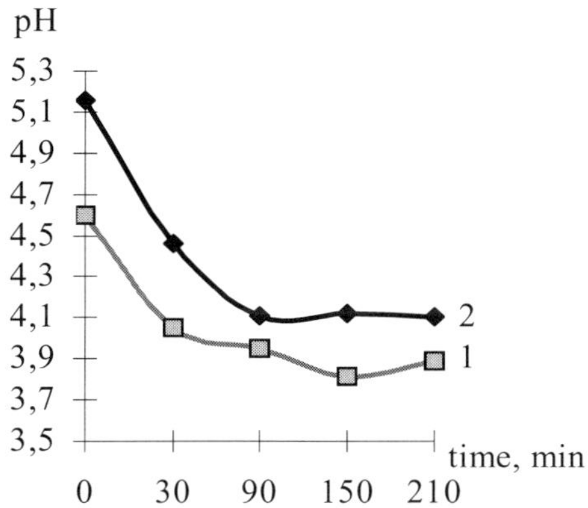

Figure 3. Yeast cells acidification power after action of EMF: (1) - control population, (2) - exposed population.

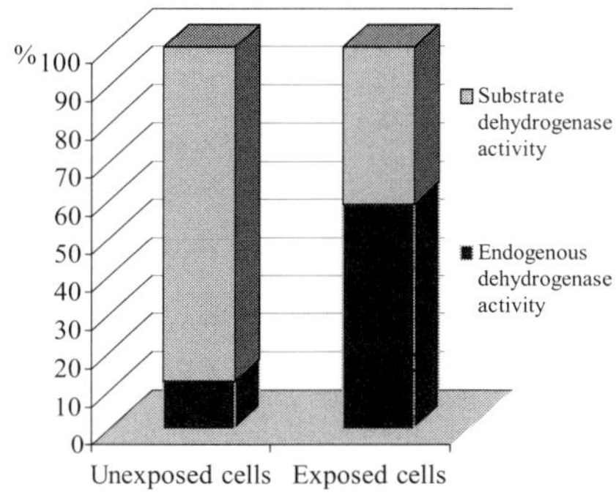

Figure 4. Dehydrogenase activity changes under EMF exposure.

yeast *S. cerevisiae* [25]. It was marked the statistically significant (95%) increase of the specific growth rate (by 7-15%) of the irradiated yeast population rather to control one (not exposured). But the revealed effect of the EMF was significant only in 16% of the experiments. In all other the cases, the rate of growth was the same as in control or difference was statistically insignificant. It is interesting to note that Pogorelov et al. [18] marked the same kind of of specific growth rate increase by 10.3% of yeast *S. cerevisiae*, when the cell suspension was exposed to microwave EMF (41.752 GHz). According to the literature data, instability of effects of the EMF action on different organisms was observed by many investigators [8]. Between the reasons of such sporadically character of the EMF action on the biological systems the researchers call: (1) an individual sensitivity of organisms and (2) the influence of helio-physics factors, geomagnetic field, daily and seasonal cycles.

To study the individual sensitivity of yeast cells to the RF EMF exposure, we investigated cell-cycle kinetics of yeast population after irradiation. It was marked that amount of cells in the stage of mitosis (M-phase) significantly increased after exposure to RF EMF (Fig. 5). The same increase was characteristic for populations in lag-phase and log-phase of growth after 3.5 and 0.5 hours. Irradiated population had higher speed of growth that correlates with the observed 15% decrease of G1-phase duration. It was calculated that on average 10% of cells in the population, which were in the G1-phase of cell-cycle, were the factor of perception and realization of effects of EMF exposure on the physiological level. From the literature data, it is known about the sensitivity of cells in G1-phase to the influence of microwave EMF [18]. But according to our results, we can say about the higher sensitivity to RF EMF only of a small part of cells that are at the beginning stages of their development (presumably these are the cells of G1-phase of first generation). According to the literature data the changes in the duration of budding-cycle can be explained by occurrence of self-organizing on the population level as the result of activation of intercellular interaction mechanisms [6].

Correlation analyses between studied processes and factors of cosmophysical nature for the period of investigations allowed to reveal the significant (95%) dependence of specific growth rate and sensitivity of the yeast population to fungicidal antibiotics with parameters of helio-cosmic activity (Fig. 6). Nevertheless, we did not reveal correlation between the effects of EMF action on the population of yeast and parameters of solar and geomagnetic activity. Thus we marked the dependence of yeast life on parameters of solar activity. At the same time, we did not reveal the correlation of stochastisity of EMF action effects with parameters of helio- and geophysical factors activity. It is possible that anthropogenous (not natural) EMF is completely independent from natural fields but, probably, it imitates them and uses the same ways of influence on the organisms.

In our researches with use of *Saccharomyces cerevisiae* it was shown that the influence of anthropogenous EMF radiofrequency, in some cases, results in effects that are known from the literature about the influence of EMFs of others frequency ranges on biological objects. Additionally, it was suggested that the solar activity renders the essential influence on growth and development of organisms, resulting to effects similar to that arise under the action of unnatural EMFs.

Our data repeat only some parts of effects characteristic for one kind of an organism and do not duplicate them. In spite of that we can suggest the existence of the universal mechanism of an action of different EMFs on biological systems. The differences which are marked in effects of EMF influence carry the "individual" character and are determined only by properties of organisms at a given moment of time and the level of evolution development. Thus the way of

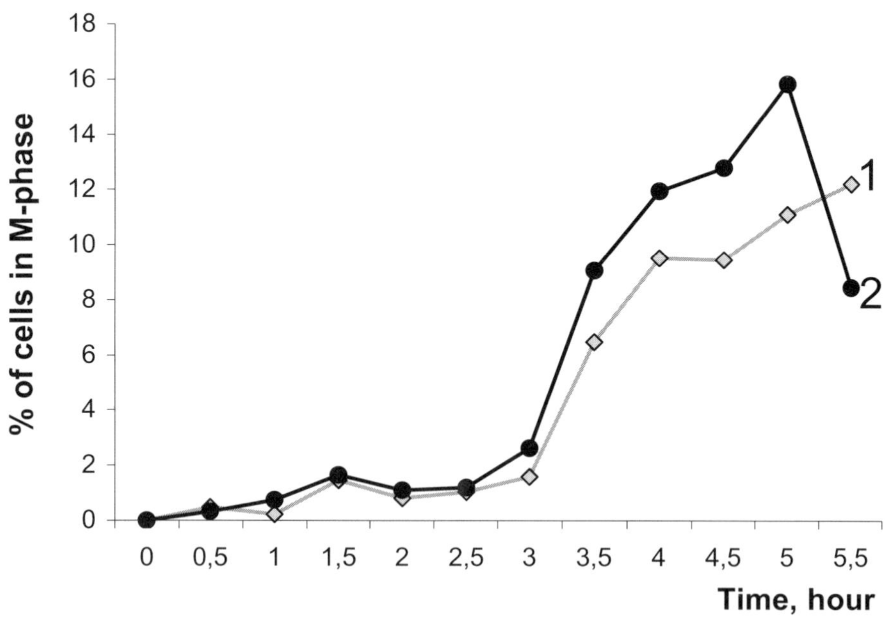

Figure 5. Increasing of mitosis after EMF exposure: (1) - control population, (2) - exposed population.

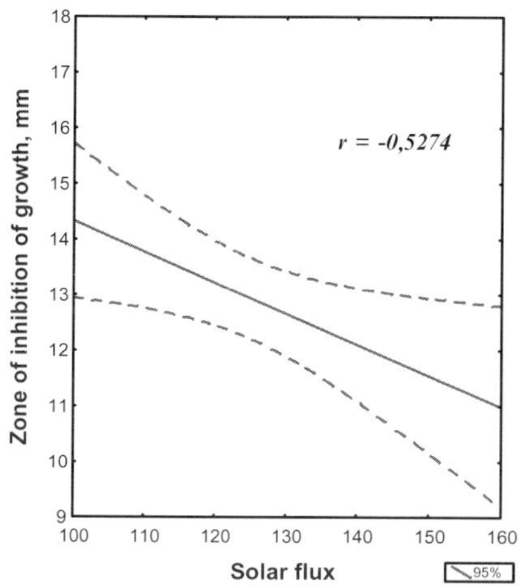

Figure 6. Dependence of yeast cells sensitivity to clotrimazol upon solar flux.

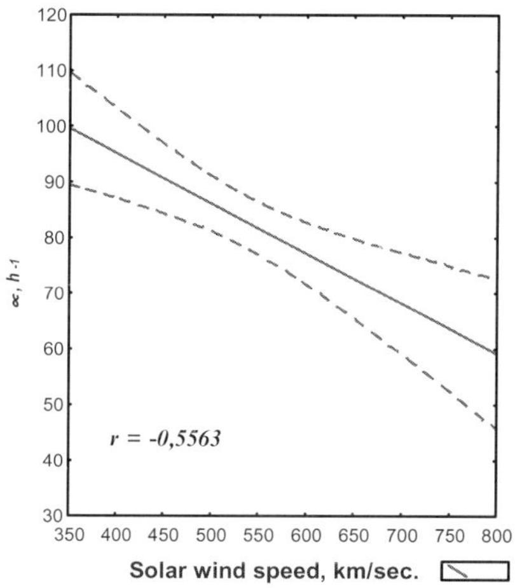

Figure 7. Dependence of growth rate upon the solar wind speed.

EMF influence on biological objects is one for any organism, the reactions of organisms on the same influence are various only.

Yeast *Saccharomyces cerevisiae* is a popular "model" organism in the laboratory. In Table 2 you can find some recent studies where this yeast was used as model organism. This yeast is been studied for a long time and investigated full enough at all levels of living matter organization. That is why the differences and similarities in effects, which arise in character of yeast physiology or biochemistry under action of EMFs of different frequency ranges, could be compared with the data received on other organisms, and can form the basis for revealing direct/indirect ways of EMF action on biological systems.

Thus we recommend using yeast *Saccharomyces cerevisiae* as modeling organism for research of EMFs influence on biological systems that not excluded researches with other organisms and carrying them in parallel. This will let to escape "dark spots" which necessarily arise at discussion of results received in researches with application of one organism independently of the level of its evolutionary development.

Table 2. Some scientific research directions where *Saccharomyces cerevisiae* used as a model eukaryote

Directions of investigations	Background of a model	References
The most intensively studied eukaryotic model organisms in molecular and cell biology. A model organism for genomic and postgenomic studies	Many essential cellular processes are conserved between yeast and humans. Almost half of genes of *S. cerevisiae* are the same as genes of the human. About 20 per cent of human disease genes have counterparts in yeast.	Ursula et al., 2001 [26] Barr, 2003 [2]
A model for studying nonthermal biological effects of extremely high frequency EMF at low power density on the division of cells.	The *S. cerevisiae* cell cycle is very similar to the cell cycle in humans and regulated by homologous proteins.	Gos et al., 1997 [7]
A model for studying single protein, including its biochemical function, role in the cell and in the whole organism, localization, mutant phenotype and genetic interactions, regulation, domains and motifs, interactions with other proteins and other relevant data.	A lot of *S. cerevisiae* proteins have similarity to human proteins involved in disease. Interacting proteins often function in conserved complexes or pathways. A pathway found in yeast might therefore exist in humans.	Costanzo et al., 2001 [3]
A model for studying the interaction of peptide hormones and G protein-coupled receptors.	Many features of this yeast recognition system are relevant to and have counterparts in mammalian cells.	Naider and Becker, 2004 [15]
A model for studying transport and homeostasis of alkali metal cations.	*S. cerevisiae* is an efficient tool for a molecular study of alkali-metal-cation transporters from higher eukaryotes.	Sychrovà, 2004 [22]

4. REFERENCES

1. S. Banik, S. Bandyopadhyay, S. Ganguly Bioeffects of microwave – a brief review, *Bioresour Technol.* Apr., **87**(2), 155–159 (2003).
2. M.M. Barr Super models, *Physiol Genomics.*, March., **3**(1), 15–24 (2003).
3. M.C. Costanzo, M.E. Crawford, J.E. Hirschman et al. YPD, PombePD and WormPD: model organism volumes of the BioKnowledge library, an integrated resource for protein information, *Nucleic Acids Res.*, Jan., **29**(1), 75–79 (2001).
4. J.I. Garrels Yeast genomic databases and the challenge of the post-genomic era, *Funct Integr Genomics.* Sep., **2**(4-5), 212–237 (2002).
5. J. Gartzke, K. Lange Cellular target of weak magnetic fields: ionic conduction along actin filaments of microvilli, *Am. J. Physiol.Cell Physiol.*, Nov., **283**(5), 1333–1346 (2002).
6. M.B. Golant, A.P. Kuznetsov, T.P. Bozjanova About the mechanism of synchronization of yeast cells culture by EHF-radiation, *Biophysics,* **39**(3), 490–495 (1994).
7. P.B. Gos, Eicher, J. Kohli, W.-D. Heyer Extremely High Frequency Electromagnetic Fields at Low Power Density Do Not Affect the Division of Exponential Phase Saccharomyces cerevisiae Cells, *Bioelectromagnetics,* **18,** 142–155 (1997).
8. M. Havas Biological effects of non-ionizing electromagnetic energy: A critical review of the reports by the US National Research Council and the US National Institute of Environmental Health Sciences as they relate to the broad realm of EMF bioeffects, *Environ. Rev.*, **8**, 173–253 (2000).
9. V.N. Ivanov, A.I. Rapoport, A.A. Pindrus et al. Fasospecificity of damages and reparation of yeast cell damages under drying-rehydratation and freezing-thawing, *Microbiol.*, **56**(2), 341-346 (1987).
10. A. Klapwuk, I. Drekt, J. Steenvoorden A modified procedure for the TTX-dehydrogenase – test in activated sludge, *Water Research.*, **8**(2), 121-127 (1974).
11. T.V. Kulakovskaya, R.N. Matyashova, V.V. Petrov, E.V. Kuranova ATPase of cytoplasmic membrane does not take part in the process of efflux of citric acid from the yeast cells of Yarrowia lipolytica, *Microbiol.*, **63**(1), 23-28 (1994).
12. P. Mackiewicz, M. Kowalczuk, D. Mackiewicz, A. Nowicka et al. How many protein-coding genes are there in the Saccharomyces cerevisiae genome?, *Yeast.*, May, **19**(7), 619–629 (2002).
13. K. Mileva, B. Georgieva, N. Radicheva About the biological effects of high and extremely high frequency electromagnetic fields, *Acta Physiol Pharmacol Bulg.*, **27**(2-3), 89–100 (2003).
14. D.J. Muehsam, A.A. Pilla The sensitivity of cells and tissues to exogenous fields: effects of target system initial state, *Bioelectrochem Bioenerg.*, Feb., **48**(1), 35– 42 (1999).
15. F. Naider, J.M. Becker The alpha-factor mating pheromone of Saccharomyces cerevisiae: a model for studying the interaction of peptide hormones and G protein-coupled receptors, *Peptides,* Sep., **25**(9), 1441 – 1463 (2004).
16. A. Niemistő, T. Aho, H. Thesleff et al. Estimation of population effects in synchronized budding yeast experiments, *Image Processing: Algorithms and Systems II, Proceedings of SPIE*, **5014,** 1–12 (2003).
17. V.G. Petin, I.I. Morozov, N.M. Kabakova, T.A. Gorshkova Some effects of radiating gormesis of bacterial and yeast cells, *Radiat. biol. Radioecol.*, **43**(2), 176–178 (2003).
18. V.V. Pogorelov et al. Sensitivity of G1-phase of yeast Saccharomyces cerevisiae cell-cycle to the action of millimeter frequency range electromagnetic waves, *Ukr. Gov. University of Food Technol.*, Kiev, 1– 8 (1996).
19. V.S. Podgorsky, S.I. Voychuk, Ye.N. Gromozova, A.S. Gordienko Protective action of electromagnetic radiation (40.68 MHz) on Saccharomyces cerevisiae UCM Y-517, *J. Microbiol.*, **66**(5), 48–56 (2004).
20. S.D. Shestakov, S.V. Elchiz Methodical instructions to laboratorial work ("Vysshaya Shkola" Publisher, Kiev, 1971).
21. A.S. Spirin Spectrophotometric determination of the quantity of nucleic acids, *Biochem.*, **23**(5), 656–662 (1958).
22. H. Sychrovà Yeast as a Model Organism to Study Transport and Homeostasis of Alkali Metal Cations, *Physiol. Res.*, **53**(1), 91–98 (2004).
23. A.H. Tambiev, N.N. Kirikova, O.A. Lyabusheva Change of microelement composition in cyanobacteria cells under influence of EHF-radiation (13 Russian symposium with the international participation "Millimeter waves in biology and medicine", Moscow, 1 - 3 Dec., 2003: Proceedings book, Moscow, 89 – 91, 2003).
24. U.B. Vahtin Development of biological models to study weak and extremely weak fields and radiations (II International Congress «Weak and extremely weak fields and radiation in biology and medicine», St.-Petersburg, 3-7.07.2000: Abstr., St.-Petersburg, 4, 2000).
25. S.I. Voychuk, V.S. Podgorsky, Ye.N. Gromozova Effect of radio-frequency electromagnetic radiation on physiological peculiarities of Saccharomyces cerevisiae strain UCM Y-517, *J. Microbiol.*, **66**(3), 51–57 (2004).
26. B. Ursula, S. Campbell, C.J. Tharappel A Model Organism for Genomic and Post-Genomic Studies (IEEE Engineering in Med. and Biol., 2001).

14
SPATIAL CHARACTERIZATION OF HUMAN ULTRA-WEAK PHOTON EMISSION

Roeland Van Wijk, Masaki Kobayashi, and Eduard P.A. Van Wijk*

1. INTRODUCTION

Weak light, spontaneously emitted from humans without any external excitation or stimulation, is commonly referred to as "human biophoton emission". The intensity of this emission in the range 200-650 nm is estimated to be the order less than $\sim 10^2$ photons/per cm^2 body surface.[1] It is thus, even though in the visible spectrum, not visible to the naked eye and cannot be captured with commonly used optical detectors. To study human biophoton emission and to clarify its basic mechanisms, one must use highly sensitive measuring instruments that record in a non-invasive and non-destructive manner. An early study using photomultipliers capable of single photon counting[2,3] reported topographical variation in emission intensity from two subjects for five different areas of the body. However, photomultipliers had not yet evolved to low-noise systems with high stability of the signal. Such a system was finally constructed by Popp and colleagues to research human biophoton emission.[4] The apparatus was hung on runners in a light-tight dark room such that the detector head could be moved over a subject lying on a bed below. The device was utilized for recording biophoton emission of 80 healthy and diseased subjects. The study confirmed differences in emission between subjects as well as between body locations.[4] However, only a few anatomic sites were recorded for each subject and a systematic measurement schedule was not followed.

Van Wijk and Van Wijk described a protocol for multi-site recording of subjects.[1,5] Anatomic sites were selected such that the distribution in photon emission could be studied as right-left symmetry, dorsal-ventral symmetry, and the ratio between the central part of the body and extremities. Although data again demonstrated the variability in patterns between subjects, some generic features were observed: (a.) the fluctuation of photon counts over the body was lower in the morning than in the afternoon; (b.) the thorax-abdomen region emits the lowest and most constant emission; (c.) the upper extremities and the head region emit the

* Roeland Van Wijk, International Institute of Biophysics, 41472 Neuss, Germany; Faculty of Biology, Utrecht University, 3584 CH Utrecht, The Netherlands. Masaki Kobayashi, Department of Electronics, Tohoku Institute of Technology, Sendai 982-8577, Japan. Eduard Van Wijk, International Institute of Biophysics, 41472 Neuss, Germany.

highest levels and increase during the day. The data suggested that a "common" human biophoton emission pattern exists in addition to individual emission patterns and dynamics.

A second system to fundamentally characterize spatial distribution of ultra-weak photon emission utilizes a highly sensitive charge-coupled device (CCD) imaging system. This technique was originally developed to study plants.[6] Biophoton emission from larger human body sites was recently examined using this CCD imaging system.[7] This imaging system highlights interesting anatomic locations that can then be quantified by, for example, a highly sensitive photomultiplier system that can be positioned over any part of the subject.[1,4]

Here, we present novel findings regarding the human biophoton emission pattern. This chapter is divided into two parts. The first section illustrates the recording of biophoton emission from the upper frontal torso, head and neck and upper extremities of a single subject utilizing these two highly advanced techniques. The final data illustrate the detection of high and low emission anatomic locations. The second section presents the systematic study of biophoton emission over specific high- and low-emission anatomic locations that include the entire axis of the body and hands. This study was carried out with a group of 32 healthy males. The data illustrate, for the first time, the existence of a "common" human body emission pattern amongst the existence of individual differences.

2. MATERIALS AND METHODS

2.1. Subjects

The study included 32 males. They were selected by posting a flyer on different Internet news groups. The subjects ranged in age from 20 to 65 years and by self-report were healthy and free of medications. They were then also interviewed to exclude any physical or emotional disorder. Exclusion criteria included the use of any antioxidant (i.e., vitamins E and C). Written consent to participate in the study was obtained after they were thoroughly informed about the research. Each subject was measured only once.

2.2. Imaging Human Body with the Highly Sensitive Charge-Coupled Device Camera

A cryogenically cooled CCD camera system that incorporates a back-thinned large area CCD sensor having full-frame architecture (CCD42-40, e2v technologies, UK) was used for imaging of human biophoton emission. Operating temperature of the CCD sensor is -100°C, resulting in the dark signal of 0.1 e^-/pixel/h. Spectral response of the CCD is ranging over 400-900 nm with quantum efficiency of >90% at the peak wavelength of 550 nm. The measurement was carried out in binning mode, resulting in the imaging format of 256 x 256 pixels. The lens system was used for imaging with magnification of approximately 3×10^{-2}. The CCD camera system was placed in a darkroom whose walls, ceiling, and floor were covered with non-fluorescent black cloths, and it was controlled from the laboratory located in juxtaposition with the darkroom. The darkroom had a chair inside; subjects were measured in sitting position after dark adaptation. The duration of a measurement inside the darkroom was 30 min.

2.3. Recording Human Photon Emission with the Photomultiplier

The photomultiplier (EMI 9235 QB, selected type) with a range of 200-650 nm was designed for manipulation in three directions. It was mounted in a sealed housing under vacuum with a 52 mm diameter quartz window maintained at –25°C to reduce the dark current. The dark current was measured before and after each experiment. During the experimental period the average background was 6.2 ± 0.3 cps (counts per second). A spacer (a ring 7 cm high) at the front of the photomultiplier tube allowed the measurement of a 9 cm diameter body area at a fixed distance. The front ring is vented inside, avoiding the condensation of moisture in the quartz window.

The photomultiplier is situated in a special dark room juxtaposition with the control room that housed the computer system. The walls and ceiling of the dark room were covered with mat black paint. The inner size of the dark room had the following dimensions: 2 m x 1.5 m x 2 m with an average temperature of 20°C. The room could be vented; the resulting small fluctuations in room temperature gave negligible change in the dark current (electronic noise) of the photon-counting device. A bed was positioned in the dark room.

Subjects were commonly recorded between 11 a.m. and 2 p.m. Before measurement, subjects were shielded from ambient light for at least one hour. Subjects remained during this period in the red dim light of the control room. Subjects then walked into the dark room and were positioned on the bed for at least 10 min. The photomultiplier tube was placed above the body, the ring at the front port of the photomultiplier touching the body. The duration of each recording was 200 s consisting of 4000 time intervals of 50 ms. Maximum duration of the measurement cycle inside the darkroom was 45 min.

2.4. Data Analysis

Statistical analysis of photon count data was performed with Statistica 6.0.

3. RESULTS

3.1. Biophoton Imaging and Biophoton Counting of Human Subjects

CCD imaging of one dark-adapted subject with the present technique was able to reveal the topography of spontaneous photon emission. The image specifies the distribution of the intensity of emission over the surface of the body.

Differences in biophoton emission are present in the ventral and dorsal images both of the torso and upper extremities (Figures 1 and 2). The set of large anatomic part CCD images of this individual subject were obtained in the Japanese laboratory by recording continuously for 30 min with cryogenic cooled

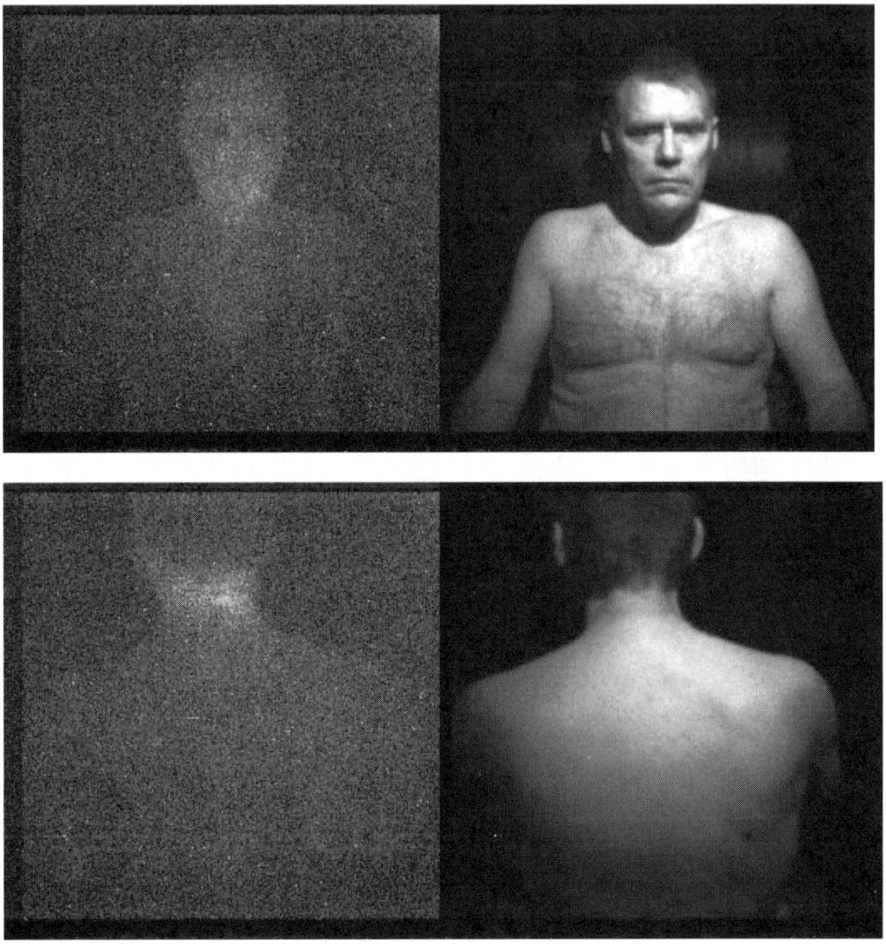

Figure 1. Biophoton emission of a human subject. Biophoton image of ventral torso (upper left panel) and dorsal torso (lower left panel) measured with the CCD imaging system. Biophoton images were taken with observation time of 30 min. Upper and lower right panels show corresponding photographs under weak illumination.

CCD camera at a distance of 100 cm. The images were displayed with the subject's corresponding photograph under weak illumination.

As illustrated in the ventral image of the superior part of the body, photon emission intensity around the face and neck is highest and gradually decreases over the torso and subsequently the abdomen (Figure 1, upper panel). There also exists a gradual decrease in intensity from the superior central torso to its lateral dimensions.

Dorsally, the highest intensity was emitted from the neck (Figure 1, lower panel). The image of arm and hand of the same subject illustrates that the low intensity of the body is extended over a large part of the arm, but increased over the hand (Figure 2).

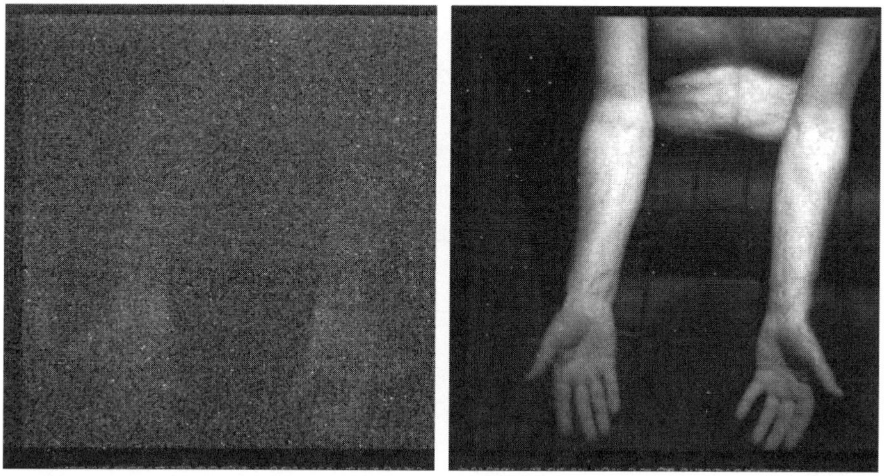

Figure 2. Biophoton emission of a human subject. Left panel: biophoton image of arm and hand of the same subject as in Figure 1 measured with the CCD imaging system. Biophoton images were taken with observation time of 30 min. Right panel shows corresponding photographs under weak illumination.

The images are compared with photon emission recordings utilizing the moveable photomultiplier device at the German collaborating institute. Data were collected from sequential anatomic locations along the ventral and dorsal longitudinal axis from head to abdomen (Figure 3).

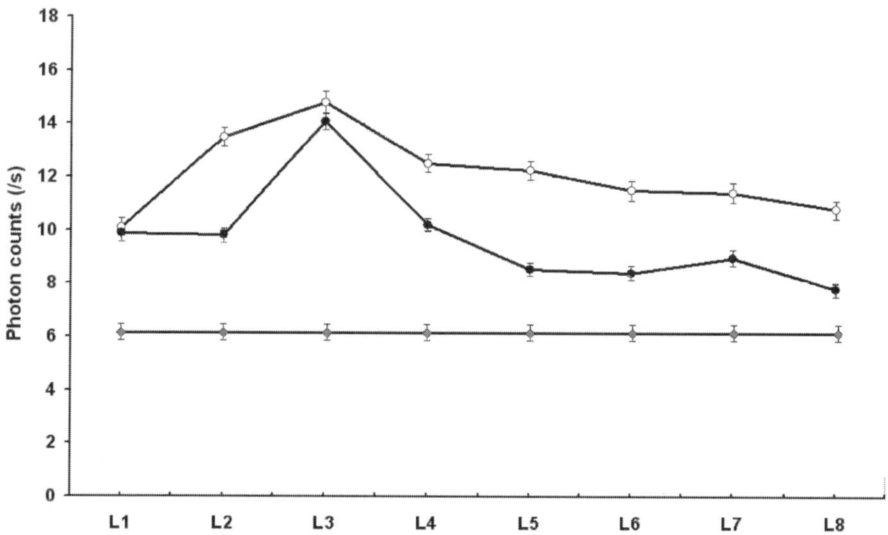

Figure 3. Photon emission recordings of the longitudinal axis of a human subject utilizing the moveable photomultiplier device. L1-8 indicate anatomical locations along the ventral (open circles) and dorsal (closed circles) longitudinal axis from head to abdomen. L1-2: head; L3: neck; L4-8: torso. Anatomical locations were recorded in series. Diameter of each location was 9 cm. Duration of each recording was 200s. Values include background electronic noise of the photomultiplier. Lowest curve represents background electronic noise.

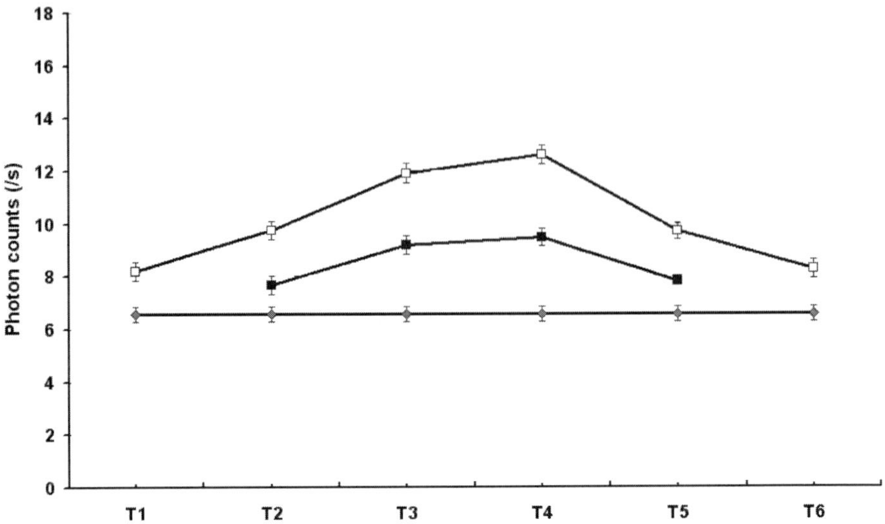

Figure 4. Photon emission recordings of a human subject transversally utilizing the moveable photomultiplier device. T1-3 indicate anatomical locations on the left side of the longitudinal axis; T4-6 indicate locations on the right side. Transversal recordings were made over the breast (open circles) and the abdomen (closed circles). Anatomical locations were recorded in series. Diameter of each location was 9 cm. Duration of each recording was 200s. Values include background electronic noise of the photomultiplier. Lowest curve represents background electronic noise.

This data also demonstrates that ventral emission decreases along the ventral longitudinal axis from head to abdomen (Figure 3, open circles). The dorsal longitudinal axis illustrates lower intensities compared with the ventral. An exception is the high intensity around the neck (Figure 3, closed circles).

Another set of data was collected transversely to illustrate emission from both left and right anterior of the longitudinal axis (Figure 4). Data could be controlled over an area of 54 cm over the breasts with the device touching the skin areas (Figure 4, open circles). Over the abdomen, data were reliably collected over a smaller (36 cm) transverse area (Figure 4, closed circles).

The data demonstrate that the central area has higher emission compared to the left and right sites. It was concluded that the data obtained with the moveable photomultiplier corresponds with the pattern of the CCD image over these parts of the body (Figure 1).

Figure 5 illustrates the recording of photon emission along arm and hand using the moveable photomultiplier device. Data were collected both dorsally and ventrally. Both right and left illustrate low emission along the arm and a strong increase over the hands. The data also correspond with the CCD image over the arm and hand (Figure 2).

The above patterns do not reflect delayed luminescence after exposure to light prior to recording. Such is excluded by sufficient adaptation to dark room conditions prior to measurements. The intensity pattern did not change during subsequent recordings.

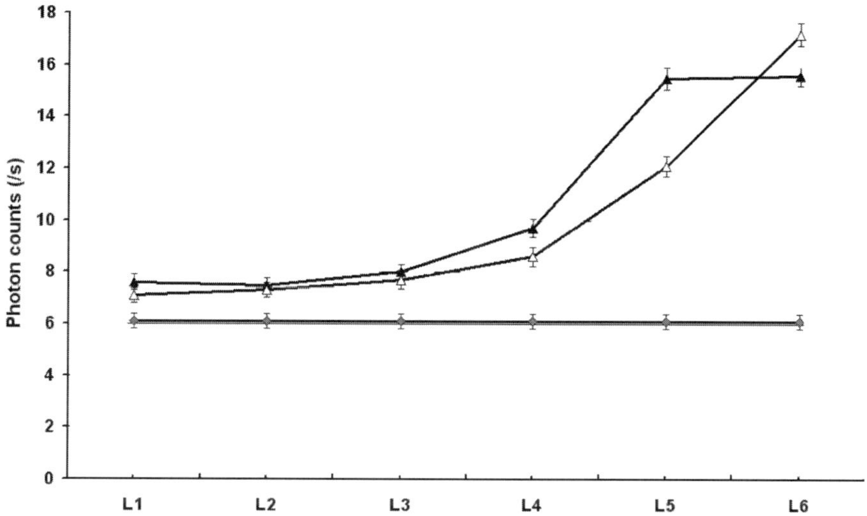

Figure 5. Photon emission recordings along arm and hand of a human subject utilizing the moveable photomultiplier device. L1-6 indicate anatomical locations of the ventral (closed triangles) and dorsal (open triangles) of the left arm and hand. L1-4 indicate locations on the arm, from elbow to wrist. L5-6 are locations on the hand (palm and fingers respectively). Anatomical locations were recorded in series. Diameter of each location was 9 cm. Duration of each recording was 200s. Values include background electronic noise of the photomultiplier. Lowest curve represents background electronic noise.

3.2. Multi-site Registration of Spontaneous Emission of a Group of Male Subjects

To study in a quantitative manner whether there is a "common" human body emission pattern, a multi-site registration of photon emission was performed with 32 healthy male subjects, utilizing the moveable photomultiplier at the German laboratory.

The anatomic locations used for recording are presented in Figure 6. The locations are selected in such a way that the distribution of emission along the longitudinal ventral axis and the left and right hands over both palm and dorsal sides were recorded. Exceptions were made at the mouth and navel areas. Both left and right sides were measured to provide total skin assessment.

Emission from each of the 12 anatomic locations of 32 male subjects were utilized to calculate average intensities of the specific anatomic locations (Figure 7). Photon emission from the abdomen was the lowest: values increased along the central axis rostally to the throat. Highest values were observed over the cheeks; emission again decreased at the forehead. Average emission over the body's longitudinal axis ranged between 4 and 10 cps.

Average emission from dorsal hand locations ranged between 6 and 12 cps. Photon emission from the dorsal sides was the lowest ($p<0.001$). Data suggest a difference in emission between left and right hands, but not statistically significant.

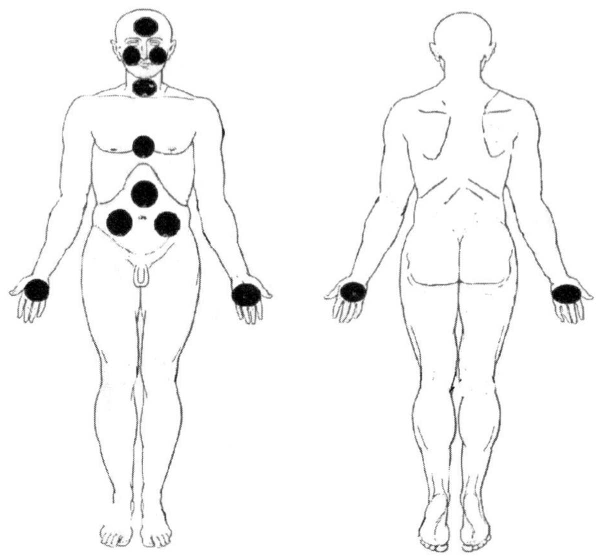

Figure 6. Anatomic locations used for multi-site registration of spontaneous emission of a group of male subjects

Average emission intensities of this male population reflect data very similar to that presented by the single subject and suggests the presence of a typical "common" human body emission pattern.

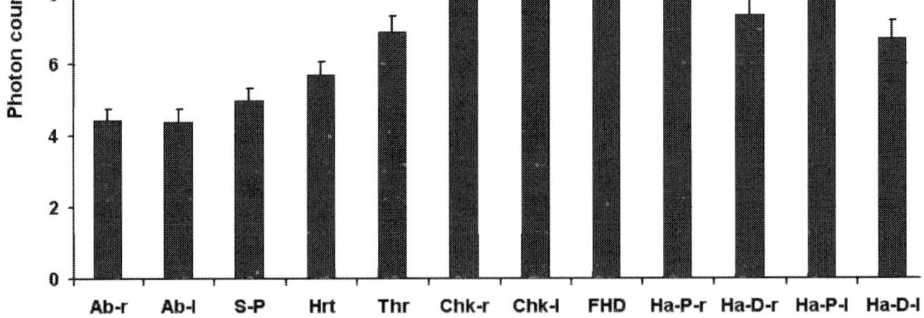

Figure 7. Average emission from anatomic locations of 32 male subjects. Anatomic locations are presented in Figure 6. Their abbreviations are: Abdomen-right (Ab-r). Abdomen-left (Ab-l), Solar plexus (S-P), Heart (Hrt), Throat (Thr), Cheek-right (Chk-r), Cheek-left (Chk-l), Forehead (FHD), Hand palm-right (Ha-P-r), Hand dorsal-right (Ha-D-r), Hand palm-left (Ha-P-l), Hand dorsal-left (Ha-D-l). Photon emission data are presented in the sequence of measurement. Diameter of each location was 9 cm. Duration of each recording was 200s. Background values were subtracted for each subject before average values were calculated.

3.3. A Typical Human Body Emission Pattern.

The data indicate that when two individual subjects differ in emission from a specific anatomic location, it reflects either an increased emission at that location without any relationship to intensities at other locations, or it reflects an increased emission at all anatomic locations. In the latter case, each anatomic location should contribute in a proportional manner to the total emission; intensity differences between subjects are then observed at all locations.

Therefore, the question whether there is a "common" human body emission pattern can also be answered when the emission contribution of each anatomic part of each subject to the total sum of emission (indicated as total emission) is calculated for each subject. Figure 8 portrays the contribution of each anatomic location to total emission for each subject.

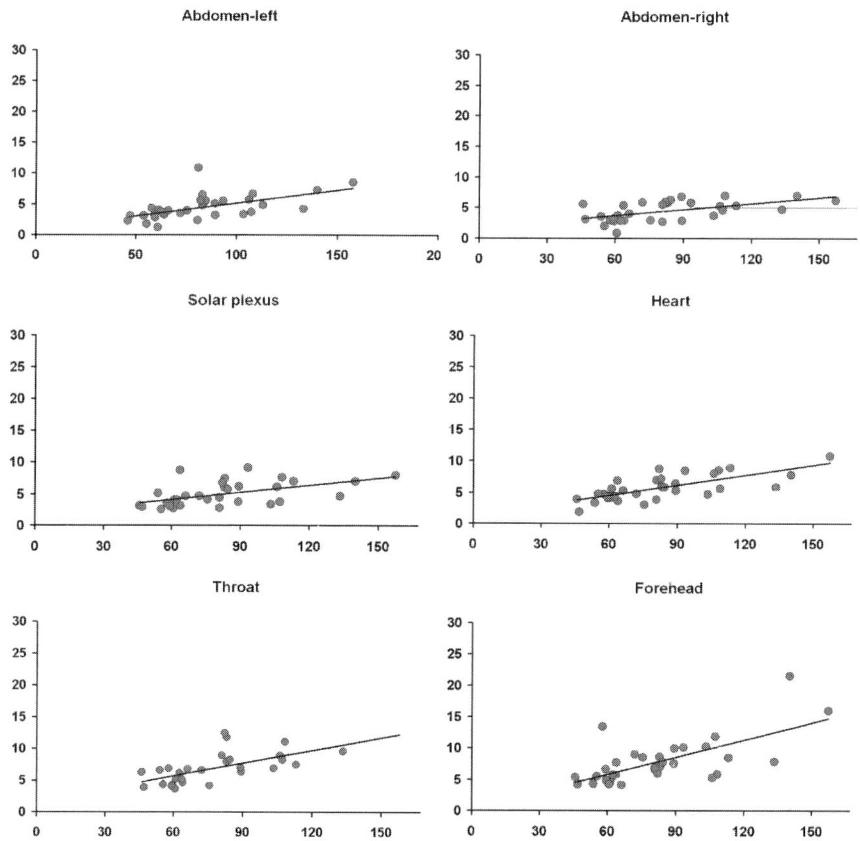

Figure 8 part 1. Contribution of photon emission from individual anatomic locations to total emission for each subject. X-axis indicates total photon emission (counts/ s); Y-axis indicates photon emission (counts/s) for each anatomic location. Each dot in a panel represents one subject, the presented line represents the best fit. The figure is continued in part 2 (next page).

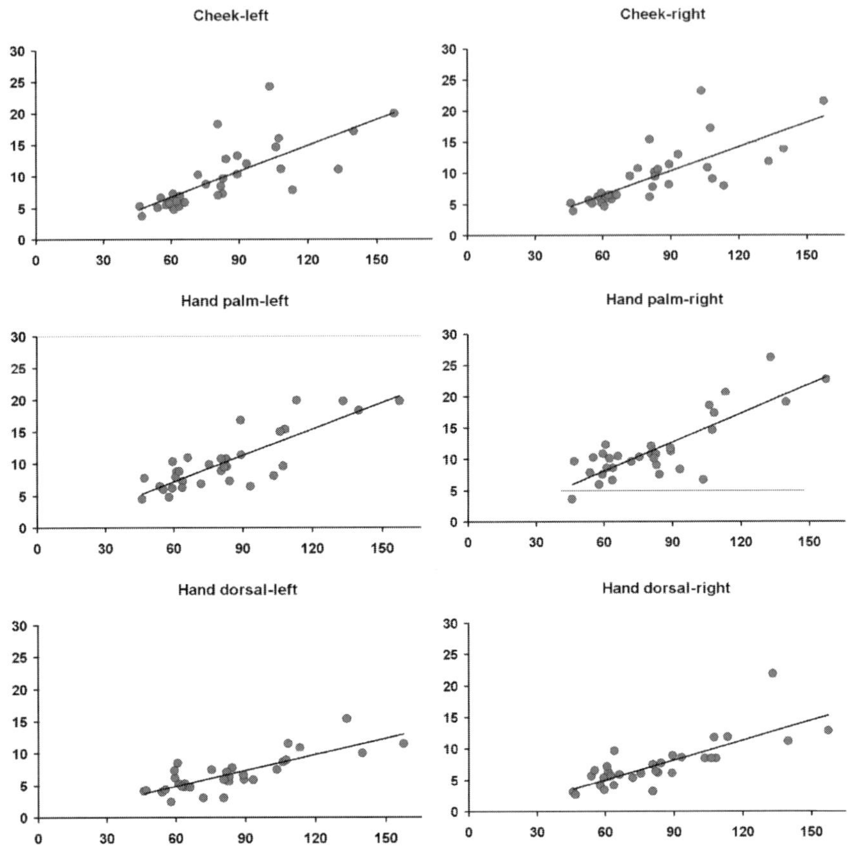

Figure 8 part 2. Contribution of photon emission from individual anatomic locations to total emission for each subject. Explanations in Figure 8 part 1.

Data demonstrate that the sum of emissions from 12 anatomic locations of each subject could differ almost 4 times between subjects; total emission can fluctuate between 45 and 160 cps. For all subjects each anatomic location's emission for each subject usually, statistically significant ($p<0.05$), correlates with the total emission for each subject.

Extrapolation of each regression line approaches zero values. It suggests that, in principle, each individual anatomic part participates in total emission with a constant percentage, independent of total emission. The contribution of each anatomic location to total emission is not equal. The slope represents the percentage of contribution to total emission values (Table 1). The individual percent contributions from each anatomic area correspond with the average emission of the anatomic locations (Figure 7).

Data suggest that the locations recorded symmetrically, i.e., abdomen, cheeks, and the back and palm of the hands, demonstrate a high degree of symmetry in photon emission. Yet, as portrayed in Figure 8, correlations are not always perfect. In some cases, a subject demonstrates a highly increased percent

Table 1. Contribution of each anatomic location to total emission

Anatomic location	Contribution (%)	Correlation coefficient
Abdomen-left	4.20	0.58 ($p<0.05$)
Abdomen-right	3.25	0.55 ($p<0.05$)
Solar plexus	3.70	0.53 ($p<0.05$)
Heart	5.36	0.72 ($p<0.05$)
Throat	6.62	0.62 ($p<0.05$)
Forehead	9.12	0.67 ($p<0.05$)
Cheek-left	13.63	0.75 ($p<0.05$)
Cheek-right	12.89	0.75 ($p<0.05$)
Hand palm-left	13.65	0.83 ($p<0.05$)
Hand palm-right	15.29	0.82 ($p<0.05$)
Hand dorsal-left	8.36	0.80 ($p<0.05$)
Hand dorsal-right	10.54	0.78 ($p<0.05$)

of emission from one specific location compared to what the percent of emission is expected from the other subjects. Those data suggests that superimposed on the "common" human emission pattern, fluctuations occur for each individual subject.

4. DISCUSSION

This study presents evidence that human subjects have a "common" pattern of ultra-weak photon emission. This conclusion was derived from recordings from the superior anterior (including the head and neck) anatomic parts of the body including upper extremities utilizing highly sensitive CCD imaging, as well as multi-site recording by a hanging and moveable photomultiplier system.

The data confirm another recent study specifying the characteristic distribution of intensity of biophoton emission from the human body surface[7]; namely, that the abdomen emits the lowest intensity which gradually increases rostally and is the highest around the face. Intensity is commonly higher along the central body line with a gradual decrease laterally. A transition in intensity was also documented from the arms being relatively low compared to the higher intensity of the hand.

Earlier studies reported two-dimensional photon counting images from smaller human anatomic areas. It is necessary to distinguish between emission studies on patients who have skin wounds and those patients with specific (chronic) diseases. The human body does demonstrate emission from wounds and other injuries affecting skin surface.[8] Other studies presented data regarding a change in intensity of photon emission from patients with chronic diseases without affection of the skin.[9-11] Two-dimensional photon emission from index and middle fingers was utilized to differentiate hypothyroidism; ultra-weak photon emission was always lower than normal. Lower emission was also reported from patients whose thyroid glands had been removed.

Photon emission features from specific anatomic areas, which might characterize states of health versus disease have also been studied using the moveable photomultiplier system. Cohen and Popp reported that a multiple sclerosis patient emitted more biophotons than ordinary healthy subjects.[12-14] They introduced a second parameter for disease, e.g., percentage of difference in emission between left and right hand. They suggested that in case of certain diseases, the left-right symmetry of emission from hands disappears. Jung et al. studied left-right photon emission symmetry from the palm and the dorsum of hands from 7 Korean hemiparesis patients, and compared the data with similar data from the hands of twenty self-reported healthy subjects.[15] The authors stated that the variation among healthy people was not large; the largest deviation was about 25% of the average value. For the hemiparesis patients, the left and right differences of biophoton emission rates were reported very large in 4 out of 7 patients. In the 3 other patients, the differences were in the normal range.

Recently, the authors studied healthy subjects regarding their differences in photon emission.[1,5] The objective was to describe a protocol for the management of subjects utilizing multi-site recording at different times of the day and different annual seasons. It was executed with the above-mentioned specially selected low noise end window photomultiplier. A 29 anatomic site recording was accomplished by manipulation in three directions of the photomultiplier hanging in a dark room. Sites were selected such that the distribution in ultra-weak photon emission could be studied as right-left symmetry, dorsal-ventral symmetry, and the ratio between the central part of the body and extremities. Data demonstrated the variability in patterns between subjects. Fluctuation of photon counts over the body was lower in the morning than in the afternoon. The thorax-abdomen region emitted the least and most constant emission. The upper extremities and the head region emitted the highest levels and increased during the day. When large fluctuations occurred, right-left symmetry remained but dorsal-ventral symmetry was not observed. Data confirm the present conclusion of the existence of a "common" human emission pattern amongst the existence of individual differences in total intensity.

At the present it is unclear if the common photon emission pattern is based on anatomical features or whether the changes in intensity might also, in part, be due to different dietary habits, psychological issues, etc. Further research is needed to clarify such issues.

REFERENCES

1. R. Van Wijk and E. P. A. Van Wijk, Human biophoton emission, *Recent Res. Devel. Photochem. Photobiol.* **7**, 139-173 (2004).
2. R. Edwards, M. C. Ibison, J. Jessel-Kenyon and R. B. Taylor, Light emission from the human body, *Complementary Medical Research* **3**, 16-19 (1989).
3. R. Edwards, M. C. Ibison, J. Jessel-Kenyon and R. B. Taylor, Measurements of human bioluminescence, *Acupuncture & Electro-Therapeutics Res. Int. J.* **15**, 85-94 (1990).
4. S. Cohen and F. A. Popp, Whole-body counting of biophotons and its relation to biological rhythms, in: *Biophotons*, J. J. Chang, J. Fisch J and F. A. Popp eds., Kluwer Academic, Dordrecht, pp. 183-191 (1998).
5. E. P. A. Van Wijk and R. Van Wijk, Multi-site recording and spectral analysis of human body spontaneous photon emission, *Research in Complementary and Classical Natural Medicine*, in press (2005).

6. M. Kobayashi, B. Devaraj, M. Usa, Y. Tanno, M. Takeda and H. Inaba, Two-dimensional imaging of ultra-weak photon emission from germinating soybean seedlings with a highly sensitive CCD camera, *Photochem. Photobiol.* **65**, 535-537 (1997).
7. M. Kobayashi, Modern technology on physical analysis of biophoton emission and its potential extracting the physiological information, in: *Energy and Information Transfer in Biological Systems*, F. Musumeci, L. S. Brizhik and M. W. Ho eds., World Scientific Publ, New Jersey, London, pp. 157-187 (2003).
8. T. Yanagawa, H. Sakaguchi, M. Ueno and K. Nitta: Sustaining faculty of living functions and its biophoton observation, *J. Intl. Soc. Life Info. Sci.* **18**, 423-437 (2000).
9. M. Usa, M. Kobayashi, S. Suzuki, H. Ito and H. Inaba, *ITEJ Technical Report* **15**, 1 (1991) [in Japanese].
10. M. Usa, B. Devaraj, M. Kobayashi, M. Takeda, H. Ito, M. Jin and H. Inaba, Detection and characterization of ultra-weak biophotons from life processes, in: *Optical Methods in Biomedical and Environmental Sciences*, H. Ohzu and S. Komatsu eds., Elsevier Science, Amsterdam, pp. 3-6 (1994).
11. M. Usa and H. Inaba, Spontaneous photon emission from human body, *Medical Imaging Technology* **13**, 47-54 (1995).
12. S. Cohen and F. A. Popp, Biophoton emission of the human body, *J. Photochem. Photobiol. B,* **40**, 187-189 (1997).
13. S. Cohen and F. A. Popp, Low-level luminescence of the human skin, *Skin Research and Technology* **3**, 177-180 (1997).
14. S. Cohen and F. A. Popp, Biophoton emission of human body, *Indian J. Exp. Biol.,* **41**, 440-445 (2003).
15. H. H. Jung, W. M. Woo, J. M. Yang, C. Choi, J. Lee, G. Yoon, J. S. Yang, S. Lee amd K. S. Soh, Left-right asymmetry of biophoton emission from hemiparesis patients, *Indian J. Exp. Biol.*, **41**, 452-456 (2003).

15
INFLUENCE OF ELECTROMAGNETIC FIELDS OF EXTREMELY DIFFERENT FREQUENCY DIAPASON ON INFRADIAN RHYTHMS OF PHYSIOLOGICAL PROCESSES

N.A. Temuryants[*], V.S. Martynyuk[†], and E.N. Chuyan[‡]

1. INTRODUCTION

The beginning of the modern electromagnetic (EM) field theories in biology appeared only in 1970, with A.M. Presman`s report of the pioneering work of Soviet Bioelectromagnetics researchers, which also contained the first outline of a holistic EM field theory of the organism and its relationships with the environment. Since then, there is ample evidence of bioeffects of EM fields and endogenous EM fields. It is now established that organisms react sensitively to the impact of electromagnetic fields, including very weak ones; effects of various types of endogenous physical fields on cellular organization and morphogenesis are very similar. We also know that several kinds of electromagnetic fields, including microwaves and optical frequencies (biophotons), are emitted from living beings.

It is known that weak extremely low frequency (ELF) dominates in the spectrum of variable magnetic fields that can be registered on the surface of the earth and, perhaps, are utilized by living organisms as means of information exchange, as an internal clock time gauge in a wide spectrum of periods[1,2]. In particular, it has been proven that 8 Hz, 5 μT EMR alters infradian rhythmicity of the studied parameters in epiphysectomized[2, 3] and hypokinetic[4] animals.

Unlike ELF electromagnetic radiation (EMR), ambient low intensity, ultra-high frequency (UHF) EMR reaches minimal value due to increased absorption in the upper atmosphere layers. This leads to a decrease in the background levels 100- and 1000-fold and prevented evolutionary adaptation to EMR in this frequency range[5]. Therefore, it is precisely in this frequency range that one would expect to see increased biological and therapeutical activity, and this has been confirmed in numerous experimental and clinical inventions[6]. However, the

* Vladimir Vernadsky Taurida National University, Simferopol, Vernadsky ave, 4, 95007, Ukraine.
† Crimean Scientific Center of National Academy of Science of Ukraine and Ministry of Education and Science of Ukraine, Vernadsky ave, 2, Simferopol, 95007, Ukraine.
‡ Vladimir Vernadsky Taurida National University, Simferopol, Vernadsky ave, 4, 95007, Ukraine.

ability of UHF EMF to affect temporal organization and, in particular, infradian rhythmicity of biological systems, remains practically not studied.

It has been established that the degree of synchronization of physiological parameters is not the same under different circumstances. Thus it could be used to assess stability of physiological systems affected by various factors[7]. Adaptation to stress-factors is accompanied by changes in the rhythmicity of function of various systems, and desynchronization, i.e., discordance in nervous and endocrine regulatory mechanisms, remains an essential component of a general adaptation syndrome[8] and leads to the weakening of an organism's resistance to noxious factors[9,10]. Therefore, one of the most important tasks of contemporary physiology is to search for the effective methods of optimization of a biorhythmic state of an organism, and of promoting its adaptation capacities.

The search for the adequate tests for performing such a task is not easy. Physiological systems of an increased sensitivity to any internal and external changes include, first and foremost, blood and, in particular, bactericidal, hydrolytic and energy systems in neutrophils. Thus, by studying these systems one could adequately evaluate the reaction that an organism may produce in response to factors of various origins.

The goal of this project was to study the effects of week UHF and ULF EMR on infradian rhythmicity of neutrophils' functional activity.

2. MATERIALS AND METHODS

The study was conducted on 120 wild-type male rats of 180-200 g weight. The individual characteristics were identified by the "open field" test that allowed for quick and confident selection of animals with similar constitutional characteristics[11]. On the basis of this test, we identified animals with low mobile activity (LMA), medium (MMA) and high mobile activity (HMA), which displayed distinct differences in the horizontal and vertical mobility.

Rats selected into each mobility category were further divided into two equal sub-groups of 20. The first sub-group included animals housed in the usual conditions of our vivarium (biological control group, C). The second sub-group consisted of animals that were exposed for 3 hours daily to 8 Hz, 5 μT VMF for 32 days.

In the present study, the selection of VMF parameters was based upon assessment of their physiological and geophysical importance. The basis for this selection and methodology is described in Ref. 12.

This study was conducted on 80 wild-type, white male rats weighing 200-250 g. We selected animals of the same age, with a medium level of mobile activity and low emotionality determined by the "open field" test. This provided the selection of homogenous groups of animals, that developed a typical reaction in response to extraneous stimuli, including EMR.

In further experiments, all animals were distributed among four equal groups. The first group was composed of rats maintained under the usual vivarium conditions (biological control, C). The second group was composed of the rats maintained under stress conditions due to imposed limitations on mobility (hypokinesia, HK)[4]. HK was produced by placing the rats in the specially constructed Plexiglass cassettes for 22 hours per day[13]. The third group was composed of animals exposed to UHF EMR. The fourth group was composed of animals that were kept in HK conditions and were exposed to UHF EMR (HK+UHF). The exposure to UHF EMR was daily effected for 30 min exposures on the occipital area during 45 days. The "Luch KVCh-01" generator that was used for this purpose was set for wavelength of 7.1mm at 0.1mW/cm^2.

A blood taken for analysis was obtained during 45 days at the same time once per day by means of a phlebotomy of the tail vein. Blood smears were studied by cytochemistry for bactericidal agents (peroxidase, PO), cation proteins (CP), hydrolytic enzymes (acid phosphatase, AP) and proteases (PR)[14]. Qualitative assessment of the studied parameters was performed by calculating the cytochemical index of content (CIC) according to the L. Kaplow's principle[14]. Average content of oxidative-reducing enzymes in neutrophiles (succinate- and α-glycerophosphatedehydrogenase, SDG and α-GPDG) was determined according to Nartsissov[15].

Phase-amplitude characteristics of the studied processes were calculated by cosinor-analysis, which completely represents the structure of physiological rhythms. The computer program for cosinor-analysis was developed by V. Martynyuk, P. Grigoriev and N. Zuyev in the Crimean Scientific Center of the Academy of Sciences of Ukraine. The using of cosinor-analysis affords an opportunity to reveal periodic components in number rows, as well as to obtain and amplitude and fluctuation phase of any periods that form the total rhythmicity of a particular parameter. For each animal, each parameter was plotted as a periodgram (spectrum) covering from 2.2 to 30.0 days periods divided to 0.5 day steps. By this technique we identified the periods (local maxima) that contribute to the rhythmicity of animals in the corresponding group. Latent co-relations between the studied parameters were established by cluster analysis, which is an adequate tool of assessing multifactorial reactions of an organism[16]. The use of agglomerative strategy of analysis allows one to construct a dendrogram of all parameters by means of hierarchical combination of them into groups (clusters) of greater commonality based on the minimal distance criterion in the spectrum of variables that describe given parameters.

3. RESULTS AND DISCUSSION

3.1. Specific Characteristics of Infradian Rhythmicity of Physiological Parameters in Rats with Different Individual Characteristics

Human beings are known to be divided into three types of biological rhythms based upon different diapasons and this effect would be commensurated with our understanding of the endogenous origins of biological rhythms. As will be shown by our data, the extent of changes in the parameters of infradian rhythms in rats expoused to different noxious stimuli is rather variable (Table 1).

Table 1. Specific characteristic of infradian rhythmicity of physiological parameters in rats with different individual characteristics

Parameter	Groups of rats		
	Low Mobile Activity	Moderate Mobile Activity	High Mobile Activity
Vertical motility			
Range of periods identification	2.2 – 11.8 days	2.2 – 20.6 days	2.2 – 22.8 days
Number of periods in the spectrum	7	10	13
No periods	$\approx 7^d$	–	–
High amplitude periods	$\approx 10^d$	–	$\approx 19^d$
Horizontal mobility			
Range of periods identification	2.7 – 14.7 days	2.2 – 22.1 days	2.7 – 21.8 days
Number of periods in the spectrum	9	13	12
No periods	$\approx 7^d$,6	$\approx 14^d$,7	–
High amplitude periods	$\approx 11^d$,5	–	$\approx 19^d$
Neutrophils acid phosphatase content			
Range of period identification	2.4 – 15.2 days	2.4 – 18 days	2.4 – 21.9 days
Number of periods in the spectrum	11	13	12
No periods	–	–	$\approx 15^d$, $\approx 18^d$,4
High amplitude periods	–	–	–
Low amplitude periods	9^d	–	2^d,4
Lymphocyte α-GPDG			
Range of period identification	2.4 – 22.3 days	2.4 – 14.1 days	2.4 – 14.4 days
Number of periods in spectrum	7	6	7
No periods	4^d,0	4^d,0	–
High amplitude periods	–	–	–
Low amplitude periods	–	–	–

3.2. Changes in Blood Lymphocyte Dehydrogenase Infradian Activity in Rats with Different Individual Characteristics Exposed to Extremely Low Frequency Magnetic Fields

The results of this study show that the exposure of rats with different mobile activity to 8 Hz, 5 μT MF leads to a shift in infradian rhythmic functional activity of blood lymphocyte and neutrophils.

ELF EMF causes polydirectional changes of infradian rhythmicity of the lymphocyte functional state in rats with different individual characteristics, i.e., the effect of VMF on rhythmic processes depends on the baseline state (Fig. 1). This data is corroborated by the previously reported results about the effects of ELF MF upon animals with different kinds of desynchronosis. These effects are accompanied by the restoration of the baseline temporal organization of physiological systems[17, 18].

3.3. Modification of Infradian Rhythmicity of Neutrophil Functional Activity by Low Intensity Ultra-High Frequency Electromagnetic Fields

UHF EMR is also capable to change the temporal organization of physiological systems. However, the character and direction of these changes depend upon the functional state of an organism.

The most expressive reconstructive processes were noticed in LMA rats. Perhaps, rats with LMA are more susceptible to ELF VMF. The results pertaining to individual sensitivity of animals to ELF VMF are in accordance with those from[19] which demonstrated polydirectional changes of non-specific resistivity in LMA and HMA rats.

Intact animals exposed to UHF EMR exhibit just a slight trend towards changes in the infradian activity of the studied parameters. For instance, no

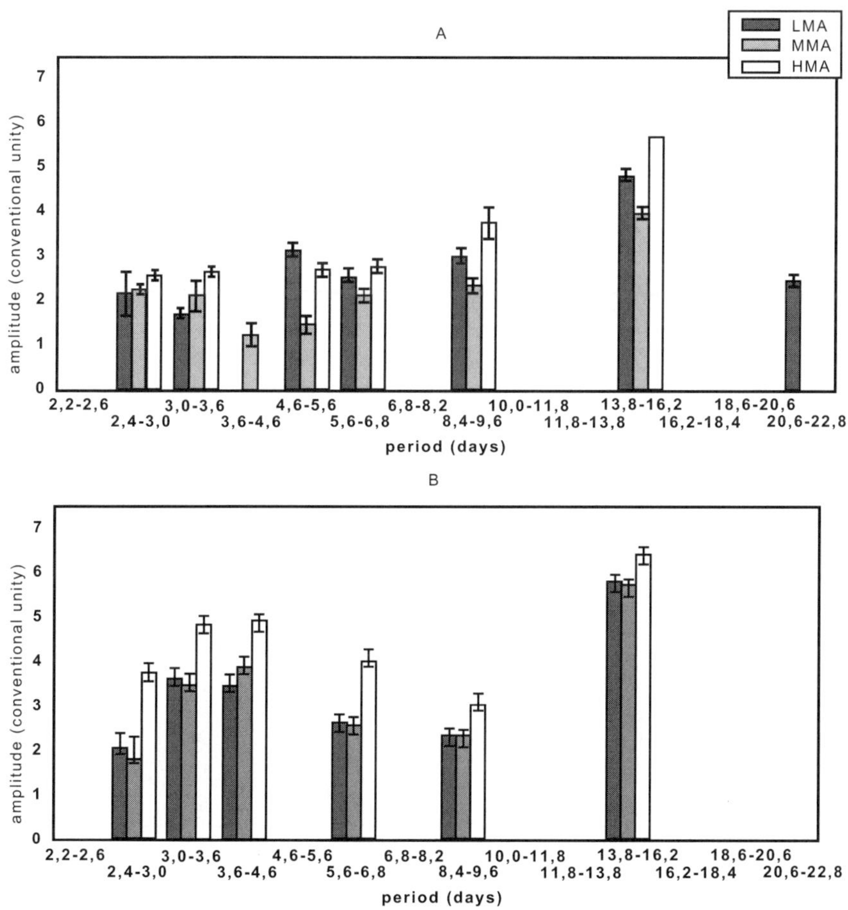

Figure 1. Average α-GPDG activity power spectra of blood lymphocytes in intact rats (A) and rats exposed to ELF VMF (B) exhibiting different mobile activity.

changes were revealed in the phase-amplitude characteristics of CIC PO and CP and the average activity of SDG and α-GPDG. At the same time, in ≈7-day rhythm an obvious increase in CIC AP ($p < 0.05$), in ≈14.0-day period no less obvious decrease in the amplitude of protease activity ($p<0.05$) (Fig. 2) and in ≈22-day period a 178° phase shift in the α-GPDG/SDG ($p<0.05$) were registered. Such changes in the rhythms of the studied parameters are not accompanied with either intra- or inter-functional synchronization (Fig. 3, and 6). Thus, the analysis of interphase correlations revealed the decrease in the difference between the phases of the average activity of SDG and α-GPDG, CIC CP and PO, CIC AP and PR, as compared to the data from cosinor-analysis of animals subjected to hypokinesia. However, in all the periods there were stable baseline phase correlations between the rhythms of the studied systems, indicating the increase of synchronization. The general distribution order of the parameters turned out to be clustered and the extent of connections was the same as in the control group (Fig. 4).

Therefore, UHF EMR induced changes in the structure of infradian rhythmicity considerably differ from the changes observed under ELF EMF exposure.

A stress reaction performed by hypokinesia leads to considerable alterations of the infradian activity of all the physiological parameters of the neutrophiles, as expressed by an extensive amplitude changes and the phase shifts, as compared with the control animals (Figs. 2, 5, 6). It was also found that hypokinesia leads to the establishment of new correlations between the functional parameters, as demonstrated by the breakdown of phase correlations between average activities of SDG, α-GPDG, CIC CP and PO, CIC AP and PR in the neutrophils of hypokinetic animals. Moreover, along with the changes in

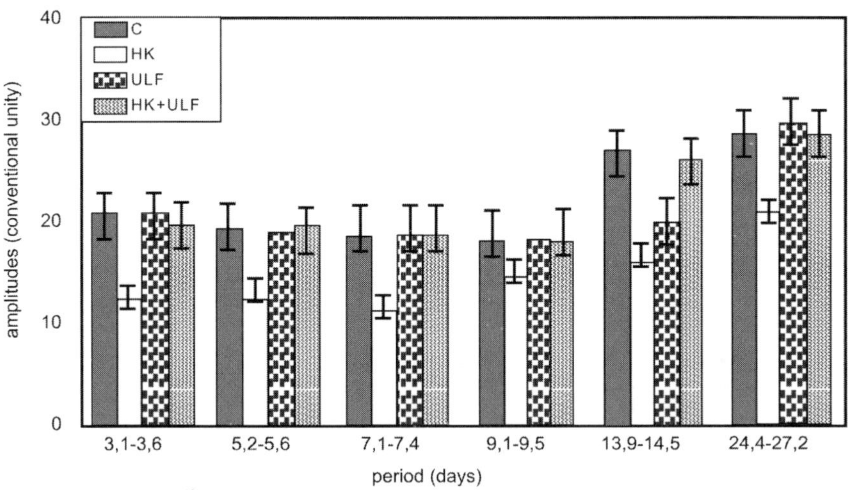

Figure 2. Amplitudes of integral rhythms of rat blood neutrophils protease activity under various conditions. C – control; HK – hypokinesia; ULF – ULF EMR; HK+ULF EMR – hypokinesia combined with ULF EMR.

intrafunctional correlations, cosinor-analysis revealed also a change in inter-functional co-relations (Fig. 3). With the help of cluster analysis (Fig. 4), inter-functional changes were also registered, as expressed by a rearrangement of the studied parameters in neutrophils, increase in the number of clusters (up to 4) and increase in the correlations lengths not only within individual clusters, but also between them. Such rearrangements demonstrate a hypokinesia-induced functional strain exerted upon the studied systems.

Therefore, a hypokinesia-induced stress reaction yields considerable influence on the temporal organization of the studied systems of neutrophils as reflected by a considerable change of amplitudes of the studied rhythms, by the phase shifts in all the identified periods, and by disturbances in the intra- and inter-functional correlations. All of this demonstrates the weakening of the coordination of enzymatic activity [20] and a progressive desynchronosis.

The analysis of results of the action of UHF EMR on the rats subjected to the limitations in mobility revealed considerable changes in the infradian activity

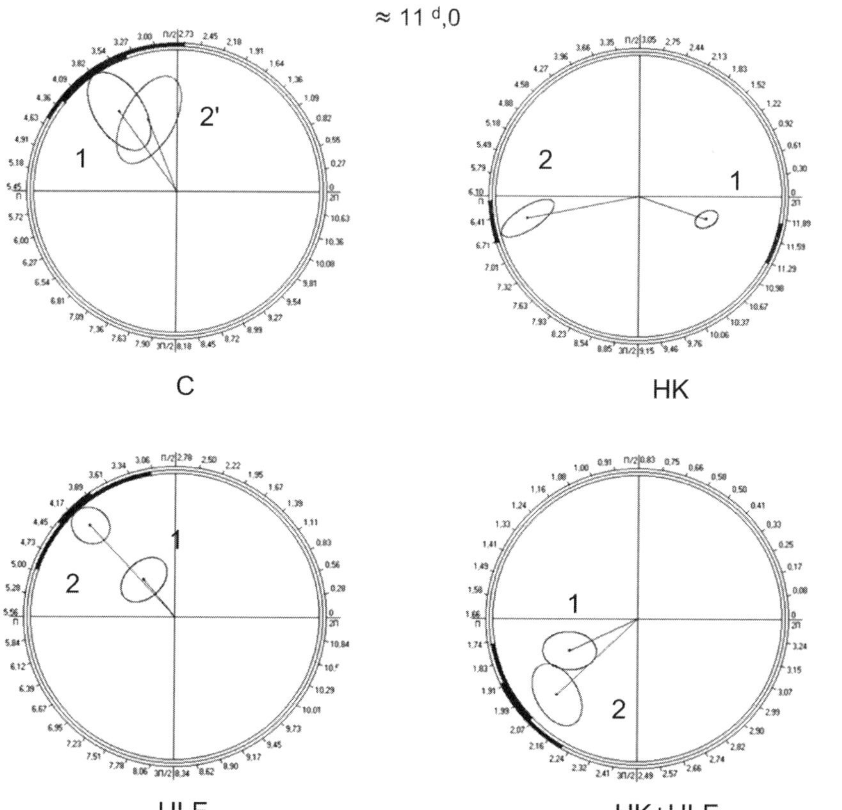

Figure 3. Cosinorograms of integral rhythms of hydrolytic enzyme ratios (CPC AP to CPC PR) and bactericidal enzymes (CPC PO to CPC CP) in blood neutrophiles of control rats (C) and in rats exposed to various stimuli: HK – hypokinesia; ULF – ULF EMR; HK+ULF EMR – hypokinesia combined with ULF EMR в in ≈11d.0 day period, where 1 is co-efficient of CPC AP to CPC PR, and 2 is co-efficient of CPC PO to CPC CP.

of the studied parameters comparing to hypokinetic rats. Thus, the exposure to UHF EMR in animals already subjected to hypokinesia had a reconstructive effect on the amplitudes of the identified rhythms in all the studied parameters, although their values did not always reach the control level (Fig. 2). Moreover, the action of UHF EMR on animals with limited mobility led to a substantial phase shift of the identified rhythms, bringing them closer to the values typical for the control group. Thus, it seems that a reconstruction of baseline phase/amplitude characteristics of hydrolytic, bactericidal and redox systems of neutrophils indicates normalization of the intra- and inter-functional correlations disturbed by hypokinesia (Fig. 2, 3, 5, 6). These data are corroborated by the results of cluster analysis (Fig. 4).

It should be noted that the combined action of UHF EMR and hypokinesia revealed changes in infradian rhythmicity of the studies parameters that were more significant that those produced under the influence of the latter factor alone. This is another proof of the correlation between the efficacy of UHF EMR exposure and the baseline state of an organism.

Thus, under conditions of desynchronosis caused by stress-reaction to limitations in mobility, a daily exposure to UHF EMR produces corrective, synchronizing action, which leads to the normalization of infradian rhythmicity of the studied parameters. These data are in agreement with previous works[21], in which it was shown that UHF EMR assists in the normalization of conditions

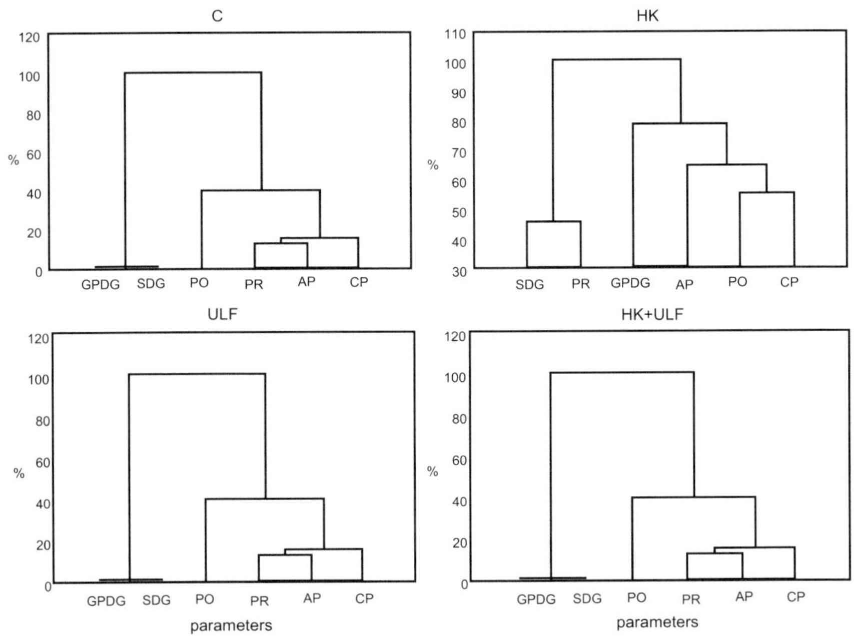

Figure 4. Cluster analysis dendrograms of blood neutrophil activity in control rats (C) and in rats subjected to various stimuli: HK – hypokinesia; ULF – ULF EMR; HK+ULF EMR – hypokinesia combined with ULF EMR, where PO – peroxidase, CP – cation proteins, AP – acid phosphatase, PR – protease, SDG – succinatedehydrogenase, α-GPDG – α-glycerophosphatedehydrogenase in blood neutrophils.

that accompany hypokinetic stress. This also provides the evidence of UHF EMR's ability to synchronize physiological processes, which may be one of the mechanisms of the anti-stress action.

According to the current understanding, a healthy organism maintains strict, but flexible correlation between different processes that comprise homeostasis, whereas various pathological processes reveal some degree of desynchronosis[9, 22]. In order to achieve synchronization of the endogenous processes, a very weak signal would suffice, such as UHF EMR, and this would bring about a "narrowing" of the near frequency[7, 23]. This phenomenon agrees with the concept of the "biological action of microdoses" of various physical and chemical agents[24].

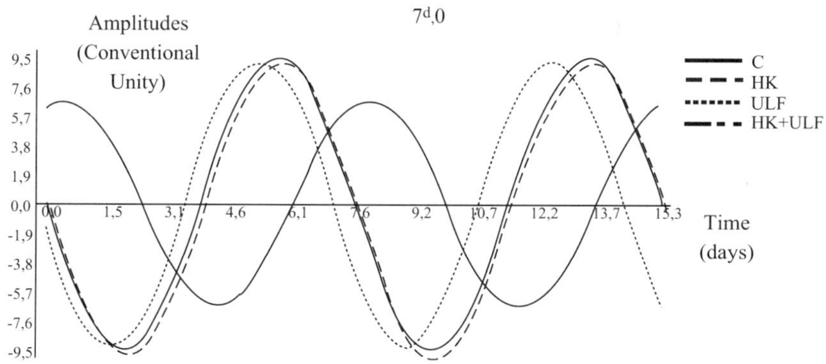

Figure 5. Phase co-relations of CPC PO in blood neutrophils in rats subjected to various stimuli: HK – hypokinesia; ULF – ULF EMR; HK+ULF EMR – hypokinesia combined with ULF EMR in $\approx 7^d.0$ period.

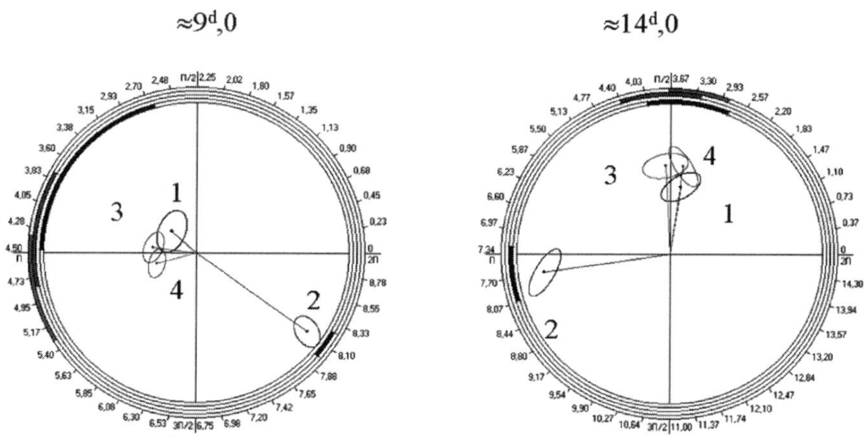

Figure 6. Cosinor.grams of integral rhythms of CPC AP in blood neutrophils in rats subjected to various stimuli: 1 – control (C); 2 – hypokinesia (HK); 3 – exposure to UHF EMR (UHF); 4 – combined action of UHF EMR and hypokinesia (HK+UHF EMR) in $\approx 9^d.0$ and $\approx 14^d.0$ day periods.

Our data allow to conclude that among the mechanisms of biological activity of UHF EMR an important role is played by its ability to modify the temporal organization of physiological systems. It is supposed that APUD-system elements localized in the skin play an important role in the mechanisms of receptivity and response to UHF EMR[25]. APUD-system structures that are located in various organs and tissues of an organism and synthesize biogenic amines and peptide hormones, including melatonin and serotonin[26], may also possess pace-maker properties and can play a role of potentially independent oscillators that are interconnected between themselves and the biosphere. These facts male it possible to use UHF EMR of the above described parameters as an external synchronizer, or a "time gauge", in the different types of desynchronoses.

REFERENCES

1. V. M. Vladimirskiy, V. Ya. Narmanskiy, N. A. Temuryants, Cosmic rhythms in magnetosphere, atmosphere, magnetsosphere, atmosbiosphere, noosphere. (Heliozhythm Publ. House Simferopol, 1994).
2. N. A. Temuryants, A. V. Shekhotkin, Chronobiological analysis of behavior in intact and epyphisectomized rats in the open field test, *Zhurnal Vyshei Nervnoi Deyatelnosti (Journal of Higher Nervous Activity)*, **49**(5), 839-846 (1999).
3. N. A. Temuryants, A. V. Shekhotkin, Change of infradian rhythmicity of blood lymphocyte dehydrogenases following epyphisectomy and exposure to weak variable magnetic fields, *Aviakosmicheskaya I Ekologichesjaya Meditsina (Aviation, Space and Ecological Medicine)*, **29**(3), 39-43(1995).
4. N. A. Temuryants, Nervous and humoral mechanisms of adaptation to the action of non-ionizing radiation, *Summary of the Doctor of Sciences Dissertation*, (Moscow, 1989).
5. L. G. Gassanov, B. I. Pyasetskiy, O.I. Pisanko, The role of ecological factors in the interactions of low intensity, ultra-high frequency electromagnetic fields with the human organism,*Vestnik Akademii Nauk USSR (Annals of the Academy of Sciences of Ukraine)*, **10**, 33-38 (1988).
6. N. N. Lebedeva, T. I. Kotrovskaya, Experimental and clinical research in the areas of biological effects of millimeter waves, *Millimetrovie Volny V Biologii I Meditsine (Millimeter Waves in Biology and Medicine)*, **3**, 3-14 (1999).
7. B. M. Vladimirskiy, V. G. Sidyakin, N. A. Temuryants, V. B. Makeev, V. P. Samokhvalov, Space and biological rhythms (Taurida Public House Simferopol, 1995).
8. B. S. Alyakrinskiy, Adaptation in the aspect of biorhythmology. Problems of temporal organization of the living systems (Nauka Moscow, 1979).
9. T. K. Breus, S. M. Chibisov, P. M. Baevskiy, K. B. Shebsukhov, Chronostructure of hearts rhythms and factors of environmental factors (Nauka Moscow, 2002).
10. S. I. Stepanova, Biorhythmological aspects of adaptation ("Nauka" Publisher, Moscow, 1986).
11. A. L. Markel, To assessment of the main characteristics of rats behaviour in "open field" test, *Journal of Higher Nervous Activity*, **31**(2), 301-307 (1981).
12. N. A. Temuryants, E. N. Chuyan, A. V. Shekhotkin, Infradian rhythmics of the functionsl state of neutrophils and lymphocytes of rats with different constitutional features, *Biophysics*, **40**(5), 1121-1125 (1995).
13. E. A. Kovalenko, N. N. Guroskiy, Hypokinesia ("Meditsina" Publisher, Moscow, 1980).
14. P. Lilli, Techniques in pathology and practical histochemistry ("Mir" Publisher, Moscow, 1969).
15. R. P. Nartsissov, Diagnostic and prognostic value of cytochemical determination of lymphocyte dehydrigenases, *Vestnik AMN USSR (Annals of the Academy of Medical Science if the USSR)*, **7**, 71-74 (1978).
16. B. S. Briskin, Z. I. Savchenko, V.N. Bukatko et al, Peculiarities of the immunologic response in patients with acute pancreatitis exposed to millimeter waves of various modifications, *Biomeditsinskie Tekhnologii I. radioelectronika. (Biomedical technologies and radioelectronics)*, **12**, 3-10 (2002).
17. N. A. Temuryants, N. A. Vladimirsky, O. G. Tishkin, Extremely low frequency electromagnetic signals in the living world ("Naukova Dumka" Publisher, Kiev, 1992).
18. N. A. Temuryants, Ye. Yu. Grabovskaya, Reaction of rats with various constitutional features to the action on weak variable magnetic fields of the extremely low frequency, *Biophysics*, **37**(4), 715-718 (1992).
19. E. Yu. Graboskaya, Reaction of rats of various constitutional peculiarities to the action of weak ELF VMF. *Biophysics*, **37**(4), 817-820 (1992).
20. L.K. Garkavi, E. B. Kvakina, T. S. Kuzmenko, Antistress reactions and activation therapy. Activation reaction as a path to health via processes of self-organization ("IMMEDIS" Publisher, Moscow, 1998).

21. N. A. Temuryants, O. M. Chuyan, N. A. Verko et al., Changes in responses of rats to hypokinetic conditions following exposure to ultra-high frequency electromagnetic field, *Fiziologicheskii zhurnal (Physiological journal)*, **49**(1), 87-93 (2003).
22. U. Ashoff, Free-flowing and restrained circadian rhythms, in: *Biologicheskie ritmy. (Biological rhythms)*. ("Mir" Publisher, Moscow, 1984), **1**, pp. 53-54.
23. L. P. Agulova, N. V. Udaltsova, S. E. Shnol, Correlation between macrospopic fluctuations in biological and physico-chemical processes and cosmo-physical factors, in: *Electromagnitnye polya v biosfere (Electromagnetic fields in biosphere)* ("Nauka" Publisher, Moscow, 1984), **1**, pp. 220-224.
24. E. B. Burlakova, A. A. Kondradov, I. V. Khudyakov, Influence of ultra-small doses of chemical agents on biological objects, *Izvestiya AN USSR (News of the USSR Academy of Sciences)*, 2, 184-193 (1990).
25. E. N. Chuyan Influence of millimeter waves of non-heat intensity on the development of hypokinetic stress in rats with different individual characteristics (Summary of Candidate of Biological Sciences Dissertation, 1992).
26. I. M. Kvetnoi, APUD-system (structural and functional organization, biological importance in norm and pathology, *Uspekhi fiziologicheskikh nauk (Achievements of physiological sciences*, **18**(1), 84-102 (1987).

16
ABSORPTION AND EMISSION OF PHOTONS BY COLLAGEN SAMPLES

T.G. Troshina, N.N. Loochinskaia, E. Van Wijk, R. Van Wijk, and L.V. Beloussov*

Abstract This study explored the integral intensity, slopes, and dynamics of the spectral content of delayed luminescence (DL) emitted from freshly prepared collagen gels, dried collagen samples, and collagen fibrils extirpated from rat tails after UV irradiation. It confirms conclusions by Ho et al. (2002) that the DL intensity is inversely proportional to the water content and it documents that DL slopes emitted from dry collagen samples are more exponential, whereas water-containing gels exhibit slopes that are more hyperbolic. While a DL timeframe persists, maximal spectral intensity shifts to the green-yellow range. In addition, optical interactions between germinating seeds and bundles of collagen fibrils induce a change of the emission rate from the seeds.

1. INTRODUCTION

Collagen proteins are significant components of extracellular matrix in all animal species. Besides their mechanical functions, collagen fibers possess unique properties that attract increased interest of modern biophysics and cell biologists (Ho, 2003). They contain a huge amount of water bound with amino acid residues (Hoeve & Tata, 1978; Peto et al., 1990), which possibly support proton conductivity (Sasaki, 1984; Ho, 2003). Also interesting are the optical properties of collagen fibrils. They are able to transmit for several centimeters "secondary radiation" from very weak biological sources (Gurwitsch and Gurwitsch, 1948). Collagen also has the capacity to simultaneously absorb two low energy photons (yellow-green range) and subsequently generate a frequency-doubled UV light (Zipfel et al., 2003). Recently, Ho et al. (2002) documented delayed luminescence (DL) of different collagen samples after monochromatic UV irradiation. Their team reported that DL intensity is inversely proportional to the amount of water in collagen samples and speculated that the increased amount of water promotes a non-radiative transfer of energy via protonic conductivity. They documented that the DL characteristics largely depend upon structural organization.

Objectives of this study were to:

1. study DL kinetics of different collagen samples during longer timeframes (>5 min).

---

* T.G. Troshina, Department of Embryology, Faculty of Biology, Moscow State University, Moscow 119899 Russia. E. van Wijk, International Institute of Biophysics, Neuss, Germany. R. van Wijk, International Institute of Biophysics, Neuss, Germany. L.V. Beloussov, Department of Embryology, Faculty of Biology, Moscow State University, Moscow 119899 Russia.

2. study changes in DL spectral content as a function of time.
3. explore the capacity of collagen fibrils to conduct the photon emission of germinating seeds.

2. MATERIALS AND METHODS

2.1. Preparation of Collagen Samples

2.1.1. Preparation of dry collagen samples from rat tails

Rat tails frozen at –2°C were maintained in 70% alcohol for 2 h. Subsequent handling was under sterile conditions. After skin removal, tendons were extracted, placed into Petri dishes with 70% alcohol, dissected and transferred into a flask with 0.1% acetic acid and then placed into a refrigerator. After several weeks, the solution was centrifuged at +4°C for 30 min at 3000 rotations per min. Supernatant was carefully removed and collagen content was determined by lyophylic drying at 37°C between 24 h to 4 days.

2.1.2. Preparation of the collagen gel

On completion of the drying, NaOH (0.34 N, 133 μl), Na_2CO_3 (70 μl), Hepes (40 μl), and PBS (200 μl) were added in succession per 2 ml of the collagen solution; the flask was then placed into a refrigerator for a few minutes. Another flask containing 1,596 ml of the collagen solution in acetic acid was placed into a refrigerator. Further handling was performed on ice. Collagen solution was then poured into a flask with alkaline and buffer solution, carefully mixed and then poured into Petri dishes or narrow quartz cuvettes. Those in Petri dishes lost water more rapidly than those in the cuvettes.

2.1.3. Preparation of "collagen bundles" from rat tails

This was completed under non-sterile conditions. Rat tails were defrosted, skin removed, tendon bundles separated from bones and muscles and detached into single fibrils of about 0.1 mm diameter and subsequently cut into 1.5 cm pieces. Several dozen fibrils were arranged parallel to each other on the bottom of a plastic Petri dish in preparation for UV radiation. The fibrils were packed as cylindrical bundles approximately 0.5 cm in diameter and 1.5 cm in length wrapped in a sheet of black photographic paper and foil for the seed-optical interaction studies. This preparation was labeled a "collagen bundle".

2.2. Irradiation

Collagen gel samples were freshly prepared and then dried for different time periods, from several hours to several days. The samples which contained 2 mg or 4 mg of a pure collagen were irradiated by a UV lamp mounted at 10, 5, or 2 cm from a sample. In all the cases, irradiation lasted for 1 min. At 15 s after the end of irradiation, the samples were put into the photomultiplier chamber for measurements.

2.3. Photon Counts

All the measurements were performed with the use of a photomultiplier with cathode EMI 9558QA, selected type, sensitive in the range from 200 to 800 nm (for technical details see Mieg, et al., 1992). Dwell time was 0.2 s. Cutoff filters RG610, GG495, and GG375 were utilized transparent only to the wavelengths greater than 610 nm, 495 nm and 375 nm, respectively.

2.4. Procedure for Testing Seeds – Collagen Bundle Optical Interactions

Two layers of germinating (maintained overnight on a wet filter paper) barley seeds (20-30 samples) were put into a quartz cuvette near the base of a hollow tube exposed at its open end facing a PMS cathode (Fig. 1 A, B). Seeds were covered on the opposite side with a sheet of wet filter paper. The entire cuvette and tube surface except the cathode-exposed end was carefully wrapped with black photographic paper. Each experiment utilized two separate collagen bundle positions in relation to seeds. The "off" position bundle lay on a horizontal wall of the cuvette exposed to the PMS cathode at its open end without optical contact with the seeds. The tube opening was covered with black paper. Thus, the seeds were optically isolated from the PMS cathode; only photon emission from the end of a collagen bundle was measured (Fig. 1A). The "on" position bundle was tightly inserted into a dual open-ended tube such that optical contacts between seeds and the bundle were maintained and only the end of a collagen bundle was exposed to the PMS cathode (Fig. 1B). Each position was maintained for 10 min (3000 measurements).

In each experiment, photon emission rates were measured under both "on" and "off" positions and the differences between them were calculated. For a succession of "off " – "on" – "off", to eliminate a possible impact of time-dependent events, the difference between the second and first positions ("on"

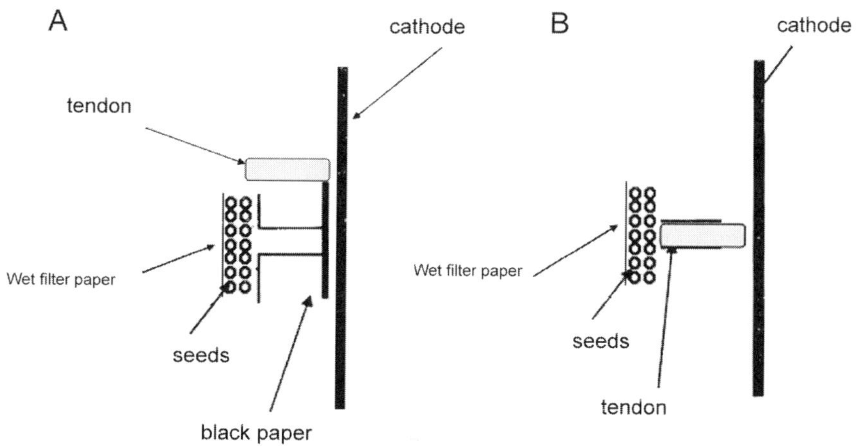

Figure 1. Experimental mount for exploring seeds-collagen bundle optical interactions. A: "off" position, no optical interactions between seeds and collagen bundle. B: "on" position, optical interactions are taking place.

differences) and that between the third and the second positions ("off" differences) were measured. In a different series, the photon emission of a seed sample with background subtracted was compared with those of the previously measured differences. For some sets of the seeds – collagen bundles optical interactions we have made also Fourier analysis corresponding to 6 min (2000 measurements) time intervals.

3. RESULTS

3.1. Integral Characteristics of DL from Different Collagen Samples

As displayed in Fig. 2 and Tables 1-4, all the samples (primarily those containing dried collagen) demonstrated considerable DL, directly proportional both to the amount of a collagen and the time of drying (Fig. 2A) and inversely proportional to the distance between a sample and the UV lamp. The correlation of the total DL intensity with the duration of drying was very high for 4 mg samples, a bit smaller for 2 mg samples, and negligible for the controls (empty plastic dishes), emphasizing that the effect is really due to the presence and the amount of collagen (Fig. 2B). Irradiation by visible (non-UV) light did not produce any significant DL.

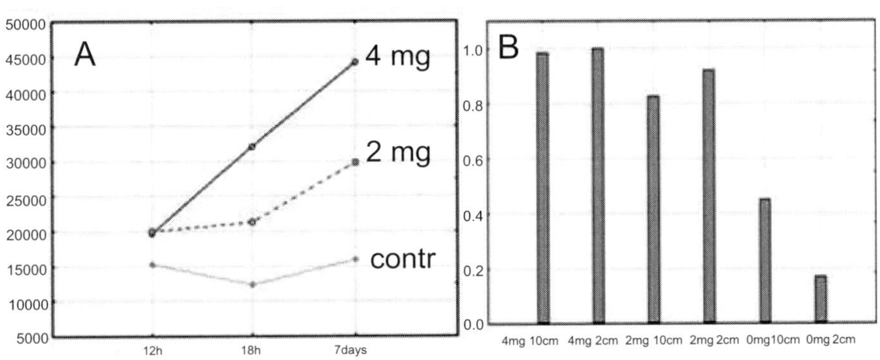

Figure 2. DL characteristics of the pure collagen samples, as compared with controls (empty plastic dishes). A: DL dependence (vertical axis) upon the duration of drying (horizontal axis: 12 h, 18 h and 7 days resp. Amount of collagen is shown). In all the cases the distance from UV lamp was 2 cm. B: correlation coefficients between DL intensity and the duration of drying. Two left columns correspond to 4 mg containing samples, next two columns to 2 mg containing ones, and two right columns which are below the significance limits to controls.

Table 1. UV-decays of a freshly prepared (1.5 h) collagen gel, as compared with the control values (UV-decay of an empty quartz cuvette). In all the tables given are the sums of decay values per 1000 measurements (200 s)

Distance from UV source, cm	Collagen gel, 5 mg	Control
1	15256	16080
5	19482	17601
2	25056	23157

Table 2. UV-decays of a dried collagen gel

Distance from UV source, cm	Collagen gel, 5 mg	Control
10	21232	11649
5	25477	12789
2	32779	14746

Table 3. UV-decays of the pure collagen samples dried for 18 h

Distance from UV source, cm / Amount of collagen, mg	10	5	2
4	23669	29990	32016
2	15182	19931	21331
Empty (control)	9618	No data	12314

Table 4. UV-decays of the pure collagen samples dried for 7 days

Distance from UV source, cm / Amount of collagen, mg	10	5	2
4	38878	39783	44169
2	29152	27522	29713
Empty (control)	17907	15307	15933

3.2. Slopes and the Dynamics of Spectral Content During the Different DL Samples

DL slopes of freshly prepared collagen gels were rather monotonous and close to hyperbolic curves (Fig. 3A). Slopes from more dry gel samples (those maintained in Petri dishes) (Fig. 2B) consisted of two sharply different parts: (1) those taking place within 100 s after counting emissions commenced and were much closer to exponential rather than hyperbolic curves; (2) those taking place after 100 sec were hyperbolic. DL slopes from dry collagen samples (Fig. 3C) and from collagen bundles (Fig. 3D) were very close to exponential. Results were reproduced during two experimental seasons.

Concerning the dynamics of the DL spectral content: the following properties were of interest (Fig. 3 F-G): (1) both fresh and dried gels during the first 100 sec exhibited a substantial UV component gradually decreasing later.

However, no UV component was detected in the fibril bundles' DL (Fig. 3G); (2) In none of the samples, as expected according to Stock's law, was there a progressive shift of the DL spectral content towards the greatest wave lengths measurable by PMS (up to 800 nm). Instead, in most cases, the maximal emission values fit the yellow-green range (495-610 nm). It is also worth mentioning, that in the course of fresh gel samples DL we detected some bursts within the red spectral range which reached the total emission level typical for the given DL period.

3.3. Seed-Collagen Bundle Optical Interaction

As observed from Table 5 and Fig. 4A, most of the "on" differences were positive and most of the "off " were negative such that the differences between the sets were highly significant (p = 0.000001). The "on" differences were of the same value order as the average photon emission of the seeds themselves (that is, without inserting the bundle) (Fig. 4A). Moreover, in several individual seed samples, the "on" differences largely exceeded the same seed's photon emission (Fig. 4B). These results reject any suggestion that the positive values of the "on"

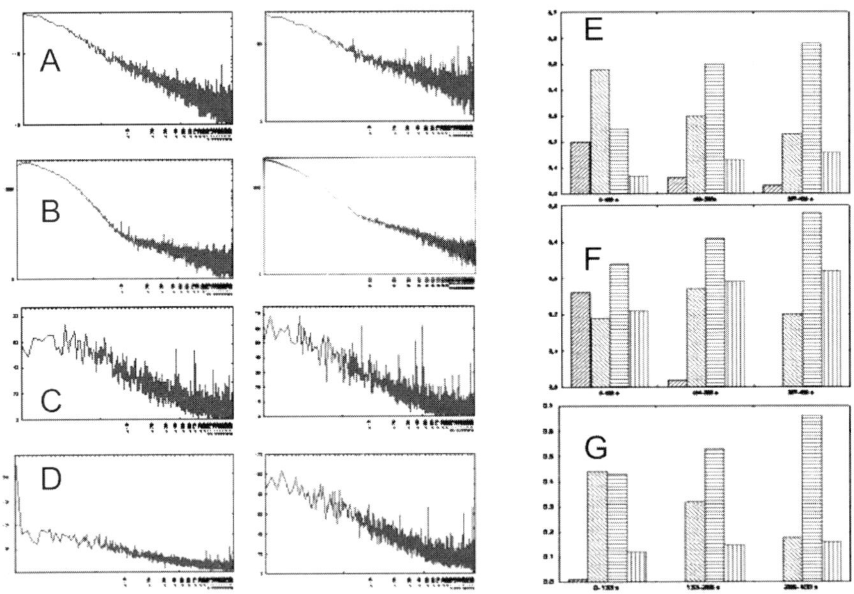

Figure 3. DL slopes and dynamics of spectral content from the different collagen samples. DL slopes from: A: freshly prepared water-containing collagen gels maintained in quartz cuvettes. B: gel samples maintained in Petri dishes. C: dry collagen samples. D: collagen bundles. A, B are plotted in double logarithmic coordinates (straight line corresponds to hyperbola) and C, D in semi-logarithmic coordinates (straight line corresponds to exponent). E-F: DL spectral content during three successive 100 s periods for dried collagen gel (E), fresh (water-containing) gel (F) and a collagen bundle (G). In each column, groups from left to right display emission intensities within: < 320 nm; 320-495 nm; 495-610 nm; 610-800 nm ranges. < 320 nm range is absent in F sample during 200-300 s period and in G sample during 100-300 s period. In all the cases at the end of DL 495-610 spectral range is dominated.

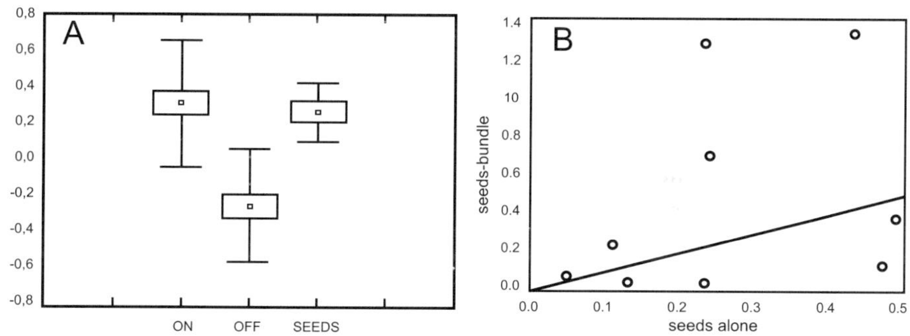

Figure 4. Quantitative estimations of the seeds-collagen bundle optical interactions. A: Box & Whisker plot. "On" is the difference in photon counts per 0.2 s between "on" position and a preceded "off" position. "Off" is the difference between the next "off" position and the preceded "on" position. "Seeds" are photon counts per 0.2 s from non-interacting seeds minus background values. B: Correlation graph of photon emission rates per 0.2 s between 9 sets of seeds while being in direct contacts with PMS and same sets of seeds while separated from PMS by collagen bundles. Straight line is the bisector of the graph angle. Circles above the bisector belong to the sets of seeds which emission rates via collagen bundles exceed those of directly contacting with PMS. As seen from the graph, in four cases such an excess was 2-fold and more.

differences were due to a leakage of seed emission via slits in the fibrils' samples; a leakage cannot produce more photons than the emitting sample itself.

Optical interaction between seeds and collagen bundles (Fig. 5) never resulted in a simple summing up of their emission values. No abrupt or stable change of emission rates were observed immediately after establishment of optical contacts. Rather, the emission rates either gradually increased (Fig. 5A) or, on the contrary, reached the highest values soon after the start of the optical contacts; subsequently, these gradually decreased (Fig. 5B).

Fourier analysis revealed considerable, although variable changes effecting both collagen bundles and seeds in the frequency patterns not only during, but also after the end of the optical interactions (Fig. 6). Typical was the enhancement of several frequency maxims in seeds after optical interaction with collagen bundles (cf E1 and A1; E2 and A2). At least some of enhanced maxims looked as if being "borrowed" from collagen samples (cf B1 and E1: 28 → 26; 135 → 133; B2 and E2: 20 → 20) while others could be traced still in pre-interacting seeds (cf A2 and E2: 7 → 7; 66 → 66). On the other hand, the spectral maxims of the collagen samples in most cases were diminished after optical interactions (cf B1 and D1; B2 and D2).

4. DISCUSSION

The study confirms the conclusion of Ho et al. (2002) that the integral DL value is inversely proportional to the amount of water in the collagen samples. In addition, the greater the water content, the closer were DL slopes to hyperbolic curves, whereas in dry samples, the slopes were close to exponential. It can be suggested that the presence of water supports a coherent interaction of the elementary photon emitters.

Table 5. Protocols of seeds-bundles optical interactions. Given are "on" and "off" differences per 0.2 s and the rates of photon emission from 9 samples of seeds, background rates subtracted

	ON	OFF	SEEDS
1	1,330	-1,100	0,430
2	1,280	-0,310	0,230
3	0,370		0,490
4	0,700		0,240
5	0,040		0,130
6	0,240	-0,090	0,110
7	0,070	0,120	0,050
8	0,040	-0,010	0,230
9	0,130		0,470
10	0,370		
11	-0,090		
12	-0,160	0,250	
13	0,200	0,240	
14	0,470		
15	0,280	-0,410	
16	0,470	-0,600	
17	0,040	-0,580	
18	0,660	-0,380	
19	0,140	-0,140	
20	0,330	-0,350	
21	0,250	-0,210	
22	0,260	-0,210	
23	0,280	-0,300	
24	0,090	-0,250	
25	0,110	-0,010	
26	0,260	-0,360	
27	0,280	-0,190	

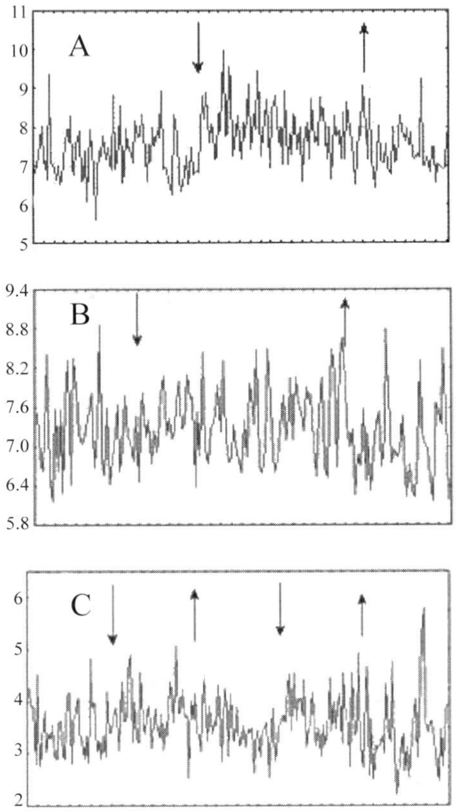

Figure 5. Some records of photon emission rates during seeds-bundles optical interactions. Downwards pointing arrows indicate the start and upwards pointed arrows the end of optical interactions. In A and C in the absence of optical interactions recorded were collagen bundles while in B the seeds. Horizontal axis: time. Total duration in A and C was 33 min. while in B 26 min. Vertical axis: photon counts per 0.2 sec.

Considering the dynamics of the DL spectral content, it is remarkable that 495-610 nm spectral range dominates (contrary to Stock's law) rather than 600-800 nm (to which a photomultiplier is also sensitive). The dominated spectral range exactly fits the doubled wave lengths of the UV range. One may speculate that the concentration of a DL after UV irradiation just within this spectral range is a process reverse to the frequency doubling described by Zipfel et al. (2003).

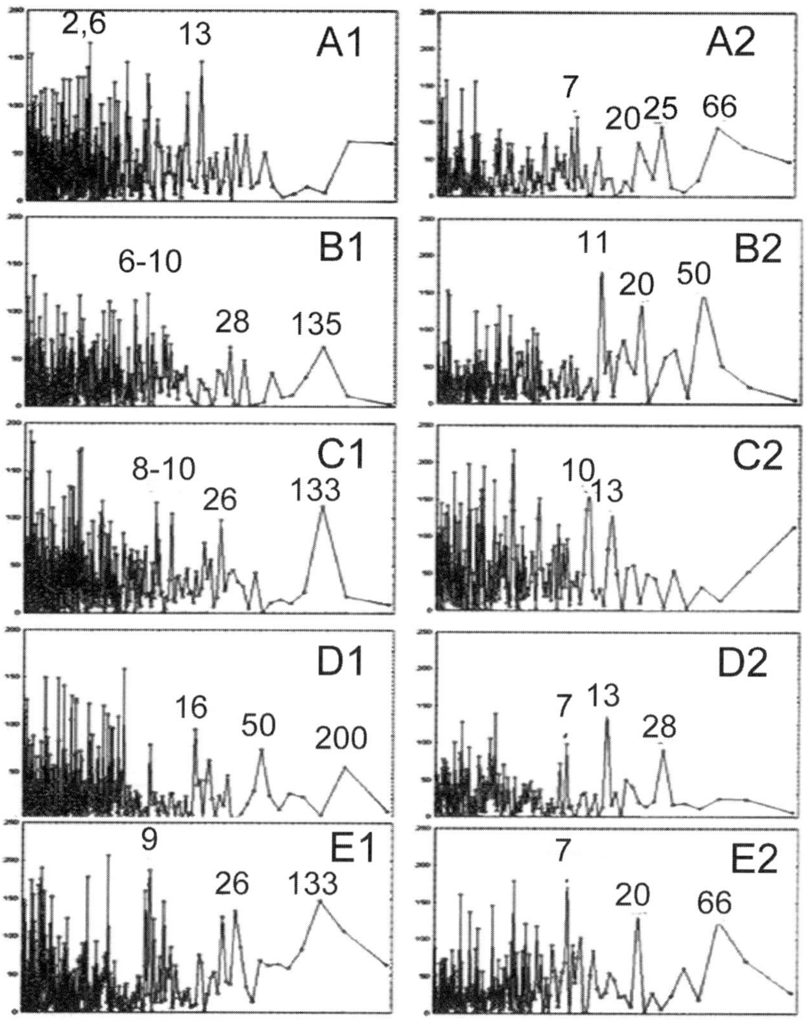

Figure 6. Fourier spectra (displayed as periodograms) for two sets of seeds-collagen bundles optical interactions. A1, A2: seeds before optical interactions. B1, B2: collagen bundles before optical interactions. C1, C2: optical interactions; D1, D2: collagen bundles after optical interactions; E1, E2: seeds after optical interactions. Figures on the graphs display the periods (in seconds) of some mostly pronounced and reproduced spectral maxims.

Whatever the mechanisms of these transformations, such a yellow-green concentration of light might permit its further transformation into UV light providing a UV relay along the collagen structures. This may have a relation to the above-mentioned Gurwitsch and Gurwitsch (1948) data on the transmission of very weak UV radiation along a collagen fiber. Meanwhile, during a few time moments were there some instantaneous releases of energy within the red spectral range.

The lack of UV component even at the beginning of DL of the collagen fibers confirms Ho et al. (2002) suggestion that in these organized samples, UV energy is more effectively used for a radiation-less transfer than in the less organized samples (gels and dried collagen).

The study demonstrates the existence of optical interactions between seeds and collagen bundles which considerably modify the rate of the total photon emission. More detailed analysis demonstrates that in no way the optical interactions could be reduced to a passive "transmission" of the seed photon emission by a collagen bundle. Instead, the interactions of both components are mutual and essentially non-monotonous. They are associated with (a) complicated dynamics of photon emission during the interactions and (b) changes of frequency patterns of both components (especially of seeds) after the end of interactions. This is typical for other kinds of optical interactions between living samples (see Beloussov, this volume). One can suggest that similar interactions, which might be of great physiological importance, are taking place between the native sources of biophoton emission in living organisms (for example, muscles) and collagen tendons.

REFERENCES

Gurwitsch, A.G., Gurwitsch, L.D. (1948). Introduction to a study of mitogenesis. Moskva, Izdatelstvo IEM 114 p. (in Russian).

Ho, M.W. (2003). The Rainbow and the Worm. World Scientific. Singapore.

Ho, M.W., Musumeci, F., Scordino, A., Triglia, A. and Privitera, G. (2002). Delayed luminescence from bovine Achilles' tendon and its dependence on collagen structure. J. Photochem. Photobiol. B: Biol. 66, 165-70.

Hoeve, C.A.J. and Tata, A.S. (1978). The structure of water absorbed in collagen. J. Phys. Chem. 82, 1661-1663.

Peto, S., Gillis, P and Henri, V.P. (1990). Structure and dynamics of water in tendon from NMR relaxation measurements. Biophys. J. 57, 71-84.

Sasaki, N. (1984). Dielectric properties of slightly hydrated collagen: Time – water content superposition analysis. Biopolymers 23, 1725-1734.

Zipfel, W.E., Williams, R.M., Christie, R., Nikitin, A.Y., Hyman, B.T. and Webb, W.W/ (2003). Live tissue intrinsic emission microscopy using multiphoton-excited native fluorescence and second harmonic generation. PNAS 100, 7075-7080.

17
INTERACTION OF PLANT SHOOTS AND ROOTS: DYNAMICS AND STABILITY

N.V. Budagovskaya*

Abstract: The use of a highly sensitive laser interference auxanometer (accuracy 0.07 μm) made it possible to register rapid (sec, min) and slow (h) response reactions of shoots (leaves, stems) under changing conditions of mineral nutrition in the root zone. Experiments were carried out with oat, barley, wheat, rice, and buckwheat plants. Disturbances in mineral nutrition of plants destabilized shoot-root interactions and caused decrease in growth rate of shoots (leaves, stems). Addition of NaCl in increased concentration to the root zone of plants caused a two-phase response reaction of leaves: decrease and the following increase in their growth rate in each phase. Duration of the 1st phase was shorter than of the 2nd. Growth rate of leaves was restored by the end of the 2nd phase (few hours after addition of NaCl). The 1st phase may be related to rapid adaptive reactions and changes in leaf turgor, the 2nd – to slower adaptive processes – *de novo* synthesis of protectors. Introduction of NaCl in high concentration caused stoppage in leaf and stem growth and shrinking of their tissues as result of dehydration. Addition of water to the root zone decreasing NaCl concentration or washing the roots of NaCl restored the turgor of leaves and increased their growth rate. Typical temporal intervals (temporal invariants) in response to reactions of shoots of plants of different species under changes in mineral nutrition of roots have been revealed. Different stationary regimes of plant growth and transitional processes of plant from one stationary state to another are characterized. Growth and turgor oscillations of about min and about hour periods were registered for all experimental plants. Relation of growth oscillations to oscillatory processes of water exchange is discussed. It was noted that coordination of rhythms of oscillation processes of different hierarchy levels was the basis for dynamic stability of plant organism and is responsible for existence of structure-functional invariants including "golden section" in whole plant composition of shoots and roots. The display of the symmetry principle in structural and functional organization of plants is considered.

1. INTRODUCTION

Higher plants consist of two main structural and functional parts: photoautotrophic (shoots) and heterotrophic (roots). Shoots and roots are functionally coordinated in the whole plant system. Earlier we demonstrated that the ratios of functional and structural parameters of shoots and roots are constant and can be expressed in the harmonic proportion "golden section" under normal growing conditions and good adaptation of plants to stress factors

*Institute of Plant Physiology of the Russian Academy of Sciences, Botanicheskaya 35, Moscow 127276, Russia. E-mail: mbeloz@genebee.msu.ru.

(Budagovskaya 1996, 2000). In the present paper, the main attention will be attributed to the temporal aspect of shoot/root interrelations. The use of a highly sensitive laser interference auxanometer made it possible to register rapid (sec, min) and slow (h) response reactions of shoots (leaves, stems) to the changing conditions of mineral nutrition (changes in NaCl concentration at the root zone).

2. MATERIALS AND METHODS

Plant growth rate was measured by laser interference auxanometer LINA-EM3D (Russia), equipped with a helium-neon laser (632.8 nm). Sensitivity of the auxanometer was 0.07 μm. Growth parameters, temperature of air and substrate at the root zone were registered every 10 sec. Plants of oat cv. Horizont, barley cv. Auxenoi, rice cv. Suxianggen 1, buckwheat cv. Molva were used in experiments. Plants were grown in washed sand or in soil in Petri dishes 9.5 cm in diameter. Three days before the start of experiment solitary plants were placed in Petri dishes of 5 cm in diameter. In the course of growing, plants were watered by tap water, the distilled water was used for control watering during growth rate measuring. NaCl solution (800 mM starting concentration) was introduced in 0.5 or 1.0 ml portions to the substrate at the root zone (sand, soil, tap water). In some experiments, roots of plants were washed of NaCl. The measurement of the growth rate of leaves was performed by auxanometer at air temperature 20-28°C and humidity of 60-64%. In special experiments the air humidity had higher values.

3. RESULTS AND DISCUSSION

Plant growth as integral process reflects the functional coherence of shoots and roots. The present paper deals with the study of effect of changes in NaCl concentration at the root zone on the growth rate of shoots.

Increase or decrease in NaCl concentration at the root zone led to a change in growth rate of leaves and stems of experimental plants (oat, barley, rice, buckwheat). A biphasic response reaction of leaves of oat (Fig. 1) and barley (data not shown) plants at single addition of NaCl was observed: decrease and increase in leaf growth rate in each phase. The first phase was noticeably shorter than the second. The ratio of phase duration of leaf response reactions was similar for oat and barley plants. The first phase may be connected with the change in leaf turgor: its decrease after addition of NaCl as result of decrease in water supply to plant and following increase due, apparently, to stoma closure and other rapid adaptive reactions maintaining cell homeostasis. Decrease in growth rate in the start of the second phase may be caused by enhanced negative effect of increased NaCl concentrations in plant. Increase in growth rate in the second phase of response reaction (started more than 3 h after addition of NaCl) may be due to accumulation of osmotically active substances and other protective compounds synthesized *de novo*.

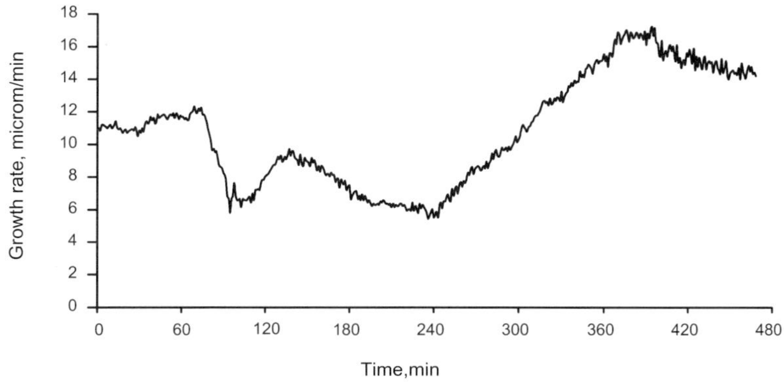

Figure 1. Effect of NaCl (50 mM) on the growth rate of the 1st leaf of 9-day-old oat plant. ↓ - the moment of addition of NaCl (0.5 ml, 800 mM) to the root zone (sand).

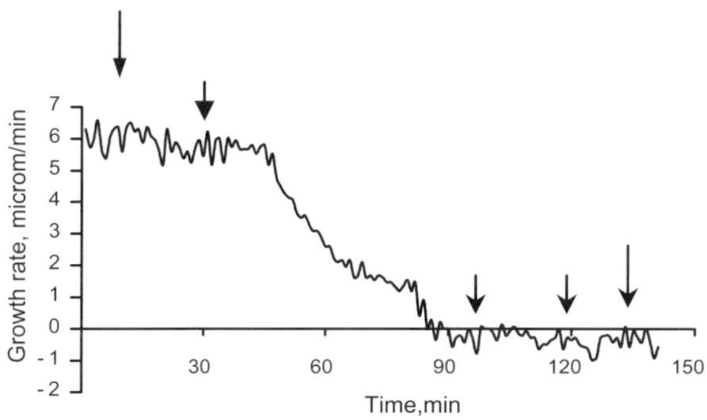

Figure 2. Effect of NaCl (50 mM) on the growth rate of the 2nd leaf of 16-day-old rice plant. ↓ – the moment of addition of NaCl (0.5 ml, 800 mM) to the root zone (sand). ↓ – the moment of addition of 0.5 ml of water, ↓ - the moment of addition of 1 ml of water.

Rice plants were more susceptible to NaCl in contrast to oat and barley plants: growth rate of their leaves decreased to zero at 50 mM NaCl concentration (Fig. 2) at which leaves of oat and barley plants continued to grow. Rice plants do not synthesize glycinebetaine under salinity, wich participates in other plants in osmotic regulation and protects membranes and proteins (Rathinasabapathi at al. 1995, Sakamoto at al. 1998). Addition of water, decreasing NaCl concentration at the root zone of rice plant, did not lead to resumption of leaf growth. Unlike to it addition of water to the root zone of oat plant, treated preliminary with NaCl (to final concentration of 48 mM), caused a rapid increase in leaf growth (Fig. 3). The following decrease in leaf growth rate of oat plant to a stationary level might be the result of stoma opening. Secondary addition of NaCl in a higher concentration than initial led to a rapid decrease in its growth rate to zero. Transition of the curve to negative region may point to shrinking of the leaf tissues as result of partial dehydration caused by addition of NaCl. The sign of response reaction changed: transfer of the leaf tissue from elongation to shrinking took place. Leaf tissue shrinking turned to its short-term elongation, after which leaf growth stopped, but pulsating shrinking (negative coordinate region) was observed followed by pulsating elongation (positive coordinate region) (Fig. 3a, fragment of Fig. 3). Leaf growth was restored for the next day.

The presence of symmetry in response reactions of shoots (elongation and shrinking) on successive addition of water and NaCl to the root zone calls attention (Figs. 3 and 3a). This symmetry of processes (temporal symmetry) reflects interelations of spatially symmetric organs: shoots and roots.

Increase in amplitude of oscillations of elongation (close to zero values) preceded restoration of growth rate and was observed repeatedly (Fig. 4). Transition of oscillation process from shrinking to elongation as well as increase in amplitude of elongation oscillations near zero values may indicate on start of restoration processes in plant. Increase in amplitude of oscillations followed by further increase in leaf growth rate occurred also at high initial growth rates of leaf. The amplitude of oscillations again decreased when the growth rate of leaves increased to a higher level. Figure 5 demonstrates substantial increase in amplitude of oscillations of leaf growth rate of barley plant. After that leaf growth rate increased from 20 μm/min to 24-25 μm/min, and oscillation amplitude decreased. Thus, increase in oscillation amplitude may point to transition of oscillation regime of the studied process from one stationary level to another, higher stationary level.

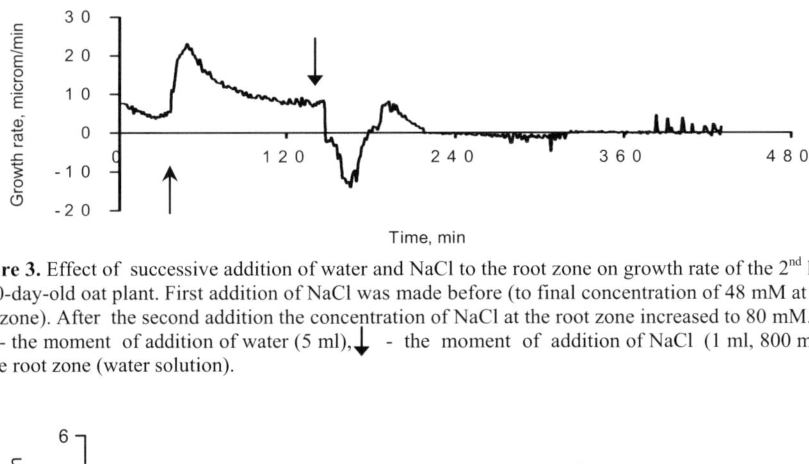

Figure 3. Effect of successive addition of water and NaCl to the root zone on growth rate of the 2nd leaf of 20-day-old oat plant. First addition of NaCl was made before (to final concentration of 48 mM at the root zone). After the second addition the concentration of NaCl at the root zone increased to 80 mM. ↑ - the moment of addition of water (5 ml), ↓ - the moment of addition of NaCl (1 ml, 800 mM) to the root zone (water solution).

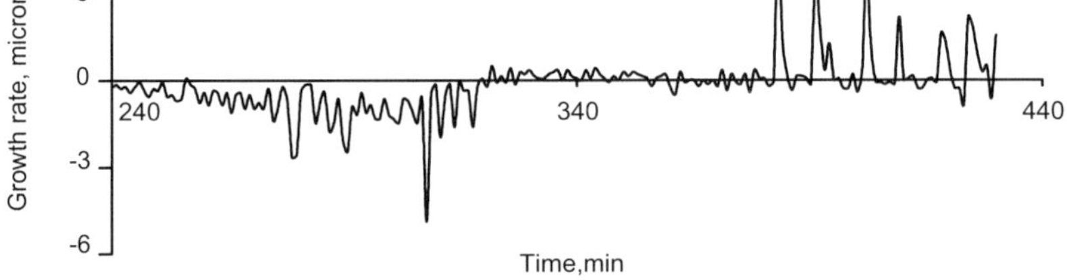

Figure 3a. Fragment of Fig. 3 (240-440 min).

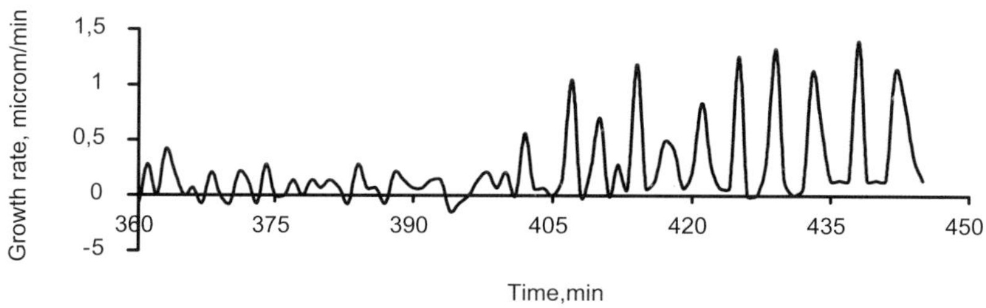

Figure 4. Oscillations of growth rate of the 1st leaf of 11-day-old oat plant. NaCl (twice in 0.5 ml, 800 mM) and water (2.0 ml) were added preliminary to the root zone (sand).

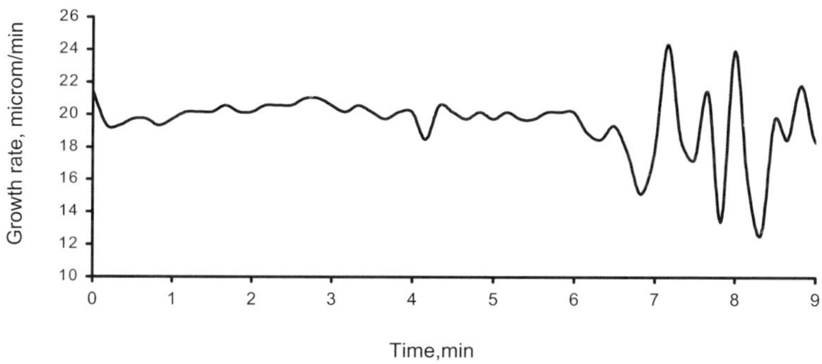

Figure 5. Oscillations of growth rate of the 2nd leaf of 12-day-old barley plant.

Figure 6 illustrates another oscillation regime of the growth rate of the same barley leaf on the next day after addition of NaCl and after 2 h after addition of water to the root zone. One can see the interchange (with a certain periodicity) of high (with 30 sec period) and low amplitude oscillations.

Oscillations with a period close to 1 min were registered by measuring growth rate of both leaves and stems of plants. Figure 7 demonstrates the curve of

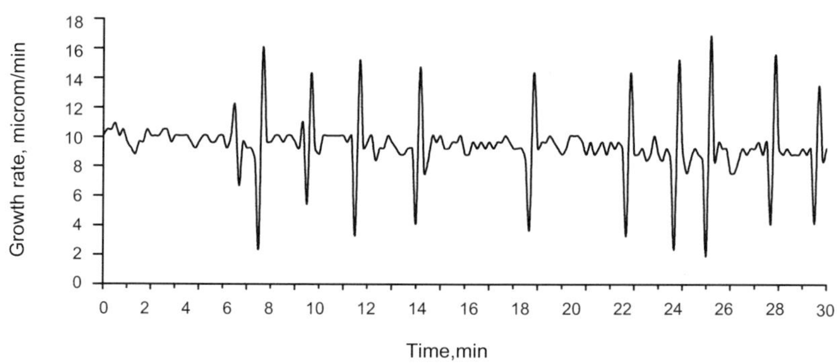

Figure 6. Oscillations of the growth rate of 13-day-old barley plant on the next day after addition of NaCl (1 ml, 800 mM) and 2 h after addition of water (1 ml) to the root zone (sand).

changes in stem growth rate of buckwheat plant at addition of NaCl to the root zone. After decrease in stem growth rate caused by NaCl, oscillations continued with a rhythmical interchange of tissue elongation and shrinking. Further they were followed by pulsating shrinking of tissue (negative coordinate region). Figure 7a (fragment of Fig. 7) shows sequential change in oscillation regimes: 2 regimes in positive and 2 in negative regions. Stepwise transitions of the curve from the positive region through zero values to the negative region first close to zero and further to higher negative values indicates on changes in the state of buckwheat plant as result of efflux of water after addition of NaCl (trigger transitions from one stationary state to another). Accordingly, the character of the oscillatory regime of growth rate may serve as diagnostics index of the state of plant.

Besides high-frequency oscillations of about 1 min period, we have also registered in our experiments low-frequency oscillations of growth rate of leaves of barley, oat, wheat, rice, and stems of buckwheat plants of about 1 h or smaller (15-30 min) period, which were occasional and dying out. These oscillations lasted for a few hours and were induced by external effect (addition of water and NaCl). Figures 8 and 9 present these oscillation regimes for barley and rice plants. The plot of oscillations of leaf growth rate of rice plant has both positive and negative values that indicates as interchange of elongation (positive coordinate region) and shrinking (negative coordinate region) of tissues. These oscillations

Figure 7. Effect of of NaCl (100 mM) on the growth rate of stem of 9-day-old buckwheat plant. - ↓ the moment of addition of NaCl (1.0 ml, 800 mM) to the root zone (sand).

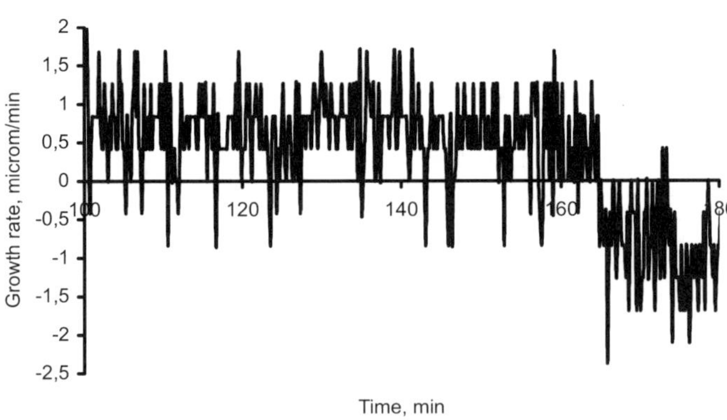

Figure 7a. Fragment of Fig. 7 (100-180 min).

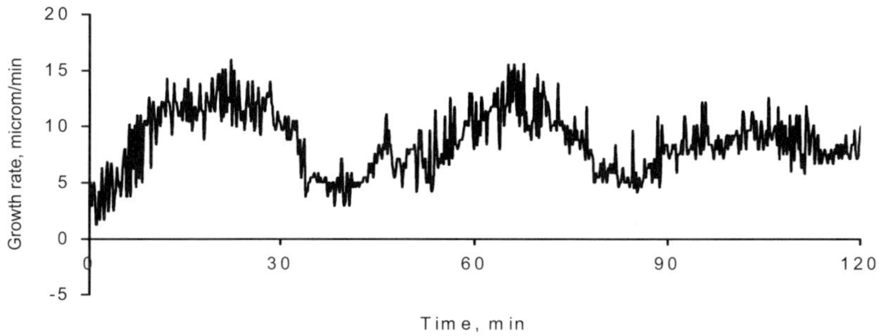

Figure 8. Oscillations of growth rate of the 2nd leaf of 15-day-old barley plant. The measurement of leaf growth rate was preceded by a double addition of NaCl (1.0 ml, 800 mM) to the root zone (sand) followed by washing the roots of NaCl.

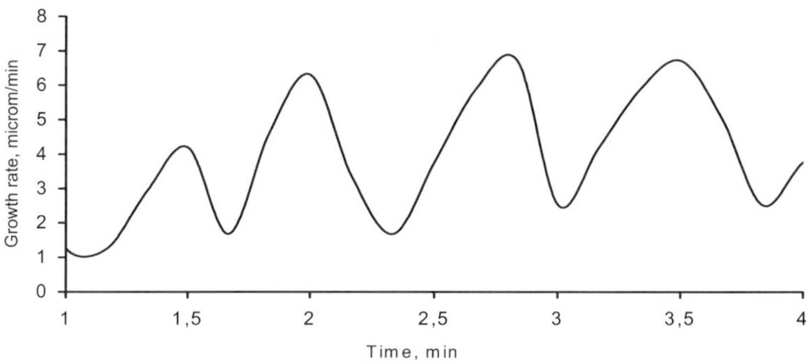

Figure 8a. Fragment of Fig. 8 (1-4 min).

were induced by sequential addition of water and NaCl to the root zone of plants. Interchange of elongation and shrinking of leaf tissues may point to existence of osmotic component in this oscillation process.

In addition to low-frequency oscillations, high-frequency oscillations can be also seen in Figs. 8 and 9. These high-frequency oscillations in barley and rice plants differ by amplitude value: for barley plants the mean oscillation amplitude values are of 4-5 μm, for rice plants 0.5-1.0 μm (Figs. 8a, 9a, fragments of Figs. 8 and 9). In this connection other peculiarities of rice plants should be also mentioned. At addition of NaCl in high concentrations to the root zone of rice plants, shrinking rate of leaf tissues as result of efflux of water was lesser than shrinking rate of leaf tissues of barley and oat plants under similar conditions. The magnitude of negative values on growth rate curves of rice plants in all experiments was lesser than for barley and oat plants in spite of lesser salt tolerance of rice plants. Besides, the leaves of rice plants dried under water deficiency were rolling into tubes and did not wilt as did the leaves of barley and oat plants at turgor decrease. Lesser amplitude of oscillations of leaf growth rate, lesser shrinking rate of leaf tissues at high salinity, preservation of vertical position of leaves at turgor decrease and leaf drying – all these events may be explained by larger inflexibility of leaf tissues of rice plants containing silicon (Erygin 1969) in contrast to more elastic leaf tissues of barley and oat plants, which are able to elongate and shrink more. Reversed proportional dependence of the amplitude value of high-frequency oscillations of growth rate and inflexibility of leaf tissue of barley, oat, and rice plants may be the confirmation of relation of these oscillations to water transport in plant. According to Zyalalov (1981), oscillations of turgor pressure of parenchyma cells and diameter of stem took place in course of water transport in plant indicating on pulsations of the cell volume. It is possible that similar pulsations of the cell volume occur also in leaf tissues. Apparently, these pulsations of cell volume have a lesser amplitude in rice plants due to larger inflexibility of containing silicon cell walls than in barley and oat plants. For the same reason, the amplitude of growth oscillations and growth rate of rice plants may be also less than of barley and oat plants as demonstrated in our experiments.

Oscillations of about 1 min period were registered permanently in the course of measuring of plant growth rate. Preservation of oscillation regime of the process at the stoppage of leaf (Figs. 3a and 4) and stem (Fig. 7) growth may also point that these are turgor oscillations connected with water exchange in plants. It was demonstrated in works of Sherif (1977), Ushakova and Koltunova (1982),

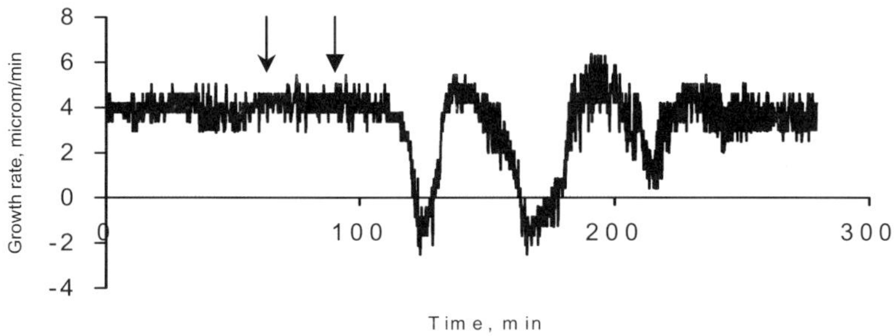

Figure 9. Oscillations of growth rate of the 3rd leaf of 16-day-old rice plant. The measurement of leaf growth rate, was preceded by a double addition of NaCl (0.5 ml, 800 mM) to the root zone (soil) followed by washing the roots of NaCl. ↓ - the moment of addition of water (1.0 ml) to the root zone. ↓ - the moment of addition of NaCl (1.0 ml, 800 mM) to the root zone.

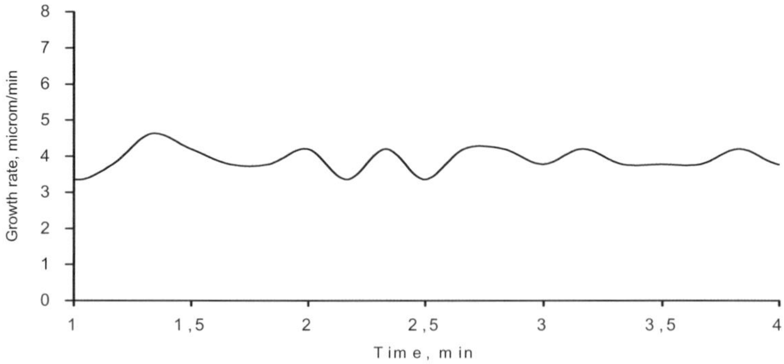

Figure 9a. Fragment of Fig. 9 (1-4 min).

and Zholkevich (1991) that absorbance, transport, and transpiration of water by plants occurs as oscillation process with a period of few min. Participation of contractile cell systems in water exchange is considered (Kushnirenko et al. 1991). It is possible that oscillations of elongation and shrinking of tissues of leaves and stems of plants mediated by oscillations in water transport also reflect cytoskeleton oscillations. Relation of growth pulsations with osmotically driven water flow was demonstrated also for animals (hydroid polip) by Belousov et al. (1993).

Growth oscillations with relatively high periods (20-40 min) were observed by Shevelukha (1992) in wheat plants and wild cereal grasses after watering of plants subjected to soil dryness. Correspondence of periods of these oscillations to oscillation periods of water transport and transpiration in plants demonstrated by Karmanov and Savin (1964) allowed Shevelukha to make a conclusion on their possible relation to water exchange. Oscillations of leaf growth rate of plants of about 1 h period demonstrated in our experiments can be also related to oscillation processes of water exchange since they involved both elongation and shrinking of leaf tissues (due to water efflux). Low-frequency oscillation character of growth rate of plants may depend also on acropetal transport of Ca^{2+} and basipetal transport of auxin occurring in oscillation regimes of about 1 h periods (Wodzicki et al. 1984, Medvedev et al. 1989). High-frequency oscillatory growth regime is possibly related to intracellular oscillations of Ca^{2+}, Ca^{2+}-dependent transformations of cytoskeleton and also water transport, synchronized at cellular ensemble level, with participation of contractile systems.

Thus, high-frequency oscillations may be due to synchronization of processes at cellular level, low-frequency – at the level of larger systems of plants and their organs: roots – stems – leaves, including stomatal apparatus. Oscillations with larger periods correspond to a higher organization level (Baevsky and Geller 1974). Oscillations of about 1 h period are induced by plant watering and addition of NaCl – factors affecting water exchange. These factors can cause changes in activity of processes of water absorbance by roots, its transport in stems and leaves and affect intensity of transpiration. As result, previous functional coordination of organs and previous stationary regime of water exchange are disturbed. Establishment of a new regime of coordinated action of all systems on the whole plant level may induce transitory oscillation processes. Changes in the rhythm of water exchange will induce changes in the rhythm of growth processes because the growth by elongation depends on water supply of plants. Relation of water status of plants to growth processes was observed in our experiments at increase in air humidity around the leaf, decreasing intensity of transpiration. Under these conditions the growth rate of the leaf changed.

Apparently, changes in oscillation regime in plant growth can be presented as following. During stationary regime of plant growth high-frequency oscillations are permanently maintained as a tool for dynamic regulation and coordination of processes on cellular level. These oscillations are functionally coordinated with processes of water transport involving contractile systems. Changes in external conditions affecting water exchange in plants (in our experiments – addition of water and NaCl to the root zone of plants and increase in air humidity) provoke transitory low-frequency oscillations. After synchronization of processes of water absorbance, transport and transpiration (on the whole plant level) a new stationary high-frequency oscillation regime is set in. In this case, low-frequency oscillations dye out. This is demonstrated in Fig. 9. Stationary regime of growth of rice plant was changed for oscillatory regime with about 1 h period. Approximately after 2 h, these oscillations died out and a stationary regime was settled again similar to the one before low-frequency oscillations started. Thus the plant is aiming to preserve relative stability of its functional activity. Plants grown in soil (Fig. 9) expressed higher salt tolerance in contrast to plants grown in washed sand (under deficiency of mineral nutrition) (Fig. 2). For plants of "soil" variant, leaf growth of rice plant was restored even at addition of NaCl in increased concentrations.

Other examples of restoration of growth rate of plants after effect of unfavorable factors (increased NaCl concentrations, calcium channel blocker verapamil) were considered above (Fig. 1) and discussed earlier (Budagovskaya 2001a, 2001b). It should be mentioned that the time interval from the moment of addition of compounds of different functional activity (NaCl and verapamil) to beginning of restoration of leaf growth rate was similar and equal to 2-3 h. It is possible that this time is necessary for realization of adaptive reactions of plants (including *de novo* synthesis of protector compounds) leading to restoration of plant growth rate. Similar time interval (2-3 h) from addition of antioxidant ambiol to the start of slow phase in increase in leaf growth rate was necessary for plants of different species (oat, rice, wheat) for transition to a higher level of functional activity (Budagovskaya 2003). It involved, apparently, *de novo* synthesis.

This 2-3 h interval may be considered as temporal invariant because it was observed in response reactions of different plant species on effect of biologically active compounds of various type of action. Another temporal invariant also revealed in pants of different species was the period of oscillation processes. In our experiments these were about min and about h periods of turgor and growth oscillations. These are relative invariants, they fluctuate around definite values. More stable values are ratios of parameters (Budagovskaya 1996, 1998, 2000) such as, for example, shoot/root ratios of structure-functional parameters, ratios of oscillation periods or frequencies, and other. They are also relatively stable and at significant variation of external conditions may change for new invariant values. In critical cases, when plant or other living organism cannot adapt to high intensity stress factors, invariance of values disappears. It may be due to functional de-synchronization. Structure-functional invariants of biological objects as dynamic systems are less stable in contrast to physical constants and may be conditionally named as "biological constants". The "biological constants" reflect certain coordination of rhythms (periods or frequencies of oscillations) of different processes in living organism. Previously mentioned correspondence of shoot/root ratios of structure-functional parameters to the "golden section" (Budagovskaya 1996, 2000) reflects harmony in rhythms of oscillation processes in different sub-systems in their hierarchical subordination in the whole plant system. Harmonic coordination of rhythmical processes in plant is determined by optimum regime of regulation, serves as integration factor and provides dynamic stability of the organism in the whole.

ACKNOWLEDGEMENTS

Author is grateful to V.I. Guliaev for help in providing experiments.

REFERENCES

1. Beloussov, L.V., Labas, Ju. A., and Kasakova N.I., 1993, Growth pulsations in hydroid polyps: kinematics, biological role and cytophysiology, in: *Oscillations and Morphogenesis*, L. Rensing, ed., Marcel Dekker Inc., New York-Basel-Hong Kong, pp. 183-194.
2. Budagovskaya, N.V., 1996, Golden section in proportions of structural and functional parameters of photoautotrophic and heterotrophic organs and whole plant as an index of harmonic development, *Plant Physiol. Biochem.* **sp. issue:** SO4-2.
3. Budagovskaya, N.V., 1998, Changes in the state of photoautotrophic and heterotrophic organs of buckwheat plants at iron deficiency and low pH, *Fagopyrum* **15**: 1-7.
4. Budagovskaya, N.V., 2000, Golden section in structure-functional organization of plants, in: *Biophotonics and Coherent Systems*, L., Belousov, F-A., Pop, V., Voeikov, and R., van Wijk, eds., Moscow University Press, Moscow, pp. 67-74.
5. Budagovskaya, N.V., and Guliaev, V.I., 2001a, Effect of calcium channel blocker on growth dynamics of plants studied by laser interference auxanometry, in: *Plant Nutrition: Food security and sustainability of agro-ecosystems through basic and applied research*, W.J., Horst, et al. eds, Kluwer Academic Publishers, Dordrecht-Boston-London, pp. 204-205.
6. Budagovskaya, N.V., and Guliaev, V.I., 2001b, The impact of calcium channel blocker on growth and development of buckwheat plants, in: *Advances in Buckwheat Research*, C.H., Park, ed., Chunchon, Korea, pp. 114-116.
7. Budagovskaya, N.V., and Guliaev, V.I., 2003, Rapid and slow response reactions of plants on effect of antioxidant ambiol, in: *Advanced Reseach on Plant Lipids*, N., Murata, et al., eds., Kluwer Academic Publishers, Dordrecht-Boston-London, pp. 323-326.
8. Baevsky, R.M., and Geller, E.S., 1974, in: *Methodological Questions of Biocybernetics*, Nauka, Moscow, pp. 162-166.
9. Erygin, P.S., 1969, Physiology of Rice, in: *Physiology of Agricultural Plants*, B.A., Rubin, ed., Moscow University Press, Moscow, pp. 266-413.
10. Karmanov, V.G., and Savin, V.N., 1964, On self-oscillation character of water exchange in bean plants, *Doklady of USSR Academy of Sciences*, **154**: 970-973.
11. Kushnirenko, M.D, Pecherskaya, S.N., Bashtovaya, S.I., Kleiman, E.I., and Zholkevich, V.N., 1991, *Doklady of USSR Academy of Sciences* **317**: 509-511.
12. Medvedev, S.S., Markova, I.V., Batov, A.Y., and Maksimov G.B., 1989, Polar flows of calcium ions and growth of plant tissues, *Fiziologia Rastenii* **36**: 990-997 [in Russian].
13. Rathinasabapathi, R., Gage, D.A., Mackill, D.J., and Hanson, A.W., 1993, Cultivated and wild rices do not accumulate glycinebetaine due to deficiences in two biosynthetic steps, *Photosynth. Res.*, **33**: 534-538.
14. Sakamoto, A., Alia, and Murata, N., Metabolic engineering of rice leading to biosynthesis of glycinebetaine and tolerance to salt and cold, *Plant Mol. Biol.*, **38**: 1011-1019.
15. Sherif, D.W., and Sinclair, R., 1973, Fluctuations in leaf water balance with a period of 1 to 10 minutes, *Planta* **113**: 215-228.
16. Shevelukha, V.S., 1992, *Plant Growth and Its Regulation in Onthogenesis*, Kolos, Moscow, 593 p.
17. Ushakov, V.Y., and Koltunova, I.R., 1982, On pulsating character of transpiration and water supply to plant leaves, *Doklady of USSR Academy of Sciences*, **266**: 766-768.
18. Zholkevich, V.N., 1991, Root pressure, in: *Plant Roots. The Hidden Half*, Y., Waisel, A., Eshel, and U., Kafkafi, eds., Marcel Dekker Inc., New York-Basel-Hong Kong, pp. 589-603.
19. Zyalalov, A.A., 1984, *Physiology-thermodynamical Aspect of Water Transport in Plant*, Nauka, Moscow, 135 p.
20. Wodzicki, T.J., Knegt, E., Wodzicki, A.B., and Bruinsma, J., 1984, Is indolyl-3-acetic acid involved in the wave-like pattern of auxin efflux from Pinus syivestris stem segments?, *Physiol. Plant.* **61**: 209-213.

18
CORRELATION OF THE "NEAR-ZONE EFFECT" AMPLITUDE DYNAMICS WITH SOLAR-GEOPHYSICAL INDICES

Tatiana A. Zenchenko[*†], Alexander A. Konradov[†], and Kirill I. Zenchenko[*]

Abstract: An array of daily values of amplitude of "near-zone effect" (NZE), the degree of manifestation of self-organization processes in a statistical population (radioactive decay), was obtained using histogram method for data analysis (Shnoll method). Significant statistical correlation between the dynamics of amplitude of "near-zone effect" and indices of solar and geomagnetic activity was revealed, so that the peaks of NZE amplitude coincide with the moments of the highest rate of changes of absolute value of interplanetary magnetic field and geomagnetic field intensities.

1. INTRODUCTION

About fifty years ago, Simon Shnoll elaborated the method of analyzing time series of quasi-stationary processes to reveal resemblances or differences in the fine structure of sample distributions (histograms) calculated according to short non-overlapping pieces of time series (1). Certain regularities in quasi-stationary processes of different nature and very wide scales of energies have been revealed. It has been found that the forms of histograms separated by definite time intervals resemble each other much more frequently than taken at random: in other words, the form of a histogram appearing in time space "t" most probably will appear in the next nearest time space 't+1', as well as in time spaces t+nT, where T is equal to a day, a month, a year and certain other easily interpreted time spaces (2). Thus it has been shown that the time series of histograms has a certain "inertia of form" (named "near-zone effect" [NZE]) and multicomponent noisy periodicity ("specific periods"). It has been also revealed that these effects may be frequently seen in the time series of histograms calculated, among other things, according to the results of measurements of alpha-radioactive decay intensity, obeying Poisson statistics and containing no important periods in corresponding Fourier spectres (3).

[*] Institute of Theoretical and Experimental Biophysics of Russian Academy of Sciences, Pushchino, Moscow region.
[†] Institute of Biochemical Physics of Russian Academy of Sciences, Moscow.

Yet, it has been also found that the degree of expression (or amplitude) of "near-zone effect" in a particular experiment varies from time to time both for the near-zone and for the specific periods. Previously it has been revealed that the dynamics of the amplitude of NZE shows a low-frequency periodical component of 84 to 110 hours for various pieces of time series (4).

The above period is characteristic of numerous solar- and geophysical indices (5). Due to the fact that though there are several alternative hypotheses, the consistent theory of Shnoll effect up to the present moment is lacking and the nature of the effect and even its phenomenology are still unclear, the aim of this work was to compare of the dynamics of NZE amplitude with the dynamics of various solar-geophysical indices.

2. MATERIALS AND METHODS

2.1. Histogram Method for the Time Series Data Analysis (Main Steps)

Figure 1 shows an array N_k of consequent values of alpha-decay intensity (counts per 10 seconds), divided into short (60 points, or 10 minutes) non-overlapping pieces. For each piece a sample distribution based on maximum possible number of bins n = max(N)-min(N) was calculated. It should be emphasized, that all examined distributions obey Poisson statistics, this fact was previously tested by standard statistical methods (6), and polymodal histogram forms obtained as a result of described procedure of histogram calculation do not mean that the general distribution should be also polymodal.

Original histograms were smoothed several times by moving average method for high-frequency discrimination. The examples of the resulting sample distributions (before and after smoothing procedure) are shown in Fig. 1, bottom.

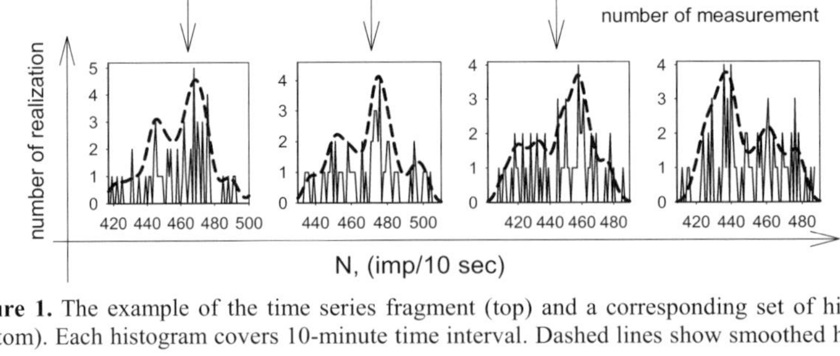

Figure 1. The example of the time series fragment (top) and a corresponding set of histograms (bottom). Each histogram covers 10-minute time interval. Dashed lines show smoothed histogram forms.

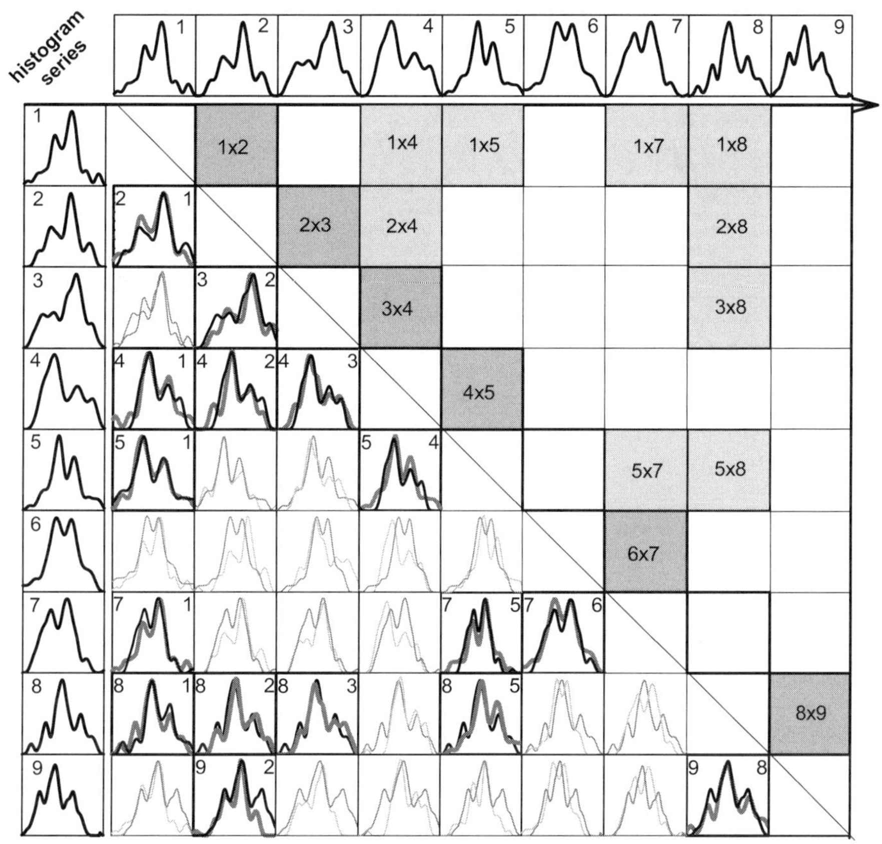

Figure 2. Illustration of a procedure of comparison of histograms – similarity matrix. See text for the details.

Figure 2 illustrates the procedure of selection of pairs of resembling histograms. Here one histogram series is shown in two dimensions, X and Y, and corresponding superposed pairs are placed at the intersection of lines (bottom part of matrix). The resembling pairs are marked by solid lines and numbers. In the top part of diagonally symmetric similarity matrix the same resembling pairs are shown as dark cells of matrix to illustrate the matter of "near-zone effect" and instability of its amplitude: the concentration of resembling pairs placed on the first diagonal in significantly higher than that on average in matrix, but this concentration is unsteady along the diagonal: in the first half four out of four possible pair turn to be resembling, in the second half – only two ones.

Figure 3 illustrates the real dynamics of number of resembling pairs corresponding to certain diagonals of similarity matrix (or certain time intervals between histograms in pair). It is seen that concentrations of resembling pairs on the first and the second diagonals synchronously vary with considerable amplitude, while this parameter remains virtually constant (16%) for fourth and

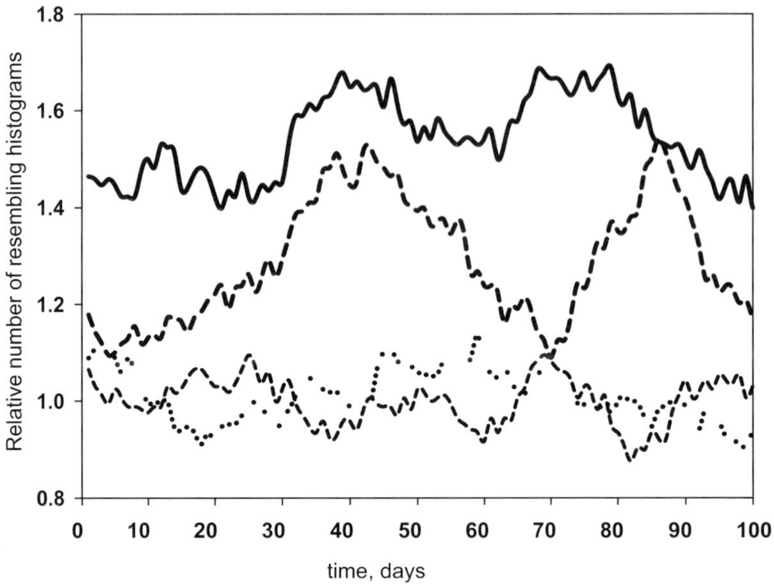

Figure 3. Illustration of daily dynamics of a relative number of resembling pairs corresponding to certain diagonals of similarity matrix in a real experiment. The horizontal axis indicates day number, the vertical axis indicates the relative quantity of resembling histogram pairs in this day. Solid line – the first diagonal ("near-zone"), dash line – the second diagonal, dotted lines – the fourth and eighth diagonals. Relative numbers of resembling pairs are calculated with respect to an average concentration equal to 16%, see text for details.

eighth diagonals. The last value can be taken as an "average concentration" ⟨A⟩ of resembling pairs for all diagonals, except for the first one. In this case the term "amplitude of NZE Ak"(for some time interval with a number k) will be the ratio of concentration of resembling histogram pairs at the first diagonal within this time interval to this average concentration ⟨A⟩.

2.2. Method for the Analysis of Dynamics of NZE Amplitude

A long time series of alpha-decay intensity of Pu-239 (common duration is 136 days, one measurement – number of counts per 10 seconds), obtained in our laboratory via semiconductor detector, were transformed into time series of 10-minute histograms (60 points per histogram, 144 histograms per day).

Expert ascertained resemblance or difference forms of adjacent histograms, t. e. placed on the first diagonal of similarity matrix. Figure 4 shows the dynamics of daily values of NZE amplitude.

For comparison with NZE dynamics, the synchronous daily averaged values of the follow solar and geophysical indices were used:

- Solar activity indices (Wolf numbers (WN) and radio flux intensity at a wavelength of 10.7cm (RF10.7));
- Interplanetary Magnetic Field total value (average of magnitudes). http://spidr.ngdc.noaa.gov/spidr/

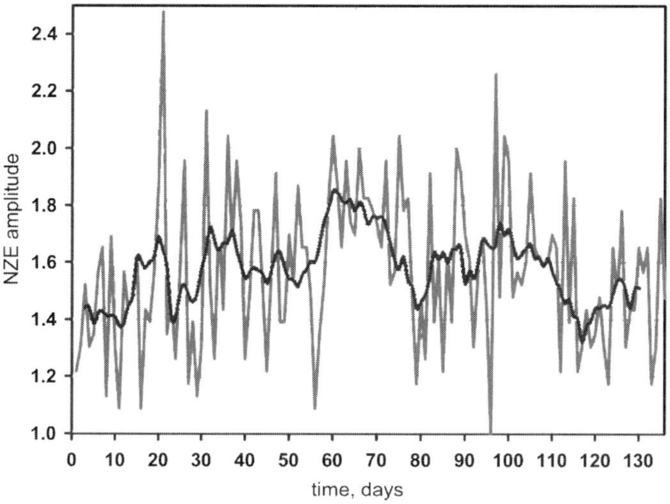

Figure 4. Daily dynamics of NZE amplitude (gray line) and its low-frequency component (black line).

- High energy proton and electron fluxes;
- Index of planetary geomagnetic activity Ap (downloaded from NOAA's National Geophysical Data Center ftp://ftp.ngdc.noaa.gov/STP/GEOMAGNETIC_DATA/INDICES/)
- Dst-variation values(downloaded from http://swdcwww.kugi.kyoto-u.ac.jp/)
- PC index (Thule) (for magnetic activity in the Polar Cap, http://www.aari.nw.ru) (7).

3. RESULTS

As it is seen from Fig. 5, Fourier density function of daily NZE amplitude contains several statistically important periods: 5.5, 7.5, 13.5, and 34 days.

Among these periods, the only 13.5-day period is presented in Wolf numbers spectrum, but we cannot claim the connection of NZE dynamics namely with WN because of the presence of this period in almost all solar-geophysical indices spectra. No significant correlation of NZE amplitude dynamics with Radio flux intensity RF10.7cm or concentrations of high-energy protons and electrons were found.

Comparison of Fourier density functions of daily NZE amplitude and Bmag IMF and Ap-index of geomagnetic activity, the coincidence of 13.5d and 7.5d periods were found, as so as the coherent increase of spectral density in 5-6d region (Fig. 5). So, three main periods of NZE dynamics are presented in Bmag spectrum, and two ones – in Ap-spectrum.

Superposition of low-frequency components of NZE amplitude and Bmag dynamics (Fig. 6) reveals very similar forms of these curves, but correlation maximum corresponds to 11-days phase displacement (NZE amplitude values

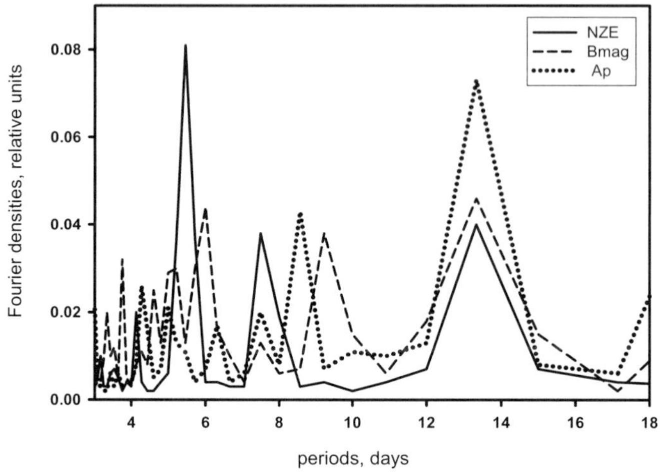

Figure 5. Fourier density functions for middle periods. Solid line – NZE amplitude, dash line – Bmag IMF, dotted line – Ap.

pass ahead of Bmag). Nevertheless in long-periods range of NZE amplitude spectrum the coincidence was not found for any investigated solar-geophysical indices. It can be explained by poor applicability of Fourier method in this range of periods.

Quite a similar result was obtained when dynamics of low-frequency components of NZE amplitude and Ap-index and Dst-variation, with the 12 days lagging (Fig. 7) were compared.

For unfiltered arrays, cross-correlation coefficients of NZE amplitude and these magnetic activity indices (Bmag, Ap and Dst-variation) are statistically important ($p < 0.01$) only under this time lag (in range from -40 to +40 days).

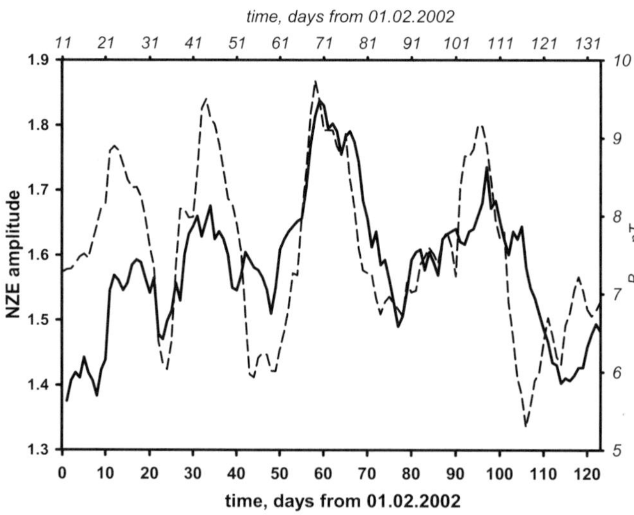

Figure 6. Comparison of dynamics of NZE amplitude (solid line) and Bmag IMF (dash line). A good agreement of these curves with 1 day time lag is seen.

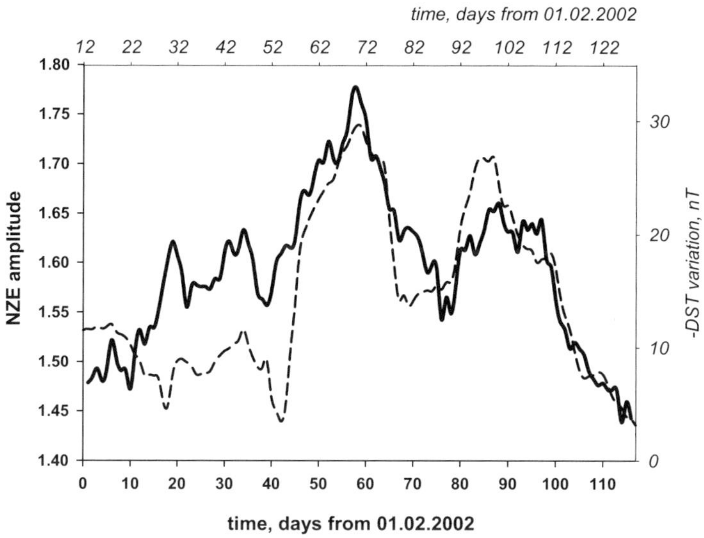

Figure 7. Comparison of dynamics of NZE amplitude (solid line) and Dst-variation with the opposite sign (dash line).

4. COMPARISON WITH TIME VARIABILITY OF DIFFERENT INDICES

Comparison of NZE amplitude dynamics with time variability (rate of changes) of Bmag also reveals significant correlation, but with zero time shift – all peaks on these curves coincide within the limits of resolution (fig 8). Daily average value of Bmag time variability (diff Bmag) was calculated as daily sums of differences of consequent hourly values of Bmag. The similar result was obtained for time variability of Dst-variations (in this case the exact anti-correlation was found). Some correlation was also found for dynamics of NZE amplitude and Ap-index variability.

5. DISCUSSION

Thus, results of this work allow us to conclude the existence of significant correlation between NZE amplitude dynamics and indices of magnetic activity (interplanetary and terrestrial). There has not been found statistical connection with solar activity and high-energy particle fluxes, except for 13.5d period in Wolf numbers spectrum, but this period is not specific for only WN spectrum.

The variability of magnitude of interplanetary magnetic field and low-latitude geomagnetic activity indices (Dst-variation) turn to be significant in this case. The connection with PC-index is much weaker. It would be reasonable to assume that observed correlation of NZE dynamics with IMF parameter does not mean direct connection, but via geomagnetic variations. It is possible, that more significant correlation with Dst-variation values, rather than PC-index or planetary Ap are somehow conditioned by location of device in latitude 54 North (Moscow). This hypothesis requires additional examination on data, obtained in high-latitudes, for example, in Antarctic.

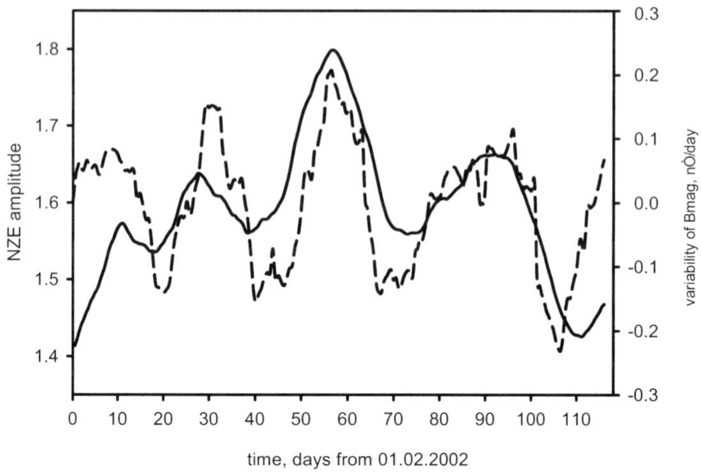

Figure 8. Comparison of low-frequency dynamics of NZE amplitude (solid line) and time variability of Bmag (dash line).

Thus, Shnoll effect ("inertia of histogram form", or the effect of nonzero relaxation time of fine structure of sample distributions) has the maximum of its appearance in the moments of maximum rate of changes of interplanetary magnetic field. The other result – that current values of Bmag correlates with NZE amplitude value realized 11 days ago – seems now absolutely inexplicable and requires further verification.

As discussed earlier (2) it is impossible that Shnoll effect is caused by direct influence of magnetic field onto radioactive decay or onto device counting only the acts of decay in "yes-no" logic. Possible theories of near-zone effect based on changes of characteristics of environmental space-time (8, 9) do not explain the correlation of NZE with magnetic field variability, but do not contradict to this result either.

So, the results obtained in this work should be treated as expansion of phenomenological picture of Shnoll effect, that can in future lead to comprehension of its nature.

ACKNOWLEDGEMENTS

We thank the National Geophysical Data Center for providing the solar and geomagnetic data and Prof. O. Troshichev for providing PC-index data.

The authors are grateful to A.A. Petrukovich for his interest in the problem and valuable discussions.

REFERENCES

1. S. E. Shnol', in Itogi nauki i tekhniki. Molekulyarnaya biologiya (Advances in science and technology. Molecular biology) Vol. 5 (Ed. V. P. Skulachev) (Moscow: VINITI, 1985) p. 130 (in Russian).
2. S. E. Shnoll, V. A. Kolombet, E. V. Pozharski, T. A. Zenchenko, I. M. Zvereva, and A. A. Konradov – Realization of discrete states during fluctuations in macroscopic processes. Uspekhi Fizicheskikh Nauk, (1998), 41 (10) 1025-1035.

3. S. E. Shnoll, T. A. Zenchenko, K. I. Zenchenko, E. V. Pozharski, V. A. Kolombet, and A. A. Konradov – Regular variation of the fine structure of statistical distributions as a consequence of cosmophysical agents. – Uspekhi Fizicheskikh Nauk, (2000), 43 (2) 205-209.
4. T. A. Zenchenko, A. A. Konradov, and K. I. Zenchenko – Macroscopic fluctuations: on the periodicity of manifestation of "nearest-zone effect". Biofizika, 2003, V. 48, No. 6, pp. 1132-1136.
5. B. M. Vladimirskii, V. Ya. Narmanskii, and N. A. Temuryants. – Cosmic rhythms: in magnetosphere-ionosphere, in atmosphere, in habitat, in biosphere-nooshpere, in the Earth crust. – Simferopol, 1994, 173 pp. (in Russian).
6. V. S. Rozhkov. Theory and methods of statistic estimation of probabilistic characteristics of random values and functions. Book 1– S-Pb, Hydrometeoizdat, 2001. pp. 53-81.
7. O. A. Troshichev, V. G. Andresen, S. Vennerstrom, and E. Friis-Christensen: Magnetic activity in the Polar Cap – a new index., Planet Space Sci. 36, 1095-1102, 1988.
8. L. A. Blumenfeld, and T. A. Zenchenko– Quantum transitions between states and cosmophysical fluctuations Biofizika, (2001), V 46(5). pp. 859-861.
9. A. A. Kirillov, and K. I. Zenchenko. On the probability of disturbance of the Poisson statistics in processes of radioactive decay type. – Biofizika. 2001. V. 46(5). pp. 841-849.

19
FREE WILL AND VIOLATION OF PHYSICAL LAWS

A new concept of volition based on A. Gurwitsch's field theory

Michael Lipkind *

> "[T]he laws of physics, applicable for inanimate matter, will have to be modified when dealing with the more general situation in which life and consciousness play significant roles."
>
> **Eugene P. Wignier**
> ***Physics and the Explanation of Life***
> *Foundations of Physics, 1970*

> "Any proposal that volition exists must necessarily involve a radical addition to the presently known physical laws.... If volition exists, then physics stands at a new frontier, in which these principles are yet to be discovered."
>
> **Jean E. Burns**
> ***Volition and Physical Laws***
> *Journal of Consciousness Studies, 1999*

> "Do we consciously cause what we do, or do our actions happen to us?"
>
> **Daniel M. Wegner**
> ***The Illusion of Conscious Will***
> *The MIT Press, 2002*

1. INTRODUCTION

The Free Will is the central philosophical problem of the mind-matter relationship that challenged human intellectual capacity since the very outset of civilization. Apart from being closely connected with the foundational problems of philosophy and natural sciences, the Free Will is the most mysterious manifestation of Consciousness. The unique importance of the Free Will problem is due to its direct connection to the problem of mental causation. This traditionally philosophical question confronting the Free Will versus physical determinism has excited, except philosophers, the most eminent physicists like N. Bohr, A. Einstein, W. Pauli, W. Heisenberg, E. Schrödinger, and E. Wignier, and since the past decade became a burning topic of neuroscience (for review, see Walter[1] and Wegner[2]).

* Unit of Molecular Virology, Kimron Veterinary Institute, P.O. Box 12, Beit Dagan, Israel 50250; International Institute of Biophysics, Neuss-Hombroich, Germany D-41472.

In the present article, the problem of the Free Will is analyzed in the view of the theory of biological field by Alexander Gurwitsch[3], especially, the part of the theory dealing with the psychic sphere[4, 5]. The theory of biological field was elaborated and developed by Gurwitsch during all his lifetime. The starting point is associated with the fact that Gurwitsch was the first to introduce the field concept into biology[6]. This fact was acknowledged in contemporary reviews[7-9] as well as in the latter-day works by biologists employing the concept of field in their theoretical considerations[10-16]. Since then, Gurwitsch's field conception underwent successive developments – from the first abstract models describing single morphogenetic phenomena[17-20] to the general theory of the vectorial cellular field embracing all the three levels of biological organization – morphological, cellular, and molecular[3-5, 21]. The initial essence of the field principle as applied to the morphogenesis was expressed by Gurwitsch as co-subjection of elements (cells) to a common morphogenetic factor, as opposed to an alternative conception considering the whole morphogenesis as a result of interactions between the elements, that was represented by the "causal-analytical" approach by W. Roux[22] culminated by famous H. Spemann's experiments[23]. In such respect, Gurwitsch's field conception stands at the crossroads of the two opposing trends of theoretical biology and, what is called now, biophilosophy[24]: the reductionist trend dominating in science since Galileo and Newton versus holistic trend which has remained unpopular and even suppressed. The final version of Gurwitsch's field theory[4, 5] has included also the psychic sphere in its interconnections with the somatic (brain) level that demonstrated unprecedented theoretical capacity of the theory describing by the same basic postulates the regularities related to different biological levels – from biochemical processes in living cells up to psychic phenomenology.

An intriguing question concerns the place of Gurwitsch's field theory in respect of the general development of theoretical biology during the past half-century period. There is no doubt that all this period was under overwhelming influence of the double-helix-triplet-code revolution that was a triumph of the reductionist trend leaving no chance to any alternative, especially holistic, approach, which has become almost complete dominion of near-scientific and esoteric activities. However, besides the modern molecular biology, during the past two decades there was significant progress in psychological and cognitive sciences that was marked by the fact that the notion of Consciousness, which was expelled from the "official science" and tabooed during the "behaviorist era", has returned in a boom-like mode.

In the latter respect, an unfortunate circumstance is that the last version of Gurwitsch's theory related to Consciousness has remained practically unknown to Western science. The last variant of the theory was elaborated by Gurwitsch in the 1940s –1950s in Soviet Russia and published in Russian[3] and in French[21]. The final version of the theory completed in 1954[4] was published only in 1991 in Russian[5], the manuscript being known only to the author's close disciples. The first comprehensive review on Gurwitsch's field theory was published in 1963 in Russian[25]. The first comprehensive review on Gurwitsch's field theory written in European language (German) was published in 1987[26, 27]. The first English review on the theory was published in 1992[28]. Some excerpts from Gurwitsch's book published in 1991 in Russian[5] were translated into English in

1994[29] with subsequent review[30]. The most comprehensive English reviews have been published just recently[32, 33]. Therefore, the recent attempts to employ Gurwitsch's field theory as scientific approach to the ***consciousness*** problem[34, 35], when reported in respective meetings[36-38], in spite of rather vivid interest, showed absence of any preliminary background from the side of the audience, that was a certain obstacle for immediate comprehension.

Another circumstance concerns the fact that as far as Gurwitschian field is autonomous, i.e., irreducible to the established physical fundamentals, it just formally can be designated as vitalistic. In fact, Gurwitsch considered himself as a vitalist and a follower of H. Driesch, the founder of the neo-vitalistic trend in biological sciences[39-41]. However, as opposed to Drieschean agnosticism, Gurwitsch considered his field theorizing as an effort to use the vitalistic principle as a powerful tool for creating working hypotheses ("practical vitalism", by Gurwitsch's expression[42]). Anyway, the vitalistic label appeared to become a serious obstacle demanding especial endeavors in order to overcome traditional dislike of scientific midst to such heretic evil as vitalism, that being due to the unlucky historical fate of the great principle.

Based on Gurwitsch's idea of irreducible biological field, a new approach to the Free Will problem is suggested below.

2. DOES FREE WILL EXIST OR IS IT AN ILLUSION?

Thus, "[d]o we consciously cause what we do, or do our actions happen to us?"[2] This alternative question reflects two alternative approaches to the Free Will problem that are so radically contrasting that one of them simply rejects the existence of the Free Will, which is declared as an illusion, while the other, defending existence of the Free Will, has to confront it versus the whole edifice of natural sciences grounded on the physical laws-based deterministic causality.

The former alternative affirms that Mind is totally reducible to the physical fundamentals. Then, according to the radical materialist formula, Mental state = Brain state, i.e., the Mind is nothing more than the functioning Brain. In neuroscience, the materialistic explanation of the Free Will is based on the behaviorist doctrine, according to which any psychic act felt as voluntary is in fact a response to a theoretically endless causal chain of the consecutive stimulus-response blocks described by the physical language. Therefore, the causality is closed ("causal closure" by Montero[43]), i.e., any cause of a mental manifestation is a physical act that was caused by another physical act, which had been caused by another physical act, etc., etc. The physical close causality is realized either by deterministic regularity (classic physics), or probabilistic randomness (quantum physics). In the view of the classic physics, an individual at every moment has only ***one*** option of acting with ***no choice***. This means that by the deterministic causality, theoretically it is possible to reveal all the causal chain in the past (up to infinity) and predict the future outcome (up to infinity). In the view of the quantum physics, an individual can behave in many different ways due to the ***blind chance mode***, i.e., again, with ***no choice*** since the probability is definitely determined. Thus (without dealing with highly disputable and controversial Copenhagen interpretation), both the ***strict regularity*** (classic physics) and the ***uncertainty*** (quantum physics) are governed by the physical laws with

no interference of the Free Will. Correspondingly, the Brain-Mind causation is realized by that the Brain is always going to do what it is caused to do by local mechanical disturbances, so that the idea that one's 'Will' is actually able to cause anything at all, is illusory. In this respect, the neuronal network is a reflection of the deterministic regularity of the Physical World in the physiologically functioning Brain. The only causes of physical events are other physical events (causal circularity). The mental events have physical effects because they, themselves (under different description), are physical events. The Brain is a material object, while the Mind is its epiphenomenal shadow like "the noise of the engine, the smoke from the fire", by W. Freeman's expression[44] as cited by S. Pockett[45]. "Physics can, in principle, predict the probability with which a human body will follow any given trajectory" (H. Putnam[46]), i.e., the Free Will as a causal agent is an illusion.

The other alternative is grounded on dualistic philosophy originating from the classical Cartesian bifurcation between ***res extensa*** and ***res cogitans***. In accordance with this, the approach rests on the following basic premises: Mind, although being intimately connected with the activity of the somatic Brain, appears to be a non-material entity, which, in turn, can make influence on the material Brain causing it to start certain physical actions (libertarian imperative). Besides such causal influence upon the material Brain, the non-material Mind has its own ***non-material activity*** ("internal life"), which is expressed in self-reflection, thinking, intentions, decision-making, and many other manifestations of consciousness. Anyhow, just formally, the causal influence of the non-material Mind on the material Brain means violation of the established physical laws[1], that topic being one of the main objectives of the suggested approach.

An approach suggested in this article is based on the vitalistic philosophy in Gurwitsch's interpretation, i.e., the vitalistic principle is realized in the form of the biological (morphogenetic) field irreducible to the established physical fundamentals.

However, before dealing with the suggested approach itself, all the pro and contra arguments about the status of the Free Will as ontological entity are to be shortly outlined. Meanwhile, it is possible to claim that in the current discussions on explanation of the Free Will, the materialistic approach has been dominating while the dualistic one is just making initial attempts to hold on scientific grounds. In the frame of the materialistic approach, scientific investigation proceeds in full accordance with the deterministic principle rejecting the existence of the Free Will as ontological entity and defining it as illusion. The Consciousness phenomenology in the whole spectrum of its manifestations, which evidently are not reducible to the established physical fundamentals, is anyway believed to be explained by the same deterministic way in a distant (perhaps, asymptotically approaching to infinity) future, that becoming a kind of a faith. As to the dualistic approach, taken as Weltanschauung, apart from philosophical classifications, like interactionism, incompatibilism, superven-

[1] The expression "the established physical laws" means the laws of the present-day physics grounded on the basic fundamentals (mass, charge, space-time), but not a notion of "physical" as an essential idea[43].

ience, etc. (for review, see Chalmers[47] and Kim[48]), the main endeavor is to make it "working" in practical research. In accordance with this, the actual task of the present study is a search for ***naturalistic*** expression of the postulated non-material Mind that is obligatory condition for unbiased scientific scrutiny and experimental investigation.

Consequently, the first step in this way is a concise consideration of the main arguments dealing with the question whether the Free Will is an illusion or it is an existing and decisive free choice-based factor determining actual behavior. Only in the case of the latter alternative, a potential applicability of the concept of irreducible field (Gurwitsch's field, in particular) for the Free Will analysis may be suitable.

2.1. The Free Will Is an Illusion

According to the main premise, any human act, except clear reflex ones, is ***accompanied*** by internal conviction and general emotional feeling that this very act is a result of realization of a previous intention, proceeding under individual's full control, and, thus, apprehended by the individual as volitional. The key word here is ***"accompanied"*** indicating that the ***real main*** act by its nature is a purely physiological (physical, in the end) process. Hence, the Free Will is an illusion, and the emotional feeling of its existence is induced by the processes, which anatomically and physiologically are ***distinct*** from those by which the Mind creates action. Such conclusion means ***denial of mental causation*** that is in the spirit of the materialist philosophy.

All the argumentation concentrates on proving that the Free Will is not a force of the Mind causing body action via somatic Brain. The Free Will as a psychological phenomenon is subject's ***experience*** of consciously willing and intentionally performing an action that has nothing to do with the causative connection of the action with the person's conscious Mind. This has been perfectly expressed by David Hume who defined the Will as "nothing but the internal impression we feel and are conscious of, when we knowingly give rise to any new motion of our body, or new perception of our mind"[49].

Apart from philosophical argumentation, scientific evidence (to be shortly accounted here) is based mainly on the neuropathological data gained by both clinical observations of different syndromes and experimental approaches. The former includes the 'alien hand' syndrome and phantom limb movements, while the latter includes different modes of brain stimulation. The 'alien hand' syndrome[50-52] is a neuropsychological disorder in which a person experiences one hand as operating autonomously as if with a mind of its own. This syndrome is often linked with damage to the middle of the frontal lobe on the side of the brain opposite the affected hand[53], suggesting that such autonomous "mental behavior", fully resembling the normal behavior (which is believed to be Free Will-caused) but without subjective emotional feeling of Free Will operation, is simply caused by stimulation from the affected brain part. The phantom limb movements always occur consciously, i.e., are not spontaneously made[52]; this also seems to substantiate the idea that a signal sent to the absent limb is somehow conscious.

The most elaborated experiments on brain stimulation include those by Penfield[54, 55], Delgado[56], and Brasil-Neto[57]. Without going into details, Penfield's experiments performed in the course of neurosurgical operations made under local anesthetic permitted to communicate with the patient during the brain stimulation. During stimulation of certain specific brain area, a conscious patient demonstrated the corresponding physiological response without feeling such response as volitional. In Delgado's experiments, electrical stimulation produced behavioral activity of the patient who considered such evoked activity as spontaneous while always offered worldly reasons for it. This observation suggests that there is a part of the brain that yields ***consciously willed*** action when it is electrically stimulated. The patient's confabulations and convenient stories made up to fit the moment versus the patient's refusal to admit that the action made after electric stimulation was voluntary, no matter where a particular responsible area might be, suggests that the brain structure that provides the experience of Free Will is ***separate*** from the brain source of action. It reveals the possibility to produce voluntary action through brain stimulation with or without experiencing Free Will. The experiments on transcranian magnetic stimulation[57] also were suggestive that the experience of Free Will can arise ***independently*** of actual causal forces influencing behavior.

The important impact into experimental studies on the Free Will was the discovery of electric ***readiness potentials*** for voluntary actions[58], which revealed that brain electrical activity starts increasing about 0.8 seconds ***before*** the voluntary movement. These experiments were further developed by B. Libet and colleagues[59-62] by including into experimental scheme the registration of the volitional moment expressed in experiencing conscious awareness of "wanting" to perform a given self-initiated movement. The results have shown that the experience of conscious Free Will appears at some point ***after*** the brain has already started preparing for the action that was demonstrated by appearance of the readiness potentials. This means that the brain starts ***first***, followed by the experience of conscious will and then by action (Libet[62]).

> "[T]he initiation of the voluntary act appears to be an unconscious cerebral process. Clearly, free will or free choice of whether to ***act now*** could not be the initiating agent, contrary to one widely held view. This is also contrary to each individual's own introspective feeling that he/she consciously initiates such voluntary acts this provides an important empirical example of the possibility that the subjective experience of a mental causality need not necessarily reflect the actual causative relationship between mental and brain events"[62].

An important fact in this respect is that the readiness potentials do not occur for involuntary movements such as tics or reflex actions[63].

The conclusion is expressed by the following utterance: "A microanalysis of the time interval before and after action indicates that consciousness pops in and out of the picture and doesn't really seem to do anything" (Wegner[2]). Accordingly, the Free Will is an illusion that is beautifully illustrated by The Little Prince, a personage from Antoine de Saint-Exupery's novel, who "ordered" Sun to rise and to set. Another beautiful illustration was given by A. Einstein:

> "If the moon, in the act of completing its eternal way around the earth, were gifted with self-consciousness, it would feel thoroughly convinced that it was

traveling its way of its own accord. So would a Being, endowed with higher insight and more perfect intelligence, watching man and his doings, smile about man's illusion that he was acting according to his own free will" (A. Einstein's contribution into *The Golden Book of Tagore*[64], cited from Home & Robinson[65]).

2.2. The Free Will Is Not an Illusion

One of the arguments that the Free Will is not an illusion is based on the assertion that "volition acts by selecting among the possibilities in a random event" (Burns[66]). If a certain voluntary act is suggested to occur due to the physical causality, a probability of such occurrence depending on the complication of the act can be calculated. For example, if one sees a fly sitting on a piece of cheese at a table and makes a decision instead of swinging lazily in an armchair, to stand up, take a towel, reach the table, raise one's hand with the towel but than change one's mind because together with the fly it would break a porcelain plate with cheese, etc, etc. According to calculation, the probability of actual occurrence of such a chain of the originally random processes is very low, while in reality this happens whenever one wants. Therefore, the practical realization of the voluntary acts goes in full disagreements with formally calculated probability.

Another argument was based on the above-described experiments by B. Libet and colleagues[59-62], which seemed to support the unconscious origin of the volitional act. However, although volitional process is initiated unconsciously, because the readiness potentials appear 350-400 ms ***before*** a subject becomes aware of intention to act, the awareness still appears 200 ms before the motor act. Therefore, the Free Will action cannot be excluded but its involvement is limited to control the final outcome, i.e., the Free Will can ***veto*** the act. Libet has come to conclusion that "these findings put constraints on views of how Free Will may operate: it would not initiate a voluntary act but it could ***control*** performance of the act"[67]. Libet attempted to extrapolate the original experimental data concerning spontaneous motor acts by encompassing more complex volitional acts with deliberation about what choice of action to adopt including pre-planning of when to act on such a choice (e.g., beginning to speak or to write). The results have shown that the time interval between conscious deliberation about the choice of action and the final decision to 'act now' (of about 200 ms) was similar to the time interval between the conscious awareness of the intention to act and the act itself in the case of fully spontaneous voluntary acts. Such coincidence is considered by Libet as a substantial reason for accepting seriously a non-deterministic alternative of the Free Will.

The experiments by Libet and colleagues performed during three and a half decades have stimulated hot discussions, critical considerations, and additional experiments to elucidate the matter. A particular topic designated as "timing of mental events"[68] needs an especial consideration which was concentrated in addendum to follow.

Thus, the conclusion on the Free Will existence-illusion dilemma is that the considered evidence leaves open the main question: whether the actual behavior is under deterministic prediction (i.e., the Free Will is an illusion), or the actual behavior is due to conscious intention to perform certain acts by making choice ("libertarian imperative"). The deterministic alternative, although being in harmony with the physical laws, exposes the Explanatory Gap[69] between physical determinism and the Free Will phenomenology. According to reductionist worldview identified with physicalism, the Explanatory Gap will eventually become explainable in some future by means of certain in principle new scientific regularities to be discovered that will make up the deficiencies of the established physical laws. As was stated before, such hope is a kind of faith, but even this faith may be shaken by what has come to be known as Hempel's dilemma[70], which concentrates on strictly formal investigation of what ***physical means***. Namely, if to define the term 'physical' via contemporary physics, then it would appear that physicalism straightforwardly false, since today physics is *a priori* incomplete and does not give the whole truth. However, if, instead, to try to define physicalism by reference to what physics may become in a near or distant future, any conclusion is incompetent because it is not known what form physics might take in the future. On the other hand, the above "in principle new scientific regularities" to be discovered must be based on such new laws, which, although being natural, are in a way 'supra-physical' if the notion 'physical' relates to the presently established physical laws[43, 71]. Such 'supra-physical' meaning would refer to a new basic fundamental(s) which is/are to be additional to the established physical fundamentals like mass, charge, and space-time.

Thus, if to accept the Free Will as illusion, there is no necessity to explain the mental causation which becomes also a kind of illusion. Using the utterance of Shakespearian King Lear, "nothing will come of nothing". As to the dualistic alternative in its radical form, the general view on irreducibility of an individual's Free Will is added with a "practical rule" for an individual to try always to act in accordance with the urge dictated by "libertarian imperative"[72].

However, any view based on the existence of the Free Will immediately leads to the problem of violation of the established physical laws. Therefore, the next task is to analyze in detail at which particular biological (neurophysiological) site(s) such violation could occur.

3. MECHANISM OF THE FREE WILL – CAUSED VIOLATION OF PHYSICAL LAWS

Free choice as the main attribute of the Free Will looks as a complete antithesis to the strict regularities dictated by the physical laws. Its realization would lead to immediate intrusion of chaotic components into the absolutely unshakable network of the physical circular causality. It seems that a human being by a thoughtful decision or a capricious desire undermines the causal grounds of the whole foundations of the Physical World, i.e., the realization of the Free Will leads to the violation of the existing physical laws.

However, if to consider the Free Will as a quality acquired as the result of the life evolution on Earth, one comes to conclusion about impossibility to determine a border, which separates the species "possessing" free will from those without it. S. Hameroff has claimed that observation on a protozoon (Paramecium) reveals elementary manifestations of the Free Will[73]. Hence, the Free Will is a prerogative quality of the animal kingdom spreading ubiquitously around the earth space. This means that at least at this part of the Universe, the Free Will-caused violation of the physical laws is as usual as the physical laws themselves, so that the Free Will, besides being natural, must be considered canonical in the same manner as gravitation or electromagnetism. This bears a global problem concerning causal realization and maintenance of the general reality based on the co-existence of the Physical World and the Free Will-possessing living entities. Hence, the Physical World – Free Will co-existence means ***continuous*** violation of the physical laws. However, due to heterogeneous spread of the whole plenty of the living species highly differing by qualitative complexity and variable intensity of the Free Will manifestations, such violation of the physical laws is ***irregular***. Consequently, the question is how the realization of the universal reality can be possible in conditions when the physical laws are ***continuously but irregularly*** violated.

Thus, the task is to confront the strict regularities and circular causality of the Physical World governed by the physical laws with continuous while irregular violation of these laws. Such confrontation immediately shows that the Physical World and the Free Will stand on the opposite sides of the abyss named the Psycho-Physical Gap, and the task is to make a bridge between the both. This would be achieved by means of a strict definition of the Free Will expressed in the ***physical*** language. Such a condition is obligatory for determination of the ontological status of the Free Will and its combination with the Physical World within a general architecture of the Universal Reality.

The studies on this direction defined as ***Mind-Brain interaction mechanisms***[74] have been determined by answering the following questions: Which physical laws are violated? How minimal can be such a violation? Would such violations be detectable? In this respect, there are two possibilities: (1) A non-physical mind would ***supply*** such energy during volitional acts, and (2) A non-physical Mind might ***harness***, rather than supply, such energy during volitional acts.

The starting question was as follows: What is a minimal level of energy, which is required to cause nerve cells in the brain to fire action potentials (to initiate neural stimulation)? A crucial point in this respect concerns the possibility that the claimed violation of the physical laws is so negligible that might occur ***undetected***. This means that if the required energy, coupled with the time during which it would need to be available, were low enough, it could be 'hidden' under Heisenberg principle of quantum-mechanical uncertainty. Hence, the practical calculation task was to find out what minimal energy would be required from a non-physical mind to trigger a ***single*** action potential in a ***single*** neuron. Such calculation was performed by taking into account all the possible stages of the action potentials formation (initiation of the nerve impulse), on which the energy would be needed, namely: (1) Opening sodium channels, (2) Altering voltage gradients, (3) Synaptic transmission, (4) Neuronal modulation, and (5) Self-generation of action potentials by neurons. The studies have shown the following results[74].

(1) Opening sodium channel. The simplest way of causing an action potential would occur by opening a number of voltage-sensitive sodium channels that would be sufficient to trigger an action potential directly. Such opening of sodium channels can be achieved through conformational change in the membrane protein, that allowing sodium ions to pass across the membrane. The conformational change which is normally triggered by a reduction in the magnitude of the resting potential at the neuronal membrane allows sodium ions to pass across the membrane thus triggering the action potential. Since the conformational change leading to the opening of the sodium channels requires energy and since such requirement is to be fulfilled by a ***non-physical mind,*** the needed energy must be created, and, hence, the realization of this process would violate the 1st law of thermodynamics. However, such a violation might go ***undetected*** if the energy required, coupled with the time during which it would need to be available, were low enough, i.e., it could be 'hidden' under quantum-mechanical uncertainty (Heisenberg principle).

The calculations have shown[72] that the opening of even ***one*** channel for an adequate period of time requires an energy input that is ***orders-of-magnitude higher*** than possible under quantum-mechanical uncertainty. Besides, causing an action potential would usually require the opening of a number of such channels.

(2) Altering voltage gradients. A second way to open sodium channels can be realized by altering the voltage across the membrane. This voltage change would trigger the voltage-gated sodium channels to open. Such change could occur by moving enough positive charges from the inside of the membrane towards the membrane or/and by moving enough negative charges away from the membrane, all that resulting in the membrane depolarization to the threshold which is reached through a typical generation of an action potential by a neuron. The respective area (axon hillock region) would contain a number of voltage-gated sodium channels.

The calculations have shown that just modifying the voltage gradient over a single channel, a much smaller area than an axon hillock, requires too much energy to be 'hidden' under the uncertainty principle. Namely, the maximum possible time period for such an energy increase, as allowed by the uncertainty principle, would be too brief to allow for any ion flow.

Thus, if the generation of the action potentials (either by direct opening of the sodium channels or by their indirect opening though altering voltage gradients) is caused by a ***non-physical mind***, this means the violation of the 1st law of thermodynamics.

(3) Synaptic transmission. The synaptic transmission has been considered as another possible site for intervention by the non-physical mind into the action potential formation. The main process of the synaptic transmission is the release of a chemical transmitter which is stored within membrane-bound vesicles in the pre-synaptic terminal and, after releasing, interacts with receptors on the post-synaptic neuron.

In the ***presynaptic neuron***, the sequence of the processes is as follows. An influx of calcium ions through the presynaptic membrane triggers the release of the transmitter. The opening of voltage-gated calcium channels in the presynaptic membrane results in the entry of calcium ions into the cell. The opening of a single calcium channel involves a change in conformation of the protein which forms the channel, that allowing a net flow of calcium ions into the cell. The release of the chemical transmitter from the pre-synaptic terminal leads to a linkage and fusion of the transmitter-containing vesicle with pre-synaptic membrane followed by interaction of the transmitter with receptors on the post-synaptic cell, that causing a change of the cell membrane potential. This chain of the processes results in final induction of the action potential upon the post-synaptic neuron. In the ***post-synaptic neuron***, the sequence of the processes is as follows: The synaptic transmitter molecules by binding to ligand-gated ion channels cause conformational changes in the membrane of the post-synaptic cell that result in opening the ligand-gated channels triggering the action potential.

Thus, at the level of the synaptic transmission, the intervention by the non-physical mind into the action potential formation would be based on opening of calcium channels (presynaptic neuron) or/and ligand-gated ion channels (post synaptic neuron). The processes at both the sides of synaptic transmission are of the same kind as those described in the previously considered possibilities – opening sodium channels and altering voltage gradients. Consequently, the energy needed for opening a single calcium channel or a ligand-gated channel was calculated by the same method[74] and gave the same result. Namely, the needed energy was ***orders-of-magnitude higher*** to be undetected, i.e., hidden under quantum-mechanical uncertainty, that if caused by a non-physical mind means violation of energy conservation, i.e., the 1st law of thermodynamics. Moreover, the opening of a single channel is not sufficient for triggering an action potential, and a considerable number of such channels at more than one synapse would usually be required[73], that further aggravates the violation of the 1st law of thermodynamics.

(4) Neuronal modulation. Apart from opening ion channels, there may be another point of mental influence connected with neuromodulation by means of especial kinds of neurotransmitters including norepinephrine, serotonin, dopamine, and some neuropeptides. These modulators produce biochemical changes in postsynaptic neurons, which alter the long-term sensitivity of the neurons to other synaptic input. Many of these neuro-modulatory processes act through a cascade of reactions, altering enzymatic activity and allowing for a considerable amplification of the signal within the neurons. The end products in the neurons can include modified ion channels, whose sensitivity to future synaptic inputs can be altered.

A hypothetical non-physical mental influence would occur at an early event in the cascade. The appropriate calculations[74] have shown that the magnitude of energy that would be required to initiate the cascade is much more significant to be allowable under the uncertainty principle. This means again the violation of the energy conservation (the 1st law of thermodynamics).

(5) Self-generation of action potentials by neurons. The last possible site of the hypothetical mental influence is connected with the fact that some neurons have special channels (usually potassium and calcium), that allow for the self-generation of action potentials. Many of these neurons are pacemaker cells, which have an oscillating membrane potential that periodically depolarizes the neurons to a threshold resulting in induction of an action potential.

The possibility that a non-physical mind might act through the modulation of such channels was investigated. The respective calculations[74] have confirmed that such possibility would lead to violation of the energy conservation.

(6) Harnessing energy. In all the previous cases, the calculations were based on supposition that the energy needed for initiation of a volitional act is to be ***supplied*** by a non-physical mind. The calculations demonstrated that the value of the needed energy is too high to be hidden (undetected) under quantum mechanical uncertainty principle, this meaning the violation of the 1st law of thermodynamics. However, there is a possibility that a non-physical mind might avoid the violation of the conservation energy law by means of ***harnessing*** a local energy, as if "borrowing" it from existing sources to generate the action potential.

The analysis of such possibility[74] has led to conclusion that opening of ion channels (causing the formation of action potential) might be achieved by changing (***vectorizing***) the movements of individual ions in a certain preferable direction on the account of energy of the ions themselves. Such an idea without any analogy in the physical (physiological) reality seems to be a mere speculation. If using a rather remote analogy, a harnessing factor may be imagined as a variant of a Maxwell's demon. However, demonic ability of a 'classic' Maxwell's demon consists in opening a trap door, allowing faster moving gas molecules to move in a given direction. In the considered case, the demon's ability is much more complicated: it would be capable to change the directions of a high number of ions to move in a certain predominant direction that would need a team of such demons. Such analogy would serve at a stretch to the hypothesis of harnessing, but it would not help: the general idea itself of a Maxwell's demon analyzed in detail[75] was considered as violation of the 2nd law of thermodynamics. Moreover, by further theoretical interpretation, the idea of the energy harnessing would violate both the 2nd law of thermodynamics and the principle of conservation of movement momentum[74].

Thus, a proper calculation, based on a single ion channel as a source of formation of a single action potential as a cause of a physiological neural impulse and attempted for any site of a possible influence of a non-physical Mind on physical Brain, has shown that the needed energy is by ***orders-of-magnitude higher*** than that minimal value which would be hidden under quantum mechanical uncertainty principle and remained undetected. Moreover, induction of an action potential would usually require the opening of a number of such channels. Besides, physiological evidence shows that transmitter release from a considerable number of synapses in a single neuron may be required[76], and a volitional act to be performed needs involvement of at least a few if not many neurons.

Thus, the final conclusion is that if a non-physical Mind is supposed to influence Brain-directed neural processes, this would mean ***violation of fundamental physical laws***.

In the view of the reductionist (materialistic) philosophy, the problem of violation of physical laws by a non-material Mind does not exist simply because the existence of a non-material Mind is denied. Therefore, the problem can be considered only within the frame of the dualistic philosophy, in which the notion of the non-material Mind is the basic postulation.

In accordance with this, the problem of the Free Will should be considered in connection with the general problem of Consciousness. Therefore, before dealing with the suggested here approach associated with the irreducible field notion, the place of the Free Will within the frame of the Consciousness problem is to be considered.

4. THE PLACE OF FREE WILL WITHIN THE PROBLEM OF CONSCIOUSNESS

Connection of the Free Will with Consciousness becomes explicit in the light of the main Consciousness problems – the binding problem and the recently proclaimed "Hard problem".

4.1. The Binding Problem

Historically, the binding problem was connected first with psychology being dependent on general philosophical views on spatial-temporal contiguity of mental representation of the external world (Hume[77]). The modern version of the binding problem is expressed on neurological level being based on the well established evidence on the disjunctive way of processing of visible percepts realized within different brain cortex areas. In spite of such spatially segregated processing of particular features of the object, it is perceived as unitary, i.e. the whole dynamically changing world appears as integral and coherent. Therefore, the synthesis of all the disjointed, dispersed, and separately processed elements (components, features) of the complex signals from the continually changing picture of the external world must be realized by ***binding*** together neurological states occurring in different brain areas. In particular, the visual data are processed separately within about fifty functionally segregated specialized cortical areas[76, 77], each one being responsible for a specific feature, like movement, colour, texture, size, curvature, some topological properties like height/width ratio, stereoscopic depth, orientation of lines and edges, and so on[78-85]. At the same time, multi-modal association areas in the cortex in which single perceptual features could be unified into a final perceptual image have not been found, so there is no explanation how the disjointed features of any perceived object are linked together. This binding problem looks more keen since there is ***no*** locus in the cortex, which could be called either ***"master map"*** [80], or ***multi-modal association areas***[86], or ***central cortical "information exchange"*** [87]. Nevertheless, still there is a certain (rather emotional) hope for future finding of "grandmother" neurons and convergence zones[87].

The binding problem is particularly complicated in the case of the visual system. The histological fact is that about 300 retinal rods (the 1st neuron of the visual network way) are structurally (histologically) connected via bipolar cells (the 2nd neuron) with one ganglion cell (the 3d neuron). Consequently, these 300 adjacent rods form a microarea within retina which during the vision process may contain heterogeneous picture as projected from a (micro)part of the visible object. Thus, the axon of the ganglion cell must conduct forward an impulse carrying such complex (integrated) visual information. This already is incompatible with the classic neuronal theory according to which the neuron firing means only the ***conduction*** of a signal that is realized according to the 'all-or-none' principle.

Thus, apart from the binding problem seemingly realized at the cortical level, two opposite processes associated with the same current visual signal go on simultaneously. One of them consists of the anatomically determined ***confluence*** of distinct signals related to different spatial parts of the perceived object into a ***single but complex*** signal, its complexity being based on the additivity (puzzle-like summation) principle – integration of different parts into a certain whole. The other one consists of the above-described ***splitting*** of the perceived object's image into dozens of quite different object's features, like shape, movement, color, etc., which are processed separately in distinct cortical areas. The possibility of the coexistence and simultaneous realization of these two antidromic processes within the same anatomic unit (system) is totally incomprehensible. On one hand, confluence (merging, junction, maybe synthesis) of a number (hundreds!) of axonal impulses reflecting ***parts*** (portions, pieces) of the object, and, on the other hand, disjunction (splitting, breaking, maybe analysis) of the object as a whole – ***not into parts*** (portions, pieces) – but into so drastically different (somewhat category-like) and causally disconnected features, like color, form, texture, movement, spatial relationships, etc. And after (or in parallel to) that, there is binding of such disjoined features into an integral coherent image. Such antidromic way of stimuli processing within the neural networks display the bias/variance dilemma that is incompatible with the neuronal theory, presenting an insuperable obstacle for the theory of computational mechanism of the brain functioning. According to the modern computational language, the problem is formulated as claiming that the objects or their different aspects have to be represented (to co-exist) within the same physical "hardware" (brain), that resulting in the "***superposition catastrophe***"[88, 89]. Therefore, if to define the main postulate of the neuronal theory in the way that each mental representation ("symbol") of the external objects is represented by the corresponding subsets of coactive neurons within the same brain structure, then if more than one of such "symbols" become active at a moment (that must occur within the real functioning pattern), they become superimposed by co-activation (structural "overlapping" of the activated subsets). In such a case, any information carried by the "overlapping" subsets must be lost, that being the mortal verdict for the whole foundations of the classic neuronal theory

Consequently, the binding problem, being not a purely theoretical construction but arisen from the very heart of the neurobiological and psychological reality, looks incomprehensible within the framework of the anatomic structure and physiological regularities associated with perception. An especial question concerns the relation of the binding problem *per se* (apart from the above integration of signals within the retina's neurons) to the Somato-Mental Gap, which is so fathomless in the case of the "Hard Problem". In this respect, the main question concerns the "localization" of the events associated with splitting of the signals into different components (constituents, features) and their consequent binding, or, better to say, timing of this signal processing in relation to the Somatic-Mental Gap (before or after?). A reasonable supposition is that, the integration of the initial signals within the retina level precedes their further splitting which, hence, occurs still within somatic level (thalamo-cortical framework), i.e., ***before*** the Gap, while the binding is realized within the mental level, i.e., ***after*** jumping over the Gap. The in principle possibility that ***both*** the splitting and binding take place within the mental sphere (***after*** the Gap) is hardly probable since the anatomical areas in the brain cortex correspond to the already split components/features[78-85], so that the splitting seems to occur before the signals reach the final 5th (cortical) neuron.

4.2. The "Hard Problem" and the Concept of Extra Ingredient

The psycho-physical gap has recently been expressed in the form of the "Hard Problem" of Consciousness proclaimed[90] and further developed[47, 91] by D. Chalmers who has exposed with clear acuity a rather simple question: how and why performance of any form of neural activity can give rise to subjective experience. The question has a more general meaning: how a physical system of any degree of complexity can be 'aware' of itself ? Chalmers claims that "a full theory of consciousness must build an explanatory bridge" between neurophysiology and consciousness[89]. In order to achieve this, an ***extra ingredient*** in the explanation is needed, because "no more account of the physical process will tell us why experience arises", i.e., "the emergence of experience goes beyond what can be derived from the physical theory" ("innocent dualism" by Chalmers[91]). This is perfectly illustrated by T. H. Huxley's famous utterance[92]: "How it is that anything as remarkable as a state of consciousness comes about as a result of irritating nervous tissue, is just as unaccountable as the appearance of Djin when Aladdin rubbed his lamp" (quoted by N. Humphrey[93]).

In such situation, the initial task of the approach suggested in this article is to combine an abstract concept of the Extra Ingredient, which according to definition is ***additional*** to the physical fundamentals, with any formalized notion closely associated with physical reality.

5. THE FIELD CONCEPT USED FOR EXPRESSION OF THE EXTRA INGREDIENT

In this respect, it would be tempting to analogize the still undefined Extra Ingredient with the well-established and strictly formalized field principle. Such endeavor may look like confrontation of two giants: ***Consciousness*** as the most mysterious enigma ever standing before scientific exploration *versus* ***Field*** as the deepest theoretical notion ever born by human intellectual power. In such confrontation, the Field concept is not so evidently "inferior" to the Consciousness enigma and success that the latter may be reduced to the former is not promised.

Therefore, if in principle the field notion is to be used for explanation of consciousness, then, instead of passive waiting for any progress in the development of the field-based theorizing, which has come to the era of the Ultimate Theory of Everything[94, 95], active endeavor for elaboration of the Autonomous Field Theory of Consciousness is to be undertaken. Such Autonomous Field Theory based on specific expression of Chalmers' Extra Ingredient as irreducible fundamental could enter as an integral part into a future comprehensive version of "The Elegant Universe"[94], or "Multiverse"[96]. Consequently, the problem consists in formulation of a non-material Extra Ingredient described by the strict mathematical language while associated with biological reality.

The modern theories of Consciousness using the field principle fall into two groups: (1) those based on the established physical fields, and (2) those grounded on autonomous fields irreducible to the established physical fundamentals. The theories of the 1st group are based on the reductionist approach explaining Consciousness from the physicalist point of view with no need for any Extra Ingredient, while in the theories of the 2nd group, the field concept has a potential capacity to express the very meaning of the Extra Ingredient. Hence, the theories based on the physical fields may, in principle, elucidate only neural correlates of consciousness (NCC)[97-99], leaving the "Hard Problem" unsolved, while the concept of the autonomous irreducible field looks as a promising tool for solving the "Hard Problem". Therefore, it seems that for the aim of the present article, only the irreducible field-based theories should be taken into account. However, since some of the physical field-based theories have approached to the binding problem, they may have a certain corollary contribution toward a possible solution of the "Hard Problem" and deserve consideration.

5.1. The Theories of Consciousness Based on the Established Physical Fields

The physical field theories of Consciousness include those based on the electromagnetic field (EM) and quantum mechanics (QM) implicating the field principle[2]. Due to the volume of the relevant evidence, only representative findings will be cited.

The first attempt to consider the unifying EM fields as involved in organizing integrated motor actions of the cerebral cortex was made by R. Sperry[102]. The theory of the "mind as a force field"[103] was based on the initial theory by K. Popper[104] considering deep similarities between Mind and physical forces.

According to the EM theories of Consciousness, the incredibly complicated network of the inter-neuronal connections within the brain is considered as a source of the respectively complicated continual electromagnetic field, which is considered as the physical basis of the subjective experience. The electromagnetic manifestations of the brain cortex have been recently analyzed using modern techniques, e.g. electroencephalography and magnetoencephalography as well as new highly sophisticated scanning (imaging) methods like positron emission tomography and functional magnetic resonance imaging[105]. By means of those techniques, a certain correlation between the functioning of nerve cells in cerebral cortex and the corresponding subjective experiences has been demonstrated[98, 106-109]. A "new generation" of the EM field theories of Consciousness[45, 110-117] directly accepts the EM field as the physical expression of the NCC.

The connection between QM and Consciousness is of especial origin. The initial work[118] was motivated by the question how the excitation of myriads of individual neurons results in unique coherence and integrity of conscious activity. In further studies in this direction[119, 120], the quantum model of the brain was based on the combined formalism of the many-body physics problem and spontaneous breakdown of symmetry. Such comprehensive prevalence of the physical aspects in the first QM theory of brain functioning has remained in the next generation of the QM theories of Consciousness[121-133]. The theoretical considerations included quantum measurement problem, e.g., collapse of the wave function,[127-129, 134, 135] Bose-Einstein condensates,[136, 137] adaptive resonance theory,[138] spontaneous breakdown of symmetries,[139] the extension of the quantum field to

---

[2] The 'holonomic' theory of consciousness (Pribram[100, 101]) sometimes is also considered as a field theory although formally the holography is not based on the field principle.

dissipative dynamics,[132, 133, 139-141] time entanglement between mind and matter,[142] quantum noise and chaos,[139] synchronization of homoclinic chaos[143].

Thus, in the most of the QM theories of Consciousness, the burning fundamental problems of ***quantum physics itself*** were considered, the Consciousness serving either as an inspiring enigma, or as a source of convenient analogies, e.g., wave/particle duality versus mind/brain duality. A certain exclusion from such QM-rather-than-consciousness-itself preference is the theory of orchestrated objective reduction (Orch OR) by Hameroff and Penrose[126, 127, 144-149].

As to the EM field theories, the essential finding of crucial importance concerns occurrence of the oscillations between the involved regions with ***zero time delay***[115, 150]. Such oscillations synchrony between ***distant*** brain regions unexplainable by the neuronal theory (discrete synaptic transactions) may be considered as a strong argument for involvement of a ***field-like*** factor.

As to the QM theories, an urge to connect Consciousness with quantum mechanics has been based on a deep intuitive feeling. On one hand, there is Consciousness enigma, which stands at the edge of human capacity for understanding and explanation. Being under shadow of the frightening Explanatory Gap, science, which is strongly believed to be physical in its ground, is meanwhile impotent to grab the Consciousness into its rigid formalistic framework. On the other hand, there is QM with its strictest formalism which is continually gains further complexity and universality supported by the powerful echelon of combined mathematical-physical intellectual forces racing toward the Theory of Everything[94, 95, 151] whose grounding conclusions also have reached the edges of human cognizing capacities. So, the urge is to unite the Consciousness, which has still unsteady scientific background, with the quantum physics, which is strictly formalized while embracing all the conceivable levels of the Physical World. Besides, such match-making between QM and Consciousness has another source noted by Chalmers[91]. "The attractiveness of quantum theories of consciousness may stem from a Law of Minimization of Mystery: consciousness is mysterious and quantum mechanics is mysterious, so maybe the two mysteries have a common source"[91]. "Is this conceptual similarity merely the basis to draw an attractive analogy, or might there be some physical correspondence between the two classes of phenomena, physical and neurophysiologic?" (John[115]). The question remains open. Anyhow, both the quantum phenomena and conscious phenomena have two common indications – non-determinism and non-locality. The former still remains on the level of incognizable, while the latter is within the experimental grasp.

The conclusion can be expressed by the following citation related to the question "Why are quantum theorists interested in consciousness?"

> "Orthodox quantum theory requires ***'something else'***. It is a plausible hypothesis to say that this something is a primitive ingredient of the world – that is, not reducible to other things in physics and to identify it with consciousness. It includes or perhaps we should say ***is*** the quale of experience. The model naturally shows us why this consciousness is causally effective. To identify the mechanisms in the brain where the required quantum "measurements" that magnify quantum selections into macroscopic effects take place is a problem for the future" (Squires[152, p. 617], the ***1st boldface*** is mine).

This "something else" is just that Chalmers' Extra Ingredient, which was suggested for Consciousness explanation because "no more account of the physical process will tell us why experience arises", i.e., "the emergence of experience goes beyond what can be derived from the physical theory" (Chalmers[91]).

However, the elucidation of the binding problem may be associated with the solution of the Hard Problem, especially if the "binding" concerns the processes located at the opposite shores of the strait formed by the Somato-Mental Gap. A paradox is that although a high potential power of the EM field concept for explanation of the binding problem is clearly evident, this is in contrast to the in principle impossibility of any physical field theory to jump over the Gap. However, the experimental evidence (using the EM-based measurements) indicates on the possible involvement of a ***field-like*** factor which, thus, has to be irreducible to the physical fundamentals. Such conclusion encourages the analysis of the autonomous field-based theories.

Thus, all the above considerations connected with both the "Hard Problem" and the binding problem are closely associated with some particular actual questions formulated in the sphere of the experimental neuroscience[115], e.g., the synchronization of neural units activities within a brain area, coherent activity between the brain areas, the role of coherent oscillations in binding, the functional significance of distributed coherence, spatiotemporal patterns of coherence, and the role of coherence in brain encoding of information. In this respect, the field concept looks as a very promising notion for explanation.

5.2. The Theories of Consciousness Based on the Autonomous Fields

The theorizing on the level of the autonomous fields harbors dangerous reefs. The very possibility for reflecting on a hypothetical irreducible field of Consciousness uncovers an inspiring temptation or even an urge to escape from the rigid chains of the physical postulations and get way to unlimited creative capacity. This may lead to a certain inclination which could be called a loss of mental discipline and responsibility. As a result, the field principle is degenerated acquiring purely belletristic expression.

The recently suggested theories of consciousness based on the irreducible field notion, namely, the "Mental field" by B. Libet[153-159], the "Unified conscious field" by J. Searle[160-162], and the "Morphic resonance" by R. Sheldrake[12, 163-166] have a claim to hit the main Consciousness riddle – the Somato-Mental Gap, i.e., they could be immediately connected with the "Hard Problem". The Libet's and Searle's theories are somehow related and even competing for priority[155], being rather burning topics of the day. The Sheldrake's theory usually is not considered seriously by conventional science[3], e.g.,

---

[3] Although I, myself, also belong to those who consider the Sheldrake's theory as highly speculative and confusing, the theory has been published, spread, popularised, and discussed until recently amongst different ***scientific*** circles and in the meetings involving highly worldwide reputable scientists, like Prof. H.-P. Dürr (Director of the Max-Planck Institute for Physics), Prof. A. Goswami (Physics, Oregon University), Prof. R. Abraham (Mathematics, Santa Cruz University), Prof. G. Schwartz (Psychology, Arizona University), and some others belonging to the "hard science" (Con-

it is not much referred to in the current discussions under the slogan "Toward the Science of Consciousness". Besides those three theories, there are two more, namely, "integrated field of consciousness" by Kinsbourne[168], which means multimodality (i.e., constellation of different brain areas), and "field of consciousness" by Hasker[169], where the word "field" in the expression "the field of consciousness" has idle meaning and can be easily replaced by various substitutions, like "stream", "state", "phenomenon", "process", etc.

Analysis of the above autonomous field theories shows that the employed field notion looses its specific fundamental character acquiring either esoteric meaning with little if any relation to scientific knowledge ("Morphic Resonance" by Sheldrake), or tautological definition ("Conscious mental field" by Libet), or merely metaphoric description ("Unified conscious field" by Searle), or purely belletristic meaning ("Integrated field of consciousness" by Kinsbourne) and the "Field of consciousness" by Hasker. Such vague allegoric use of the great principle leads to its emasculation and devaluation.

The above conclusion does not mean that the autonomous irreducible field principle, in general, is far away from the Consciousness problem. On the contrary, potential advantage of employing such concept for understanding Consciousness looks as attractive possibility to join to the grand theoretical edifice of physics pervaded by the field principle from microcosm to macrocosm. Such perspective is stunning, but the indispensable condition that a postulated field must be ***irreducible*** means that the urge to join to the Grand Physics is limited only to the physical glossary to be used for a formal description of a hypothesized field, apart from its physical (or metaphysical) nature. The problem could be analyzed in the frame of "naturalistic dualism"[91] if the Extra Ingredient is described by means of the abstract field vocabulary, but *hic haeret aqua*. As far as "irreducible" means "non-physical", a danger is that a postulated irreducible field would lose, together with physicality, its "field face". Then, the urge to join to the Grand Physics may be looked as a kind of mimicry when the authority of the universal scientific principle would mask theoretical void of a suggested "field" hypothesis. The Searle's Unified Conscious Field is a clear example of such mimicry.

Therefore, for Consciousness theorizing, the obligatory condition for an autonomous irreducible field is its subordination to the same formal field postulates, which are common to all the established physical fields. The task is to fill such abstract skeleton with novel ontological flesh. Then, an autonomous field theory could explain "nomological emergence" of Consciousness, when "higher level entities, properties, etc., are governed by higher-level laws that are not determined by or necessitated by the fundamental laws of physics governing the structure and behaviour of their most basic physical parts" (Silberstein[170]).

---

ference "Rupert Sheldrake in Discussion"[167]. Therefore, just formally, Sheldrake's theory cannot be omitted.

5.3. The Postulates of the Abstract Field Principle

Therefore, I have attempted to characterize the field principle in a generalized abstract form, ***irrespective of the field physical nature*** (gravitational, electromagnetic, and weak and strong nuclear), while expressed using the physical glossary. Such attempt has resulted in summarizing the following formal prerogative features of the field principle:

1. Implication of actual field source(s).
2. Action-at-a-distance from the field source upon sensitive-to-field entities (substrate) – as opposed to direct ("mechanical") contact (collisions).
3. Unboundedness – absence of any distinct boundaries of the field influence, which, hence, theoretically is infinite.
4. Field integrity, i.e., continuity "where the hole cannot be" (Campbell[171]).
5. Dependence of certain parameters of any in-field-occurring process on its coordinates within a certain whole which is under significant field influence.
6. Field 'direction', which may have different expressions from the 'classic' scalar and vectorial fields till the modern highly complicated expressions (tensor-string-twistor, etc.) associated with 'over-classic' numbers of dimensions.
7. Measurability, i.e., the possibility for quantitative estimation of certain field particulars, like intensity, decrement (distance dependence), fluctuation, and any other parameters.

From the above postulates, the 5th one seems to be the most indicative: empirically found dependence of any parameters of a process on its coordinates in a certain whole is direct indication that this process is realized within an actual field.

Any physical field-based theory includes the above postulates into its formulation and the same rule should be kept toward any version of the autonomous field theory but this is not the case. Therefore, it could seem that the task to elaborate a non-tautological autonomous field theory of Consciousness based on the above-summarized formal field postulates is destined for future efforts. However, the field theory by Alexander Gurwitsch, which was elaborated long before[3-5] the above-described irreducible field-based theories, meets all the above demands for a basic fundamental, and, accordingly, the next step is consideration of its postulates.

5.4. The Postulates of the Theory of the Vectorial Biological Field by A. Gurwitsch

In this theory, a notion of the autonomous irreducible field is neither tautological, nor metaphoric, nor allegorical, being based on strictly defined postulates deeply rooted in the biological reality. Accordingly, the basic postulates of the theory are as follows:

1. Each cell is a source of the field generated in the nucleus.

2. The field is of vectorial nature and the vectors are directed centrifugally from the field source.

3. The generation of field is associated with certain processes in the nucleus, namely those related to transformations of "chromatin". Extranuclear chromatin (cytochromatin) is also a source of the field.

4. There are ***elementary "flashes" of generated field,*** connected with certain elementary acts in the chromatin metabolism. A total number of such flashes per time unit designated as ***field intensity*** depends immediately on the intensity of the chromatin metabolism and, hence, on the general cell metabolism.

5. The elementary flashes of generated field can occur only if these acts proceed within the sphere of influence of the already existing field. Essentially, this is the expression of the succession of processes in living systems, or, in other words, the proclamation of the same principles declared by W. Harvey ("***omne vivum ex ovo***") and L. Pasteur (***denial of a spontaneous generation of life***).

6. The field vectors directed from the nucleus are the resulting values from a total statistical number of the elementary flashes of the field at any moment. Therefore, the field intensity is a completely dynamic fluctuating parameter subtly reacting to metabolic changes.

7. The elementary field is ***spatially anisotropic*** and this is the main postulate. Its meaning is that the isodynamic surface at which all the vectors are equal ***is not*** merely ***spherical*** but more complicated. However, this "complication" may be sufficiently simple, e.g., it may be considered as ***ellipsoidal***. The anisotropy of the ellipsoid can be expressed as a particular ratio between its three axes and such ratio, being species-specific, is considered as an ***invariant species constant***. An infinite number of possible axes ratios covers all the potential number of possible species.

8. The field vector has a certain decrement along with the distance from the field source. An exact function of the dependence is a matter of empirical examination. In spite of the decrement, the influence of the field is not limited to the cell boundaries

9. Field vectors exert influence upon the ***excited*** protein molecules (those which have just got a portion of metabolic energy and are in the excitation state) transforming a portion of the molecule general excitation energy into ***directed kinetic energy*** and the direction of the movement is determined by the field vector at this spatial point. This is expressed either in the directed movement of the excited protein molecules along with the vector or in specifically directed deformations of the protein molecules if they are anchored to any structures. This means that in living conditions the field "works" against the chaotic movement (agitation) of the protein molecules.

10. The intensity of the field at a certain cell point (the length of the vector at this point) determines what share of the whole molecule excitation energy is transformed into the directed kinetic one. E_d / E_t ratio, where E_d is the directed kinetic energy and E_t is the molecule total excitation energy, displays this share. The intensity of the field depends not on the amount of the chromatin but on its ***turnover***.

11. The vectors from separate field sources can be composed geometrically and the resulting vector at the point of composition will determine the direction of the kinetic constituent of the full molecular excitation energy. Therefore, in different parts of the system consisting of a number of cells and, hence, the cor-

responding number of field sources, there is an ***integral (actual) field*** resulting from the total geometrical composition of all the vectors issued from all the sources (nuclei). Evidently, in such composition, both the field intensity (being a function of metabolic activity as well as the distance of the point of the composition from the field source) and the field anisotropy (relation of the point to the nuclei' axes) make contribution into the value (length) of the resulting vector.

12. The last postulate is, essentially, the inference from the above postulates, especially the previous one concerning the notion of integral (actual) field. Geometrical configuration of the integral actual field imposing vectorization upon energetically excited protein molecules of intracellular substrate (the object of the field influence) determines dynamic configuration of the protein molecular continuum. Such dynamic associations of the energetically excited protein molecules maintained by continuous metabolic energetic influx are called ***"unbalanced (non-equilibrial) molecular constellations"***. These constellations provide dynamic conditions for steric facilitation or hindrance for certain reactions which, hence, are not due to the canonic chemical properties of the molecules – members of the constellation, but to specificity of configuration of the constellation which is determined by the geometrical configuration of the integral actual field at any considered locus. The unbalanced molecular constellations are considered as a "working substrate" of most biological manifestations.

Although the question about the nature of the biological field is not especially touched upon in the above postulates two principal comments should be added.

1. Gurwitsch's biological field cannot be reduced to any known physical field: it is an immanent property only of living objects. According to the postulate 5, the elementary flash of the biological field is induced ***only by the existing field so that the field is successive and cannot originate de novo.*** This is the full expression of the vitalistic principle.

2. Gurwitsch's biological field ***is not energetic*** which means that no special energy is focused in the field source. The field vector just transforms a portion of the metabolic energy accumulated in the excited protein molecules into directed kinetic energy moving or deforming the molecules. The energy is not supplied by the field to the spot of its action but the field vector as if ***harnesses the local energy accumulated at this spot*** [74].

Thus, the Gurwitschian field is neither tautological, nor metaphoric, while responding to all the demands for any physical field. The explanatory capacity of the theory was tested by Gurwitsch using different levels of the biological organization: molecular (metabolism), cellular (mitosis, differentiation and histogenesis), and organismic (morphogenesis, neuro-muscular system, brain cortex structure and functioning) [4, 5]. The mode of the field action expressed on the morphological level is defined as subjection of equipotential elements (cells) to integral morphogenic field causing the cells' ***spatial*** orientation or/and movement. On the molecular level, the field action is expressed as vectorization of the molecules' chaotic movements *in vivo* as opposed to the molecules' chaotic movement *in vitro*.

6. THE MAIN PREREQUISITES OF THE SUGGESTED APPROACH

The approach suggested in this article was elaborated on the basis of Gurwitsch's theory of biological field in connection with the recent development of the modern dualistic philosophy combined with the renaissance of the science of Consciousness. In such constellation, namely Gurwitsch's field theory adds an in principle new aspect that seems to mean the development of an entirely new paradigm.

6.1. The Vitalistic Principle versus Classic Dualism: the Crucial Difference

While the reductionist philosophy is based on the pure physical laws, the classic dualistic philosophy makes jump from the physical sphere directly to the sphere of Consciousness skipping off the phenomenon of Life. Classical dualism establishes a principal gap between ***conscious and non-conscious*** realms. The latter realm includes both physical and biological phenomena with conviction that the living state can be reduced to (comprehensively explained by) the same physical laws. The vitalism establishes a principal gap between ***living and non-living*** phenomena (i.e., all forms of Life versus the "dead" Physical World) with conviction that the Consciousness can be reduced to (comprehensively described by) the autonomous principles of Life. In the nowadays science, the dualistic view is considered as legitimate while the vitalistic view is totally intolerable with determined inclination to exorcise this evil from the Science.

Such simple but crucial shift in division between the realms (location of the "gap") is the most important step of the suggested approach. It means, that the primary question is "What is Life?" and the question "What is Consciousness?" becomes to be the secondary one inferred from the primary question. By other words, this shift is hoped to provide that construction of a larger formal system (Life versus Consciousness) which, by breaking the limits of the Gödel's ring[172], would make the 'Hard Problem' solvable.

Thus, the essence of the vitalistic principle is related not to the conscious ***Mind*** but to ***Life***, a species-specific living entity. Unfortunately, according to the commonplace concept of the vitalistic principle, the latter is a kind of an abstract metaphysical ***omnipotent*** "vital spirit" ("*vis vitalis*", "élan vital"), "organizing" ("animating") material components of a living system and thus determining its structural appearance and coherent functioning. Such evidently tautological character of the vitalistic concept has neither explanatory power, nor scientific value. Therefore, contrary to such vulgar understanding of the vitalistic principle, the latter is considered here as a working tool permitting formulation of working hypotheses[42], in particular, those to be applied for the analysis of specific aspects of Consciousness including the Free Will.

6.2. Formulation of Protophenomenal Fundamental

The initial aim on this way is to "naturalize" the Extra Ingredient, which will be designated as Protophenomenal Fundamental, by formulating its unequivocal definition. The immediate conclusion is that Consciousness ***itself***, as being a "hodgepodge", by O. Flanagan's expression[173], i.e., highly variable (and dynamically changing) in its attributes and manifestations, cannot in principle be in the role of a fundamental, alongside the established physical fundamentals. That is in contrast to the idea by Chalmers that "...the ***experience itself*** is accepted as a fundamental feature of the world, alongside mass, charge, and space-time"[91] (***boldface*** is mine). However, there is another Chalmers' utterance, namely:

> "We ought to be able to develop formalisms for capturing the ***'structure' of experience***, i.e. similarities and differences between experiences of related sorts and the detailed structure of something like... I don't know what exactly such a formalism would look like, but perhaps something bringing in ideas from geometry or topology or from information theory might be useful."[174] (***boldface*** is mine).

Evidently, these two utterances, the ***"experience as a fundamental"*** and the ***"structure of experience",*** totally contradict each other because if the experience is a fundamental, it must not have any structure: rather, the constituents of such a structure may be candidates for a fundamental. Therefore, the main question is formulated by the following way: May any ***attribute*** of the hodgepodge Consciousness be in the role of the Protophenomenal Fundamental? Before dwelling on this question, I have summarized the general demands, which are accounted as obligatory for any physical fundamental and, hence, must be obligatory for the Protophenomenal Fundamental to be formulated.

6.3. General Demands to the Protophenomenal Fundamental

Like the established physical fundamentals (Mass, Charge, Time-Space), the Protophenomenal Extra Ingredient must be:

1. ***Elementary*** (further unsplittable);
2. ***Axiomatic*** (further unquestionable);
3. Strictly and unequivocally ***defined***;
4. Qualitatively ***uniform*** (homogeneous);
5. ***Measurable***.

The 1st and the 2nd demands do not need any comment.

The 3rd demand, which is obligatory for any axiomatic fundamental, is the most problematic and would seem hardly realizable in the case of Consciousness that is notoriously famous for escaping any definition. The peculiarity of the situation is that, on one hand, the meaning of Consciousness is so self-evident and intuitively clear that does not need any explanation while, on the other hand, any practical attempt of "understanding", or "explaining", or, especially, "defining" Consciousness meets insuperable obstacles.

The 4th demand which is obvious for any (physical) fundamental is stipulated here because of the heterogeneous ("hodgepodge") expressions of Consciousness, which are either almost synonymous like awareness, experience, mentality, or are Consciousness's attributes (manifestations), like feeling, purposefulness, cognition, intention, volition, intuition, thought, mood, hesitation, decision-making, imagination, creativity, morality, etc., etc., or are rather metaphoric expressions, like self, "I", psyche, psi, soul, etc. However, a physical fundamental, e.g., Mass, is **constant** in its axiomatic properties in any physical processes and in every physical entity, let it be the mass of an electron, or a molecule, or the Earth, or Sun, or Galaxy, or a dead brain, or a living one, or dreaming one, etc., etc. The properties of the Mass as a fundamental are constant along with the transient time while the properties of Consciousness manifestations are changing and interchanging along with time in any combinations, e.g., feeling would be followed by imagination or thinking, then exchanged for hesitation during decision-making or volitional acts etc., etc. The Protophenomenal Fundamental to be defined must be **the basis for any manifestation** of the hodgepodge Consciousness.

The measurability (the 5th demand) can be realized at least by the simplest mode of quantitative estimation, like more – less, higher – lower, quicker – slower, more intensive – less intensive, more complicated – less complicated, etc., etc. Such demand is hardly realizable in the case of any Consciousness' manifestation, except maybe pain and some other qualia which could vary in intensity ('less hurt – more hurt' might sound, but high – low 'redness of red' … is doubtful). Probably, it is possible to make quantitative estimation in the case of feeling ('less merry – more merry'), or thinking ('less thoughtful – more thoughtful – highly thoughtful'), or creativity (less creative – more creative), or morality… although such estimations look clumsy or funny, while in the case of such manifestations like intention, volition, intuition etc. any quantitative estimation looks inappropriate. Besides, such "quantitative estimation" is totally subjective, i.e., based on the first-person perspective.

Thus, the consecutive analysis clearly shows that not only Consciousness ***as it is***, which in principle cannot be in the role of a fundamental, but ***any one*** of the Consciousness hodgepodge manifestations does not meet the above obligatory demands for an axiomatic fundamental.

Therefore, in the frame of the suggested approach, the formulation of the Protophenomenal Fundamental means reduction of Consciousness not to any of its hodgepodge features and manifestations, but to an elementary entity meeting ***all*** the above-formulated demands while ***still preserving phenomenal quality***. Consequently, the Protophenomenal Fundamental to be postulated must be axiomatically uniform in relation to ***any*** Consciousness attribute, so that the whole ***plurality*** of the Consciousness expressions could be reduced to a ***singular source***. Since such formulation has been described in detail (Lipkind[35]), the specific aim of the present article is to use the formulated Protophenomenal Fundamental specifically for the analysis of the Free Will. For such a purpose, only the main steps of the formulation of the Protophenomenal Fundamental and the respective corollaries are to be dealt with below.

6.4. Geometrical Feeling as the Protophenomenal Fundamental (Consecutive Formulation)

The suggested definition of the Protophenomenal Fundamental is based on the morphogenetic approach, according to which ***form*** **(*morpha*)** is accepted as an intrinsic species-specific property of ***any living*** entity. In accordance with this, the initial definition of the Protoconsciousness has been formulated,[34] namely:

*The **Protoconsciousness** is an embodied immanent capacity of any living cell to **feel** its own evolving dynamically fluctuating species-specific **form**.*

Such "capacity to feel its own form" is designated as ***Geometrical Feeling*** which becomes to be the central notion in this article[4]. According to this definition, which is not tautological[5], the Protoconsciousness is an inalienable attribute of Life, i.e. any cell as soon as it is alive feels its own geometry. The latter is conceived not only as an external shape characterized by morphological contours: the Geometrical Feeling transpierces through all the three-dimensional whole of a living entity, i.e. each geometrical (stereo metrical) dot within such living whole is felt, sensed, being aware of by that whole. This could be associated with mathematical descriptions of space which may be not limited to the classic Euclidean space but to more complicated space models, e.g. Riemannian space or modern string theory-based models[94, 95, 177].

However, the above definition does not cover some general aspects of the Consciousness problem, which are related especially to the topic under consideration, i.e. the Free Will. These aspects concern association of the Protoconsciousness with the material substrate of living cell. The connection between the cellular material substrate with the Free Will (if it is not accepted as an illusion) is evident: the main prerogative of the Free Will is to ***activate somatic*** (material) mass of the body, or, simply, to "move" the physical body within which the Free Will abides. In accordance with this, the Protoconsciousness as a current process is put into effect through its immediate interconnection with the cellular material substrate, which is under continuous metabolic exchanges with the external environments. Such metabolically active material substrate filling the abstract ("ideal") geometrical form as a "receptacle" and being under genetic and biochemical regularities is in the state of dynamic heterogeneity and turbulence, thus "gushing over" the external and internal geometrical "borders" of the ideal geometrical form. Just this dynamic ***non-congruence*** between the ideal geometrical form and the real spatial distributions of the material substances filling this form is continuously felt (experienced) by the cell[6]. Therefore, the above definition of the Protoconsciousness is exchanged for a more complete version that is as follows:

*The **Protoconsciousness** is the embodied immanent capacity of any living cell to **feel** any spatial **non-congruence** between the cell's evolving species-specific **"ideal" geometrical form** and the real distribution of the **material** stuff **"filling"** this form.*

In accordance with this definition, the "Geometrical Feeling" is not the feeling of the geometrical form ***as it is*** but the feeling of the non-congruence

---

[4] The initially suggested concept of Geometrical Feeling[34] was further developed[35-38, 175, 176] and described in detail[35].

[5] A tautological version of this definition would be as follows: "The Protoconsciousness is the immanent capacity of any living entity to feel its own ***state***" (instead of ***form***).

[6] Besides the 'ordinary' metabolic exchanges between the cell and environments, there are some specific physical signals coming from the environmental material sources and "switching on" some cellular reactive capacities.

between this abstract geometrical form and its physical (material) realization. The non-congruence means incompatibility (discrepancy, divergence, collision, conflict) between the Ideal Geometry and the Solid Physics within a living entity that is "felt" by the entity. The Ideal Geometry is species-specific and initially pre-existing, while the Solid Physics is actually occurring and continually fluctuating to adapt (to adjust, to accommodate, to fit, to approximate) to the Ideal Geometry. A living cell "feels" not only an external three-dimensional shape of its own body but also all the internal lines ***as soon as*** the actively fluctuating (dynamically "boiling") material stuff constituting the cell body "violates" the above geometrical abstract borders. However, since the absolute coincidence of a physical (material) framework with its abstract geometrical image is ontologically impossible (only asymptotic approximation to the never reachable coincidence can be imagined), the inevitable non-congruence between the both will never cease, and so will the feeling of the non-congruence by a living cell. Such non-congruence is, essentially, a formalized expression of the Psycho-Physical gap which, hence, is analogous to the discrepancy between the ideal geometry and its physical imitation (Abstract Mathematics versus Solid Physics), like a geographical meridian upon the globe of natural size versus the corresponding particular relief upon the Earth.

Hence, the Protoconsciousness can be imagined as current continuous awareness by a living cell about the eternal gap between the cell's ideal geometrical form and its real physical (material) framework. In the light of this idea, the living process in general can be expressed as continuous dynamic approximation of the actual physical form to its geometrical ideal.

Another postulated attribute of the Protoconsciousness is connected with the capacity of a living cell to ***preserve*** its species-specific morphology by means of the morphogenetic reaction smoothing the non-congruence between the "ideal" geometrical form and the material stuff inside it. This means that if any physical factor has an effect upon the material substrate which constitutes the species-specific form of the living cell, this upsets the balance by perturbing and thus increasing attained minimal ("background") non-congruence. The increased non-congruence causes (or may conceived as) a "tension" between the "ideal" species-specific geometry (stereometry) of the cell and the actual spatial distribution of the material stuff within this "ideal" form. This tension is "felt" by the cell which reacts ("behaves") morphologically by changing the spatial distribution of its material content directed to its better adjustment to geometrical "fetters". This results in decreasing the non-congruence and "smoothing" the disharmony between the geometry of the ideal form and its material embodiment. Therefore, the combination: ***"geometrical feeling" – "morphological reaction"*** can be considered as a ***rudimentary psychic act***[7], that must be considered as an integral part of the Protoconsciousness. Therefore, taking into account the intrinsic capacity of a living system to react on the dynamically changing non-congruence, the final comprehensive definition of the Protoconsciousness is formulated as follows:

---

[7] Such "rudimentary psychic act" has been designated as "morphological mind" (Lipkind[34]).

*The **Protoconsciousness** is the embodied immanent capacity of any living cell to **feel** and immediately **minimize** any spatial **non-congruence** between the cell's evolving species-specific "ideal" geometrical form and the real distribution of the material stuff "filling" this form.*

The advantage of this definition is that by combining two capacities (***"to feel and minimize any spatial non-congruence"***) it represents both the enigmas associated with Consciousness: its origination within the physical stuff and its further influence upon the physical stuff.

6.5. Reflections on the Final Definition of Protoconsciousness

The above-declared advantage of the final definition of Protoconsciousness may bear a certain doubt. Namely, such double-potential capacity – to feel and minimize non-congruence – may look incompatible with the 1st demand to a basic fundamental – to be elementary (further unsplittable). "To feel and to minimize" seems to relate to different mechanisms, namely, "to feel" non-congruence is immediately clear, while "to minimize" the non-congruence means to "act", i.e., to "move" physically (but on the molecular level) the material stuff to reach maximal approximation of the physical mould to its geometrical prototype. Hence, it would seem that these two aspects of the suggested concept of Protoconsciousness correspond to different fundamentals, e.g., the postulated geometrical feeling corresponds to the common capacity "to feel", while the capacity to minimize non-congruence concerns something like a reaction to feeling associated somehow with intention and realization of a volitional act.

However, by intuitive conviction, the Consciousness *per se* necessarily includes both the aspects, i.e., feeling is fraught with inevitable immediate consequences – I believe, a "pure feeling" does not exist, i.e., the feeling is immediately associated with its experience. This experience would be expressed by different states: from volition resulting in active behavior till purely internal state expressed quite differently, e.g., deep intellectual reflection, or imagination, or ecstatic delight, etc. Therefore, if on the conscious level, feeling and its ***immediate*** experience are inseparable, i.e., they are part and parcel of the same phenomenal quality, such a principle must be correct also for the level of the postulated Protoconsciousness. Then, the capacity to feel ***and*** minimize non-congruence can be prerogative qualities of ***a*** **single** Protophenomenal Fundamental, which looks like two-faced Janus.

Such situation may be illustrated by the classic ***Le Chatelier's principle*** concerning conditions of the chemical dynamic equilibrium state, which is described as follows:

> "Le Chatelier's principle states that if a system at equilibrium is subjected to a disturbance or stress that changes any of the factors determining the state of equilibrium, the system will react in such a way as ***to minimize the effect of the disturbance***" (Mahan & Myers[178], ***boldface*** is mine).

I believe this is not only an illustration but also an expression of a deep ontological analogy. The analogy, however, concerns only the "mode of (re)action": certainly there is no analogy between the dynamic state of the non-congruence between Geometrical and Physical "experienced" by a living cell and the state of the dynamic chemical equilibrium in solutions. The analogy becomes evident namely at the stage of reaction, which is realized by the minimization of the effect of any disturbance caused either by any factors changing the chemical equilibrium (Le Chatelier's principle), or by any factors upsetting the balance of the non-congruence (the concept of Protoconsciousness under consideration). However, the reaction of the chemical equilibrium system expressed "in such a way as to minimize the effect of the disturbance" proceeds as if automatically, according to the natural law.

> "The condition of a system at equilibrium represents a compromise between two opposing tendencies: the ***drive*** for molecules to assume the state of lowest energy and the ***urge*** toward molecular chaos or maximum entropy" (Mahan & Myers[178], ***boldface*** is mine).

Therefore, although the suggested definition of Protoconsciousness based on the notions of irreducible Extra Ingredient and Protophenomenal Fundamental is extravagant enough, the above analogy between the 'reactive' part of the Protophenomenal Fundamental (minimization of the "non-congruence") and Le Chatelier principle associated with the general laws of thermodynamics maintains scientific validity of the suggested definition. The anthropomorphic meaning (with a subtle psychological nuance) of the words "***drive***" and, especially, "***urge***" in relation to molecules motions in ideal solutions (given in a dry chemistry textbook!) even more emphasizes analogy between both the formulations.

Thus, the postulated Geometrical Feeling fulfills all the demands required to a basic fundamental, namely:

1. It is ***elementary***, i.e., ***unsplittable (indecomposable)*** into any components.
2. It is ***axiomatic***, i.e., further ***unquestionable*** by its essence.
3. It is strictly and unequivocally ***defined***. Consequently, Geometrical Feeling is ***neither tautological***, ***nor metaphoric***: it is not synonymous either to Consciousness itself, or to any its hodgepodge manifestations.
4. It is ***measurable***: Geometrical Feeling as feeling of the non-congruence between the "ideal"geometrical form and the material substrate filling the respective physical framework permits quantitative estimation, i.e., there may be different quantitative degrees of the non-congruence.
5. It is ***homogeneous***: Geometrical Feeling is the ***same*** axiomatic basis for any hodgepodge expressions of Consciousness, and is postulated to lie in the grounds of three proposed forms of Consciousness: ***Protoconsciousness, Primordial Consciousness***, and ***Consciousness per se***.

Thus, the task to formulate the Protophenomenal Fundamental by obeying all the above demands to basic fundamentals while ***still preserving phenomenal quality*** seems to be fulfilled. Accordingly, the postulated Protophenomenal Fundamental is axiomatically uniform in relation to ***any*** Consciousness attribute, so that the whole ***plurality*** of the Consciousness expressions could be reduced to a ***singular source***.

The important addition of theoretical significance should be taken into account. According to the suggested theory, the form (morpha) of any living entity is characterized not only by its external shape (three-dimensional contours) but also by the ***internal*** geometrical microstructure, i.e., internal spaciousness of the living entity that could be described in mathematical language. The latter gives a full scope of the existing mathematical models of space from the "simple" Euclidean geometry of Newtonian space till Riemannian geometry of Einsteinian "space warp" and the modern theoretical summits based on hyperdimensional spaces[179,180] including "reflection space" and "catastrophe structures"[181, 182] as well as all the inferences of the "superstring revolution"[183] and the Ultimate Theory of Everything[94, 95, 151]. The assertion that the internal spaciousness of the living entity is in principle different from the external physical space would express the very essence of the vitalistic view about irreducible uniqueness of Life.

In this respect, the above "internal geometrical borders", i.e., the spatial lines of the unique spaciousness of a living entity, can be considered as ***lines of force*** that may be accounted as manifestation of a ***field***-like factor. However, such hypothetical field must not be energetic like all the established physical fields, but ***harnessing*** the local energy produced as a result of living entity's metabolic activity. Such property of harnessing is compatible only with an autonomous field irreducible to the physical fundamentals, and the Gurwitsch's field is just the case.

6.6. Transition from the Defined Protoconsciousness to the Consciousness *per se*

Although there is still a gap between the postulated Geometrical Feeling as the expression of Protoconsciousness and the full phenomenological bouquet of the Consciousness *per se*, both the defined Protoconsciousness and the Consciousness *per se* to be defined belong to the same ontological level (with no "break of continuity"), being described by the same ***psychic*** vocabulary. Namely, the both have at least one common attribute – ***feeling***: either the postulated "Geometrical Feeling" as a basic fundamental for the psychic phenomenology (Protoconsciousness), or the "ordinary" feeling as a basic constituent of the full spectrum of manifestations of the Consciousness *per se*. However, it is not at all self-evident that the abstract morphic principle together with the postulated Geometrical Feeling would be sufficient for the description of the Consciousness *per se*. Accordingly, the subsequent step is to develop a definition of the Consciousness *per se* based on the suggested concept of Geometrical Feeling.

The main task in this way is to elucidate the central syntactic point of the Protoconsciousness definition: the predicate "***form***" with its main attribute "***species-specific***" ("species-specific 'ideal' geometrical form"). In this respect, the postulated "Geometrical Feeling" concerns each geometrical dot within the living stereometric morphological whole. Accordingly, the main effort is to formulate the naturalistic factor determining the species-specific morphology. In the context of the present consideration, such factor is that Extra ingredient based on the vitalistic principle whose necessity was proclaimed above. Consequently, the morphic principle that could be considered as an epistemological tool must be expressed in ontological meaning. Such fundamental can be provided by the theory of the biological field by Alexander Gurwitsch.

The first step on this way is to reformulate the final definition of Protoconsciousness concretizing the expression "the cell's evolving species-specific ideal geometrical form" by uttering it in the language of the Gurwitsch's theory of biological field, namely:

*The **Protoconsciousness** is the embodied immanent capacity of any living cell to **feel** and immediately **minimize** any spatial **non-congruence** between the species-specific **configuration** of the cell's **Gurwitschian field** and the actual distribution of the material stuff adjusting to ("squeezing into") this configuration.*

In such a form, the definition of the universal Protoconsciousness related to any living cell is a basis for further formulation of the human Consciousness *per se*. However, between those realms, there is ontogenetic development from the human fertilized zygote (considered as a possessor of the Protoconsciousness) to the human normal adult as a possessor of the Consciousness *per se*. Therefore, this can be done via definition of the Primordial Consciousness related to an early non-differentiated ("pre-cerebral") human embryo:

*The **Primordial Consciousness** is the capacity of an early non-differentiated human embryo to **feel** any non-congruence between the geometrical configuration of the evolving **integral Gurwitschian field**, on one hand, and the real spatial distribution of the **material** stuff within the field-influenced space on the other hand, and to **react** morphologically to smooth the non-congruence.*

This definition determines rudimentary psychic phenomenology during "pre-cerebral" stages of the human embryo development. Gurwitsch uses the expression of the embryo's ***knowledge*** about currently changing momentary configuration of the embryo's integral field[3, 4]. The original Gurwitsch's expression is that "embryo knows the state and changes of its own integral field and acts in order to smooth the tension appeared as a result of the changes"[4]. Consequently, the early embryogenesis can be imagined as the chain of the embryo's actions, each one being associated with the ***act of choice*** by the embryo between different possibilities. Such embryonic "Free Will" becomes evident in the case of the experimental interference that has been clearly demonstrated by the classic experiments by H. Driesch ("harmonic regulations"[39]) having led to the notion of equifinality, i.e. the development of the same final species-specific form realized by different ways[40, 41]. In the glossary of the modern theory of supervenience[182-184], the equifinality can be expressed as the absence of the isomorphic identity during the morphological development: the same final species-specific morphology can be realized via quite different processes occurring on cellular and molecular levels.

The final step concerns the conversion of the Primordial Consciousness into Consciousness *per se* that is associated with the formation of brain. Because this has been described in detail elsewhere (Lipkind[35]), only the main logical thread will be described here. In view of the Gurwitsch's field theory, the formation of the brain means the combination of two remarkable features: together with the development of highly complicated cytoarchitectonics, the brain develops as a ***morphologically compact geometrical whole.*** This leads to the formation of the brain integral field. As a result, a remarkable combination is established: a high complexity of geometrical configuration of the integral field is combined with incredible sophistication of the neural cytological infrastructure. Now, the embryonic integral field is substituted by the integral field of the Brain that is exposed to the current stream of diverse signals (impulses) coming from external and internal somatic receptors via developing highly sophisticated neural network. All these impulses are conducted by neurophysiological way, i.e., the physical energy causing specific excitation of the respective receptors is transformed into the neural impulses (described by biophysical and biochemical terms) which transfer the initial signals from the receptors towards the corresponding areas of the brain. The current waves of such stream flow into the Gurwitschian integral field of the Brain, that causing current 'disharmonies' in the Brain integral field configuration. These disharmonies are due to the same postulated ***non-congruence*** – this time between the ideal geometry (stereometry) of the whole Brain integral field continuum (internal "lines of force"), on one hand, and the actual distribution of the physical stuff filling this geometric frame, on the other hand. So, the non-congruence is a result of the disturbances in the Brain material substrate caused by the impulses coming from all the receptors and reflecting physical characteristics of the initial irritants. Since the non-congruence is related to the spatial configuration of the whole Brain's integral field, while the comprehensive current stream of the coming signals includes the whole total of all the stimuli from all the body receptors, the experience of this non-congruence reflects the whole external World as-perceived in all its integral totality.

Accordingly, the definition of the Consciousness *per se*, is as follows:

*The **Consciousness per se** is the capacity of a mature human Brain to **feel** any non-congruence between the **geometrical configuration** of the Brain's integral Gurwitschian **field**, on one hand, and the current fluctuations of the spatial distribution of the **material stuff** within the field-influenced space, on the other hand.*

The important elucidation to be added to this definition relates to the above-uttered expression *"current fluctuations of the spatial distribution of the physical stuff"*. The detailed version of this part of the definition sounds as follows: *"current fluctuations of the spatial distribution of the physical stuff that are caused by the continuous stream of afferent neural impulses initiated in the sense organs (including all the totality of the external, internal, and proprioceptive receptors) and conducted physiologically into the brain".*

In the above definition, the non-congruence means the same discrepancy between the geometrical ideal form and its material realization, i.e., "filling" ("stuffing") the geometrical frame with the physical "meat". However, as opposed to the Protoconsciousness and Primordial Consciousness, in the case of the Consciousness *per se*, the source disturbing (and thus increasing) a minimal ("background") non-congruence is the current dynamic stream of the afferent neuronal firing initiated from various receptors which are excited by different physical stimuli causing specific sensations. Namely, different mechanic stimuli cause tactile, equilibrial, and proprioceptive sensations, chemical stimuli cause olfactory and gustatory sensations, acoustic stimuli cause auditory sensations, and photonic stimuli cause visual sensations. Specificity of the conducted impulses is determined by the corresponding physical characteristics of the initial stimuli. Because the current dynamic stream of the impulses causing the non-congruence proceeds from all the totality of the receptors bombarded ceaselessly by all the possible stimuli coming from the external World as well as the own body, the non-congruence as-felt unequivocally reflects the World as-perceived. Thus, the causal chain is as follows: (1) Irritation of the receptors by the physical stimuli leading to their excited state; (2) Conduction of this excitement via neuronal network into the brain; (3) Disturbance by the neuronal firing of the relative conformity ("minimal background" non-congruence) between the field-determined geometrical frame and the molecular substrate filling this frame; (4) Feeling (experience) of the dynamically changing non-congruence which ***reflects in every detail coherent and comprehensive picture of the external world*** (including the own body) as-perceived by an individual.

6.7. The Free Will in the Light of the Gurwitsch's Field Theory

Thus, after all the above considerations, the epigraphs to the article return with their full force emphasizing the main question: "Do we consciously cause what we do, or do our actions happen to us?"[2]. "Any proposal that volition exists must necessarily involve a radical addition to the presently known physical laws. If volition exists, then physics stands at a new frontier, in which these principles are yet to be discovered"(Burns[66]). So, does Free Will exist, or it is an illusion? Both the possibilities are to be considered in the light of the Gurwitsch's field theory on the basis of the suggested notion of Geometrical Feeling, taking into account that the existence of the Free Will means violation of the established physical laws..

The comparison of the above-formulated definitions of Protoconsciousness, Primordial Consciousness, and Consciousness *per se* shows that the last one concerning the Consciousness *per se* differs in the following important detail. As opposed to the Protoconsciousness and Primordial Consciousness which are defined as "capacity to ***feel and minimize*** the non-congruence...", the Consciousness *per se* is defined as the "capacity to ***feel*** the non-congruence" without mentioning its "minimizing". This difference leads to important consequences. As it was discussed earlier, the "capacity to ***feel and minimize*** the non-congruence" means a single qualitatively inseparable phenomenal action analogical to the above-discussed Le Chatelier's principle. Accordingly, "minimizing" is realized immediately with "feeling" as if automatically.

In the case of the Consciousness *per se*, such "automatic" reaction is not postulated due to the involvement of the nervous system and centralization of conscious activity within the Brain, that providing transition from the slowly reacting "morphological mentality" through the instinctive/reflective behavior to the highest manifestations of conscious phenomenology. In contrast to the Protoconsciousness and Primordial Consciousness in which case the disturbance of the material substrate is realized directly when disturbing factors are rather "mechanical", i.e., acting directly upon the molecular material substrate of living entity, in the case of the Consciousness *per se*, the disturbance of intracellular substrate of the brain neurons is mediated via the nervous system. The latter as if "equalizes" quite different stimuli (mechanical, chemical, acoustic, photonic) by reducing them to the process of neurophysiological conduction of the neural impulses. All the enormously huge amount of continuously proceeding neural impulses are brought into the vigil Brain disturbing the neuronal material substrate and causing the non-congruence between the super-complicated web of internal lines of Gurwitsch's integral field continuum (transpiercing through the whole Brain cytoarchitectonics) and the material substrate distribution within this web. The feeling of the resulting non-congruence comprehensively reflecting the external reality is so multicolored that the postulated idea of the immediate (automatic) "minimization" of the non-congruence does not seem already valid because it cannot be realized in the vigil Brain. Consequently, reaction to such multicolor feeling, instead of unrealizable minimization of the non-congruence, must be realized in a more specific way which means that a kind of the reaction has to be somehow ***selected*** amongst a plenty of possible responses. Such selection means a ***choice*** that is part and parcel of the Free Will.

The validity of such conclusion can be examined by analyzing the human ontogenetic development leading to the exchange of the Primordial Consciousness for the Consciousness *per se*. The question is at which ontogenetic stage the initial capacity "to feel and immediately minimize" the non-congruence (compatibility with the Le Chatelier principle) abandons the "minimizing" component. The evidence proves that such event is closely connected with cell differentiation of the early embryo including development of the neural tissue forming the neural cord, brain, and peripheral neural network. But this is not sufficient for abandoning the "minimization component" unless the developing neural system becomes functionally active. Therefore, the first movement of the embryo within the maternal uterus is a signal of transition from the "pre-cerebral" slowly acting "morphological mentality" to the quickly reacting neurophysiological (instinctive/reflective) "behavior" of the embryo. This is a clear-cut case when a new quality emerges ***not*** in the entrails of the preceding stage, i.e., there is ***no*** metamorphosis of the Primordial Consciousness into the Consciousness *per se*. The latter appears as a result of independent process going ***in parallel*** to the existing Protoconsciousness, i.e., the developing morphologically and functionally highly specialized neural system reaches the ready-to-use stage and starts acting (at a particular moment!) taking the reins of governing the embryo individual psychic development and, thus, substituting the preceding slow mechanism of embryo reactivity.

Thus, contrary to the Primordial Consciousness belonging to the developing embryo as a whole, the Consciousness *per se* is born within the developing brain. However, the Consciousness *per se* is still grounded on the Geometrical Feeling of the non-congruence but without realization of its immediate minimization. So, what is an intimate mechanism, the holy of holies of the sacred kitchen of the decision-making? In terms of the suggested approach, it can be described as follows:

If to consider the non-dreaming sleep as the state of a minimal (background) Brain activity, it must correspond to the respectively minimal non-congruence between the spacious highly complicated web of the integral Gurwitschian field configuration within the whole brain continuum and the spatial distribution of the material substrate within this continuum. In a vigil state, the whole brain continuum is under current bombardment by signals coming persistently from all the receptors. These stimuli uninterruptedly perturb the material substrate of the Brain continuum, so that the high level of the non-congruence is continuously maintained while the minimizing component is not realized, being just masked by such incessant torrent of disturbances. Together with this, such continuously maintained heightened non-congruence between the integral Gurwitschian field and the Brain's material substrate fluctuates irregularly depending on random diversity of disturbing impulses coming from the external Physical World. This dynamically changing mosaic of the substrate perturbations is followed by the correspondingly changing non-congruence which is ***felt*** according to the basic definition. The feeling of such continuously changing non-congruence adequately reflects not only the external Physical World but also the internal personal "world" which includes all the "***vestigia***" (Gurwitsch's expression), i.e., everything that Brain has accumulated during the individual lifetime. The vestigia include the inherited and formed during the lifetime personal psychological inclinations, trends, habits, emotional features, etc., as well as all the memorized facts and events. So, since the immediate minimization of the non-congruence is not realized, there is a suspended state of fluctuating heightened non-congruence as if "waiting" for any reaction "easing" the "tension". Such suspended state is imagined to consist of the emotionally colored processes of reasoning, considering, analyzing, accounting, interpreting, doubting, recollecting, thinking, hesitating, choosing, etc., etc. – all that resulting in a balanced final decision that is the full expression of the Free Will. But is it justified to make such a jump from an abstract "feeling of non-congruence" to the above full spectrum of actual conscious manifestations? The logical proof of such inference is as follows.

As it was mentioned above, according to the view that the Free Will is an illusion, any human act is simply ***accompanied*** by internal personal conviction and general emotional feeling that this very act is a result of realization of a previous intention, proceeding under individual's full control, and, thus, apprehended by the individual as volitional. The main argument is that such "accompanied" processes anatomically and physiologically are ***distinct*** from the Mind-created ***real main*** act which is a purely physiological process. In the full contrast to that, in the present approach, the "emotional feeling" of the Free Will is being born in the ***same source*** that is the whole Brain's integral Gurwitschian field where the afferent impulses are coming to and where the efferent signals are delivered from. The immediate inference is that the Free Will is not localized in a particular Brain area but associated with the integral Brain cortex. Detailed consideration of the "localization problem" is beyond the format of the present article. Therefore, the Free Will is not an "accompanied" epiphenomenon but an actual factor (actor!) causing the final ***decision-making***-based effect. This leads to the immediate further inference about reality of mental causation, as opposed to the utterances asserting the opposite.[8]

The last point concerns the definition of the Free Will. Indeed, if the Protoconsciousness, Primordial Consciousness, and Consciousness *per se* have been defined, the same attempt should be made in the case of the Free Will although its meaning is intuitively quite clear. Fortunately, there is a perfect abstract definition of the Free Will used in the philosophy of mind that is as follows:

---

[8] "Mental causation…is epiphenomenal causation, that is, a causal relation that is reducible to, or explainable by, the causal processes taking place at a more basic physical level" (Kim [184]).

> "Free will is an influence on physical events that corresponds with mental intention and causes a physical change, which would ***not*** otherwise occur in ***identical physical*** circumstances" (cited from Burns[66], ***bold face*** is mine).

However, this definition is valid if the Free Will is accepted as existing phenomenon but not as illusion. Therefore, in the light of the above conclusions that the Free Will and the real mental act originate from the ***same*** source, the Free Will is not an accompanied epiphenomenon anatomically and physiologically ***distinct*** from the ***real main*** act, but a causing factor.

7. DISCUSSION

The suggested theory based on the concept of Geometrical Feeling as a capacity to feel and minimize non-congruence between the Gurwitsch's field-determined geometrical form of living entity and its material fulfillment relates to several aspects of the Free Will problem.

First, the whole consideration has resulted in conclusion that the Free Will is not an illusion but a really existing phenomenon contradicting physical determinism and violating the physical laws[66, 67] that was clearly proved and calculated[74]. It should be taken into account, that the violation concerns the established laws of contemporary physics, namely, the laws of conservation (the 1st and 2nd laws of thermodynamics, and the principle of momentum conservation), while according to Hempel's dilemma[70], the notion of physicalism itself is considered as not properly defined[43, 71], or even false[70]. In this respect, the Gurwitschian field irreducible to the established physical fundamentals can be considered as that novel law which is additional to but not contradicting the physical laws of contemporary physics.

The suggested theory is a means of overcoming the epistemic Explanatory Gap[69] represented now as ontological Psycho-Physical Gap expressing the evident irreducibility of the psychic phenomenology to the established physical laws. The definition of Protoconsciousness of a living entity is the basic formulation from which the Primordial Consciousness of human embryo and the Consciousness *per se* have been defined. In the suggested theory, the Psycho-Physical Gap, which usually is connected with ***emergence*** of Consciousness (at an indefinite non-determinable stage of the ontogenetic development) ***from*** the physical stuff and the action of the emerged Consciousness ***upon*** the physical stuff, is moved onto the ***initial basic level***. This means that the Protoconsciousness is not an emergent property but exists within the ***zygote,*** the fertilized egg (i.e., at the very start of the ontogenesis) as the Protophenomenal Fundamental additional to the known physical fundamentals. However, the alternative panpsychistic doctrine, according to which the mental quality divides down to lower components up to elemental units (atoms, electrons, quarks, etc.), meets the combination problem[185], that leading again to the emergence problem. The suggested notion of Geometrical Feeling belonging to the Life realm liberates atoms and electrons from the burden to possess elementary mentality, i.e. the physical elementary particles are totally devoid of any conscious quality. However, the material substrate in general is inalienable component of the conscious experience because the Geometrical Feeling as a basis of the Protoconsciousness is induced by the non-congruence ***between*** Gurwitsch's field-determined geometrical form of the living entity and its material substrate. The consecutive inference of Protoconsciousness, Primordial Consciousness, and Consciousness *per se* from the ***same*** basic postulate of Geometrical Feeling means overcoming of the Psycho-Physical Gap.

An intriguing point for discussion arises from comparison of the suggested theory with the problem of connection between awareness and computation pioneered by J. Lucas[186, 187] and thoroughly analyzed by R. Penrose[126-128, 130]. Their basic point concerns argumentation against computational modeling of mind, for all that the Physical World is computational by its nature. Penrose's consideration consists in classification of possibilities illustrated by the well-known classic arguments, like Gödel theorem[172], Turing test[188], Searle's "Chinese room" argument[189], and Chalmers' "dancing" and "fading" qualia argument[47]. The relationship between conscious quality and computation was formulated by Penrose in the form of four possibilities[127, 130], namely: (a) All the conscious qualities, awareness in particular, are induced merely by the carrying out of appropriate computations; (b) Awareness is induced by the brain's physical action while, although any physical action can be simulated computationally, computational simulation by itself cannot induce awareness; (c) Appropriate physical action of the brain, while inducing awareness, cannot be simulated computationally; (d) Awareness cannot be explained in either physical, or computational terms. Avoiding complicating discussion about all the above arguments, it is quite clear that the theory suggested in this article corresponds to the 4th possibility, which, to my mind, is fully compatible with the vitalistic principle. Anyhow, Penrose claims that there is a fundamental conflict between mathematical ***understanding*** and purely computational process that is revealed by Turing test together with Gödel's theorem and can be demonstrated by certain forms of mathematical induction, e.g., transfinite induction based not on the natural integers but on Cantorian transfinite ordinal numbers[128]. The conclusion is quite unequivocal: "It seems to me that whatever understanding is, it is something that requires *awareness* – whatever awareness is!"(Penrose[130]). In accordance with this line, comes a rather trivial lament: "There is the mystery of the very existence of mentality and of how physical structures of the appropriate kind seem to be able to evoke it"[130]. The final Penrose's construction consists of three intimately connected Worlds: the Physical World, the Mental World, and Platonic World of absolute mathematical truth[125]. Biological aspect considered by Penrose is connected to the natural evolution of the quality of understanding which can be the result of natural selection only in condition that "the physical laws contain a non-computational ingredient"[130].

In this respect, it is possible to say that the theory suggested here comes out of the frame of the existing paradigm, which, in accordance with the "undecidable Gödel's sentences", cannot fulfill all the three Gödelian conditions, i.e., to be (1) finitely describable, (2) consistent (free of contradictions), and (3) complete (be proven to be true or false). The dominating in contemporary science paradigm based on the parsimonious Ockham's Razor principle fulfills the 1st and the 2nd conditions (finite and consistent) and, thus, remains incomplete dealing ***only*** with constructive proofs[190]. Then whenever any unavoidable incompleteness is met, the studied problem is usually left because of the absence of constructive proofs, and the efforts are transferred to another task of the same logical sort (consistency and finiteness without completeness). The golden mean is between the finite but incomplete combination ("easy" problems) and complete but infinite combination ("hard" problems), that meaning an attempt to go beyond the Procrustean bed of dominating scientific paradigm (finite and consistent combination) by creating a larger frame. The suggested theory of the Free Will based on the concept of Geometrical Feeling and Gurwitsch's theory of biological field is an attempt in this direction. The Geometrical Feeling as Protophenomenal Fundamental is that addition to the physical fundamentals which explains and ***balances*** the evident violation of the physical laws caused by the very existence of the Free Will[66, 74]. The Gurwitsch's concept of irreducible field acting (as energy harnessing factor) according to its purely mathematical postulation (while being deeply rooted in the biological reality) is that "completing" addition to the established physical laws that tends to meet the 3rd Gödelian condition (completeness). Then the golden mean between the Gödelian combinations of ***finite but incomplete*** and ***complete but infinite*** is hoped to be achieved.

REFERENCES

1. H. Walter, Neurophilosophy of Free Will: From Libertarian Illusions to a Concept of Natural Autonomy (The MIT Press, Cambridge, MA – London, 2001).
2. D. M. Wegner, *The Illusion of Conscious Will* (The MIT Press, Cambridge, MA – London, 2002).
3. A. G. Gurwitsch, *The Theory of the Biological Field* (Sovetskaya Nauka, Moscow, 1944) [In Russian].
4. A. G. Gurwitsch, *Principles of Analytical Biology and the Theory of Cellular Fields* (Unpublished manuscript, 1954) [In Russian].
5. A. G. Gurwitsch, *Principles of Analytical Biology and the Theory of Cellular Fields* (Nauka, Moscow, 1991) [In Russian].
6. A. G. Gurwitsch, Die Vererbung als Verwirklichungsvorgang, *Biolog. Zentralbl.* **32**, 458-486 (1912).
7. L. von Bertalanffy, *Modern Theories of Development: An Introduction to Theoretical Biology* (Oxford University Press, Oxford, 1933).
8. W. McDougall, *The Riddle of Life: A Survey of Theories* (Methuen, London, 1938).
9. P. Weiss, *Principles of Development* (Holt, Rinehart, and Winston, New York,1939).
10. C. H. Waddington, Fields and gradients, in: *Major Problems in Developmental Biology*, edited by C. H. Waddington (Academic Press, New York – London, 1966), pp. 105-124.
11. D. J. Haraway, *Cristals, Fabrics, and Fields: Metaphors of Organicism in Twentieth-Century Developmental Biology* (Yale University Press, New Haven, 1976).
12. R. Sheldrake, Morphogenic fields: Nature's habits, in: *Dialogues with Scientists and Sages: The Search for Unity*, edited by R. Weber (Penguin Group, London 1986), pp. 71-88.
13. B. C. Goodwin, Is biology an historical science? In: *Science and Beyond,* edited by S. Rose and L. Appignanes Blackwell, London, 1986), pp. 47-60.
14. G. R. Welch, An analogical 'field' construct in cellular biophysics: history and present status. *Progr. Biophys. & Molec. Biol.* **57**, 71-128 (1992).
15. E. Laszlo, *The Creative Cosmos: A Unified Science of Matter, Life, and Mind* (Floris Books, Edinburgh, 1993).
16. S. F Gilbert, J. M. Opitz, and R. A Raff,. Resynthesizing evolutionary and developmental biology, *Develop. Biol.* **173**, 357-372 (1996).
17. A. G. Gurwitsch, Der Vererbungsmechanismus der Form, *W. Roux' Arch. Entwicklungsmech.* **39**, 516-577 (1914).
18. A. G. Gurwitsch, Über den Begriff des embryonalen Feldes, *W. Roux' Arch. Entwicklungsmech.* **51**, 383-415 (1922).
19. A. G. Gurwitsch, Weiterbildung und Verallgemeinerung des Feldbegrifes, *W. Roux' Arch. Entwicklungsmech.* **112**, 433-454 (1927).
20. A. G. Gurwitsch, *Die Histologischen Grundlagen der Biologie*, (G. Fischer Verlag, Jena, 1930).
21. A. G. Gurwitsch, Une theorie du champ biologique cellulaire, *Bibliotheca Biotheoretica (Leiden), ser. D* **11**, 1-149 (1947).
22. W. Roux, *Gesammelte Abhandlungen über Entwicklungsmechanik der Organismen,* vols. 1 and 2 (Leipzig, 1895).
23. H. Spemann and H. Mangold, Über induction von Embryonalanlage durch Implantation artfremden Organisatoren, *W. Roux' Arch. Entwicklungsmech.* **100**, 599-638 (1924).

24. R. Sattler, *Biophilosophy: Analytic and Holistic Perspectives,* (Springer-Verlag, Berlin – Heidelberg – New York–Tokyo, 1986).
25. L. V. Beloussov, The sources, development, and perspectives of the theory of biological field, in: *Physical and Chemical Grounds of Living Phenomena: Historical Essays* (USSR Academy of Sciences Publishing House, Moscow), pp. 59-117.
26. M. Lipkind, Gurwitschs Theorie vom biologischen Feld. Teil I, *Fusion* **8**(4), 30-49 (1987a).
27. M. Lipkind, Gurwitschs Theorie vom biologischen Feld. Teil II, *Fusion* **8**(5-6), 53-65 (1987b).
28. M. Lipkind, Can the vitalistic entelechia principle be a working instrument? (The theory of the biological field of Alexander Gurwitsch), in: *Recent Advances in Biophoton Research and Its Applications,* edited by F.-A. Popp, K. H. Li, and Q. Gu (World Scientific, Singapore – New Jersey – London – Hong Kong, 1992), pp.469-494.
29. L. V. Beloussov, Alexander Gavrilovitch Gurwitsch, *Rivista di Biologia* **86**(1), 119-126 (1994).
30. L. V. Beloussov, Life of Alexander G. Gurwitsch and his relevant contribution to the theory of morphogenetic fields, *Intern. J. Develop. Biol.* **47**, 771-779 (1997).
31. M. Lipkind, Application of the theory of biological field by A. Gurwitsch to the problem of consciousness, in: *Current Development of Biophysics*, edited by Changlin Zhang, F.-A. Popp, and M. Bischof (Hangzhou University Press, Hangzhou,1996).
32. M. Lipkind, Alexander Gurwitsch and the concept of the biological field. Part 1, *21st Century Science & Technology* **11**(2), 36-51 (1998a).
33. M. Lipkind, Alexander Gurwitsch and the concept of the biological field. Part 2, *21st Century Science & Technology* **11**(3), 34-53 (1998b).
34. M. Lipkind, The concepts of coherence and 'binding problem' as applied to life and consciousness realms: Critical consideration with a positive alternative, in: *Biophotons*, edited by J.J. Chang, J. Fisch, and F.-A. Popp (Kluwer Academic Publishers, Dordrecht, 1998c), pp. 359-373.
35. M. Lipkind, Definition of consciousness: Impossible and unnecessary? In: *Integrative Biophysics,* edited by F.-A. Popp and L. Beloussov (Kluwer Academic Publishers, Dordrecht, 2003), pp. 467-503.
36. M. Lipkind, The concept of field applied for explanation of consciousness: Attempts for naturalistic expression of extra ingredient, in: Abstracts of the Conference *Consciousness and its Place in Nature*, Skövde, Sweden, August 7-11, 2001, Abstract 141.
37. M. Lipkind, The concept of irreducible field applied for consciousness understanding, in: Abstracts to the conference *Quantum Mind 2003,* Tucson, AZ, March 15-19, 2003, Abstract C8, pp. 42-43.
38. M. Lipkind, The concept of field in consciousness theorizing: Is it a tool for solving the 'Hard Problem'? In: Consciousness Research Abstracts of the 6th Conference *Toward a Science of Consciousness 2004*, Tucson, AZ, April 7-11, 2004, Abstract 187 (C.13), pp. 77.
39. H. Driesch, Entwicklungsmechanische Studien. 1. Der Werth der beiden ersten Furchungszellen in der Echinoder menentwicklung Experimentelle Erzeugung von Theil- und Doppelbildungen. *Zeitschr. Zool.* **53**, 160-178.(1891).
40. H. Driesch, *Science and Philosophy of the Organism* (Adam & Charles Black, London, 1908).
41. H. Driesch, *Vitalism: Its History and System*, Russian Edition, the authorized translation from German into Russian by A. G. Gurwitsch (1915) [In Russion].
42. A. G. Gurwitsch, On practical vitalism, *American Naturalist* **49**(1915).
43. Barbara Montero, Varieties of causal closure, in: *Physicalism and Mental Causation: The Metaphysics of Mind and Action*, edited by S. Walter and H.-D. Heckmann (Imprint Academic, Exeter, UK, 2003).
44. W. J. Freeman, *How Brains Make Up Their Minds* (Weidenfeld and Nicolson, London, 1999).
45. Susan Pockett, Difficulties with the electromagnetic field theory of consciousness, *J. Cons. Stud.* **9**(4), 51-56 (2002).
46. H. Putnam, *Renewing philosophy* (Harward University Press, Cambridge, MA, 1992).
47. D. J. Chalmers, *The Conscious Mind: In Search of a Fundamental Theory* (Oxford University Press, New York, 1996).
48. J. Kim, *Mind in a physical world: An essay on the mind-body problem and mental causation* (Bradford book, MIT Press, Cambridge, MA, 2000).
49. D. Hume, *A treatise of human nature*, 1739, edited by L. A. Selby-Bigge (Oxford University Press, London, 1888).
50. G. Banks, P. Shot, A. J. Martinez, R. Latchaw, G. Rattcliff, and F. Boller, The alien hand syndrome: Clinical and postmortem findings, *Arch. Neurol.* **46**, 456-459 (1989).
51. P. G. Gasquone, Alien hand sign, *J. Clin. Exp. Neuropsychol.* **15**,653-667 (1993).
52. R. Leiguarda, S. Starkstein, M. Nogues, and M. Berthier, Paroxysmal alien hand syndrome, *J. Neurol. Neurosurg. Psychiatr.* **56**, 788-792 (1993).
53. L. A. Jones, What do they reveal about proprioception? *Psychol. Bull.* **103**, 72-86 (1988).
54. W. Penfield and K. Welch, The supplementary motor area of the cerebral cortex, *Arch. Neurol. Psychiatry* **66**, 289- 317 (1951).
55. W. Penfield, *The Mystery of Mind* (Princeton University Press, Princeton, NJ, 1975).
56. J. M. Delgado, *Physical Control of the Mind: Toward a Psychocivilized Society* (Harper and Row, New York, 1969).
57. J. P. Brasil-Neto, A. Pascual-Leone, J. Valls-Solé, L. G. Cohen, and M. Hallett, Focal transcralialc magnetic stimulation and response bias in a forced Choice task, *J. Neurol. Neurosurg. Psychiatr.* **56**, 964-966 (1992).
58. H. H. Kornhuber and L. Deecke, Hirnpotentialandeerungen bei Wilkurbevegungen und passiv Bewegungendes Menschen: bereitshaftspotential und reafferente Potentiale, *Pflügers Arch. Ges. Psychol.* **284**, 1-17 (1965).

59. B. Libet, C. Gleason, E. W. Write and D. K. Pearl, Time of conscious intention to act in relation to onset of cerebral activity (readiness potentials): The unconscious initiation of a freely voluntary act, *Brain* **106**, 623-642 (1983).
60. B. Libet, Unconscious cerebral initiative and the role of conscious will in voluntary action, *Behavior & Brain Sci.* **8**, 529-566 (1985).
61. B. Libet, *Neurophysiology of Consciousness* (Birkhäuser, Boston,1993).
62. B. Libet, The neural time-factor in perception, volition, and free will, *Rev. Métaphys. Morale* **97**, 255-272 (1992).
63. J. A. Obeso, J. C. Rothwell, and C. D. Marsden, Simple tics in Gilles de la Tourette syndrome are not prefaced by a normal premovement EEG potential, *J. Neurol. Neurosurg. Psychiatry* **44**, 735-738 (1981).
64. R. Tagore, *The Religion of Man* (George Allen & Unwin, London, 1931).
65. D. Home and A. Robinson, Einstein and Tagore: Man, nature and mysticism, *J. Cons. Stud.* **2**, 167-179 (1995).
66. J. Burns, Volition and physical laws, *J. Cons. Stud.* **6**(10), 27-47 (1999).
67. B. Libet, Do we have free will? *J. Cons. Stud.* **6**(8-9), 47-57 (1999).
68. B. Libet, The timing of mental events: Libet's experimental findings and their implications, Cons. Cog. **11**, 292-299 (2002).
69. J. Levine, Materialism and qualia: the explanatory gap, *Pacific Philos. Quart* **64,** 354-61 (1983).
70. C. G. Hempel, Comments on Goodman's ways of worldmaking, *Synthese* **45**, 193-194 (1980).
71. Barbara Montero, Post-Physicalism, *J. Cons. Stud.* **8**(2), 61-80 (2001).
72. U. Uus, The libertarian imperative, *J. Cons. Stud.* **6**(10), 48-64 (1999).
73. S. Hameroff, Quantum coherence in microtubules: A neural basis for an emergent consciousness? *J. Cons. Stud.* **1**(1), 91-118 (1994).
74. D. L. Wilson, Mind-brain interaction and violation of physical laws, *J. Cons. Stud.* **6**(8-9), 185-200 (1999).
75. H. S. Leff and A. F. Rex, *Maxwell's Demon: Entropy, Information, Computing* (Princeton University Press, Princeton, 1990).
76. B. Hille, *Ionic Channels of Excitable Membranes*, (Sinauer Associates, Sunderland, MA, 1992).
77. D. Hume, *Enquiries Concerning Human Understanding and Concerning the Principles of Morals.* (Oxford University Press, Oxford – New York, 1777/1975).
78. D. H. Hubel and M. S. Livingstone, Segregation of form, color, and stereopsis in primate area 18, *J. Neurosci.* **7**, 3378-3415 (1987).
79. M. S. Livingstone and D. H. Hubel, Psychophysical evidence for separate channels for the perception of form, color, movement, and depth, *J. Neurosc.* **7,** 3416-3468 (1987).
80. A. Treisman, Features and objects in visual processing, *Sci. Amer.* **254**, 114-125 (1986).
81. V. S. Ramachandran and S. M. Anstis, The perception of apparent motion, *Sci. Amer.* **254**, 101-109 (1986).
82. V. S. Ramachandran, Visual perception in people and machines, in: *AI and the Eye*, edited by A. Balke and T. Troscankopp (John Wiley and Sons, New York, 1990), pp. 21-77.
83. S. Zeki, The visual image in mind and brain, *Sci. Amer.* **254**, 68-77 (1992).
84. A. Bartels and S. Zeki, The theory of multistage integration in the visual brain, *Proc. Roy. Soc., London, Ser. B* **265**, 2327-2332 (1998).
85. S. Zeki, The disunity of consciousness. Proceedings of the 4th Conference of the Association for the Scientific Study of Consciousness, *Cons. Cog.* **9**(2), S30 (2000).
86. A. R. Damasio, The brain binds entities and events by multiregional activation from convergence zones, *Neural Comp.* **1,** 123-132 (1989).
87. Valerie Hardcastle, Psychology's binding problem and possible neurobiological solutions, *J. Cons. Stud.* **1**(1), 66-90 (1994).
88. C. Von der Malsburg and W. Schneider, A neural cocktail party processor, *Biol. Cybern.* **54**, 29-40 (1986).
89. C. Von der Malsburg, Synaptic plasticity as basis of brain organization, in: *The Neural and Molecular Bases of Learning*, edited by J.-P. Changeux and M. Konishi (John Wiley and Sons, New York, 1987), pp. 411-432.
90. D. J. Chalmers, Concluding remarks on the Conference 'Toward a Scientific Basis for Consciousness', in Conference Report by Jane Clark, *J. Cons. Stud.* **1**(1), 152-154 (1994).
91. D. J. Chalmers, Facing up to the problem of consciousness, *J. Cons. Stud.* **2**(3), 200-219 (1995).
92. T. H. Huxley, *Lessons in Elementary Physiology* Macmillan, London (1866).
93. N. Humphrey, *A History of the Mind* (Simon and Schuster, New York,1992).
94. B. Greene, *The Elegant Universe: Superstrings, Hidden Dimensions, and the Quest for the Ultimate Theory,* (W.W. Norton & Company, NewYork-London,1999).
95. B. Greene, The future of string theory, *Sci. Amer.* **289**, 48-53 (2003).
96. A. H. Guth, *The Inflationary Universe* (Addison-Wesley, Reading MA, 1997).
97. F. Crick and C. Koch, Feature article: Consciousness and neuroscience, *Cerebr. Cortex,* **8**, 97-107 (1998).
98. C. Frith, R. Perry, and E. Lumer, The neural correlates of conscious experience: an experimental framework, *Trends Cog. Sci.* **3**, 105-114 (1999).
99. *Neural Correlates of Consciousness: Empirical and Conceptual Questions,* edited by T. Metzinger (Bradford Book, MIT Press, Cambridge, MA – London, 2000).
100. K. Pribram, *Languages of the Brain: Experimental Paradoxes and Principles of Neurophysiology* (Brandon House, New York, 1971).
101. K. Pribram, *Brain and Perception: Holonomy and Structure in Figural Processing* (Lawrence Erlbaum Associates, Hillsdale, NJ – Hove – London, 1991).
102. R. W. Sperry, Cerebral regulation of motor coordination in monkeys following multiple transactions of sensorimotor cortex, *J. Neurophys.* **10**, 275-294 (1947).

103. B. I. B. Lindahl and P. Århem, Mind as a force field: Comments on a new interactionistic hypothesis, *J. Theor. Biol.* **171**, 111-122 (1994).
104. K. R. Popper, B.I. B. Lindahl, and P. Århem,. A discussion of the mind-brain problem, *Theor. Med.* **14**, 167-180 (1993).
105. Imaging Brain Structure and Function: Emerging Technologies in the Neurosciences, edited by D. S. Lester, C. C. Felder, and E. N. Lewis, *Ann. N.Y. Acad. Sci.* **820**, 1-113 (1977).
106. R. Näätänen, R. J. Ilmoniemi, and K. Alho, Magnetoencephalography in studies of human cognitive brain function, *Trends Neurosci.* **17**, 389-395 (1994).
107. M. E. Raichle, The correlates of consciousness: an analysis of cognitive skill learning, *Philos. Trans. Roy. Soc. Lond. B* **353**, 1889-1901 (1998).
108. D. L. Schacter, R. L. Buckner, and W. Koutstraal, Memory, consciousness, and neuroimaging, *Philos. Trans. Roy. Soc. Lond. B* **353**, 1861-1878 (1998).
109. M. Mölle, L. Marshall, B. Wolf, H. L. Fehm, and E. Bohm, EEG complexity and performance measures of creative thinking, *J. Psychophysiol.* **36**, 95-104 (1999).
110. Susan Pockett, Anaesthesia and the electrophysiology of auditory consciousness. *Cons. Cog.* **8**: 45-61 (1999).
111. Susan Pockett, *The Nature of Consciousness: A hypothesis* (Lincoln NE, Universe Ltd., 2000).
112. J. McFadden, *Quantum evolution* (Harper Collins, London, 2000).
113. J. McFadden, Synchronous firing and its influence on the brain's electromagnetic field: Evidence for an electromagnetic theory of consciousness. *J. Cons. Stud.* **9**(4), 23-50 (2002a).
114. J. McFadden, The conscious electromagnetic information (cemi) field theory: The hard problem made easy? *J. Cons. Stud.* **9**(8), 45-60 (2002b).
115. E. R. John, A field theory of consciousness. *Cons. Cog.* **10**, 184-213 (2001).
116. E. R. John, The neurophysics of consciousness. *Brain Res. Rev.* **39**, 1-28 (2002).
117. H. Romijn, Are virtual photons the elementary carriers of consciousness? *J. Cons. Stud.* **9**(1), 61-81 (2002).
118. L. M. Ricciardi and H. Umezava, Brain physics and many-body problems, *Kybernetik* **4**, 44-48 (1967).
119. C. I. J. Stuart, Y. Takehashi, and H. Umezava, On the stability and non-local properties of memory, *J. Theor. Biol.* **71**, 605-618 (1978).
120. C. I. J. Stuart, Y. Takehashi, and H. Umezava, Mixed system brain dynamics: neural memory as a macroscopic ordered state, *Found. Phys.* **9**, 301-327 (1979).
121. H. P. Stapp, *Mind, Matter, and Quantum Mechanics* (Springer-Verlag, Berlin – Heidelberg, 1993).
122. H. P. Stapp, Quantum theory and the role of mind in nature, *Found. Phys.* **31**, 1465-1499 (2001).
123. M. Jibu, S. Hagan, S. Hameroff, K. Pribram, and K. Yasue, Quantum optical coherence in cytoskeleton Micro-tubules: implications for brain function, *BioSyst.* **32**, 195-209 (1994).
124. M. Jibu and K. Yasue, *Quantum Brain Dynamics and Consciousness: An Introduction* (John Benjamins, Amsterdam,1995).
125. M. Jibu, K. Pribram, and K. Yasue, From conscious experience to memory storage and retrieval: The role of quantum brain dynamics and boson condensation of evanescent protons, *Intern. J. Mod. Phys. B* **10**, 1735-1754 (1996).
126. R. Penrose, *The Emperor's New Mind* (Oxford University Press, Oxford, 1989).
127. R. Penrose, *Shadows of the Mind* (Oxford University Press, 1994).
128. R. Penrose, Beyond the doubting of a shadow, *Psyche* **2**(1): 89-129 (1996a).
129. R. Penrose, On gravity's role in quantum state reduction, *Gen. Relat. Grav.* **28**(5), 581-600 (1996b).
130. R. Penrose, On understanding understanding, *Intern. Stud. Philos. Sci.* **11**(1), 7-20 (1997a).
131. R. Penrose, *The Large, the Small, and the Human Mind* (Cambridge University Press, Cambridge, UK, 1997b).
132. G. Vitiello, Dissipation and memory capacity in the quantum brain model, *Intern. J. Mod. Phys. B* **9**, 973-989 (1995).
133. G. Vitiello, *My Double Unveiled* (Benjamins, Amsterdam, 2001).
134. R. Penrose, Wave function collapse as a real gravitational effect, in: *Mathematical Physics* 2000, edited by A. Fokas, A. Grigouryan, T. Kibble, and B. Zegarlinski (Imperial College, London, 2000), 266-282.
135. D. J. Bierman, Does consciousness collapse the wave-packet? *Mind & Matter* **1**(1): 45-57 (2003).
136. I. N. Marshall, Consciousness and Bose-Einstein condensates, *New Ideas Psychol.* **7**, 73-83 (1989).
137. D. Zohar, Consciousness and Bose-Einstein condensates, in: *Toward a Science of Consciousness (The 1st Tucson Discussions and Debates)*, edited by S. Hameroff, A. Kaszniak, and A. Scott (Bradford Book, MIT Press, Cambridge, MA and London,1996), pp. 439-450.
138. S. Grossberg, The attentive brain. *Amer. Sci.* **83**, 438-449 (1995).
139. E. Pessa and G. Vitiello, Quantum noise, entanglement and chaos in the quantum field theory of mind/brain states, *Mind & Matter* **1**, 59-79 (2003).
140. E. Pessa and G. Vitiello, Quantum dissipation and neural net dynamics. *Bioelectrochem. Bioenerg.* **48**, 339-342 (1999).
141. E. Alfinita and G. Vitiello, Formation and life-time of memory domains in the dissipative quantum model of brain. *Intern. J. Mod. Phys. B* **14**, 853-868 (2000).
142. H. Primas, Time – entanglement between mind and matter, *Mind & Matter* **1**(1): 81-119 (2003).
143. F. T. Arecchi, Chaotic neuron dynamics, synchronization, and feature binding: quantum aspects, *Mind & Matter* **1**(1), 15-43 (2003).

144. S. Hameroff, Fundamental geometry: The Penrose-Hameroff Orch OR model of consciousness, in: *Geometry and the Foundations of Science: Contributions from an Oxford Conference Honoring Roger Penrose*, edited by N. Woodhouse (Oxford University Press, Oxford, 1997), pp. 103-127.
145. S. Hameroff, Funda-Mentality: is the conscious mind subtly linked to a basic level of the universe? *Trends Cog. Sci.* **2**(4), 119-124 (1998).
146. S. Hameroff, Consciousness, the brain, and space-time geometry, in: Cajal and Consciousness: Scientific Approaches to Consciousness on the Centennial of Ramón y Cajal's Textura, edited by P.C. Marijuan, *Ann. NY Acad. Sci.* **929**, 74-104 (2001).
147. R. Penrose and S. Hameroff, What gaps? Reply to Grush and Churchland, **2**(2), 99-112 (1995).
148. S. Hameroff and R. Penrose, Conscious events as orchestrated space-time selections, *J. Cons. Stud.* **3**(1), 36-53 (1996a).
149. S. Hameroff and R. Penrose, Orchestrated reduction of quantum coherence in brain microtubules: A model for consciousness, in: *Toward a Science of Consciousness (The 1st Tucson Discussions and Debates)*, edited by S. Hameroff, A. Kaszniak and A. Scott (Bradford Book, MIT Press, Cambridge, MA – London, 1996b), pp. 507-540.
150. J. D. Desmedt and C. Tomberg, Transient phase-locking of 40Hz electrical oscillations in prefrontal and parietal human cortex reflects the process of conscious somatic perception, *Neurosci. Lett.* **168**, 126-129 (1994).
151. J. D. Barrow, *Theories of Everything* (Fawcett-Columbine, New York, 1992).
152. E. J. Squires, Why are quantum theorists interested in consciousness? in: *Toward a Science of Consciousness, II,* edited by S. R. Hameroff, A.W. Kaczniak, and A.C. Scott (MIT Press, Cambridge, MA, 1998), pp. 609-618.
153. B. Libet, A testable field theory of mind-brain interaction, *J. Cons. Stud.* **1**(1), 19-126 (1994).
154. B. Libet, Conscious mind as a field, *J. Theor. Biol.* **178**: 223-224 (1996a).
155. B. Libet, Solutions to the hard problem of consciousness, *J. Cons. Stud.* **3**(1), 33-35 (1996b).
156. B. Libet, Neural processes in the production of of conscious experience, in: *The Science of Consciousness*, edited by M. Velmans (Routledge, London, 1996c), 96-117.
157. B. Libet, Consciousness, free action and the brain: Commentary on John Searle's article (with a short reply from John R. Searle), *J. Cons. Stud.* **8**(8), 59-65 (2001).
158. B. Libet, The timing of mental events: Libet's experimental findings and their implications, *Cons. Cog.* **11**(2), 291-299 (2002).
159. B. Libet, Can conscious experience affect brain activity? *J. Cons. Stud.* **10**(12), 24-28 (2003).
160. J. Searle, Consciousness and free action, in: *Toward a Science of Consciousness Tucson 2000, Consciousness Research Abstracts*, Tucson AZ, April 10-15, 2000, Abstract 123, PL9 (2000a), pp. 70.
161. J. Searle, Consciousness, *Annu. Rev. Neurosci.* **23**, 557-578 (2000b).
162. J. Searle, Consciousness, free action and the brain, *J. Cons. Stud.* **7**(10), 3-22 (2000c).
163. R. Sheldrake, *A New Science of Life: The Hypothesis of Formative Causation* (Tarcher, Los Angeles, 1981).
164. R. Sheldrake, Morphic resonance and collective memory, in: *Proceedings of the 2nd Symposium on Consciousness,* January 1992 (Athens, 1993), pp. 93-98.
165. R. Sheldrake, *Seven experiments that could change the world: A do-it-yourself guide to revolutionary science* (Fourth Estate, London, 1994a).
166. R. Sheldrake, Experiments that could change the world, *Network* **54**, 8-10 (1994b).
167. *Rupert Sheldrake in Discussion: The New Science of Life*, edited by H.-P. Dürr and F.-T. Gottwald (Scherz Verlag, Bern – München – Wien, 1997).
168. M. Kinsbourne, Integrated field theory of consciousness, in: *Consciousness in Contemporary Science,* edited by A.J. Marcel and E. Bisiash (Oxford University Press, Oxford, 1988), pp. 239-256.
169. W. Hasker, *The Emergent Self* (Cornell University Press, Ithaca, 1999).
170. M. Silberstein, Converging on emergence: Consciousness, causation, and explanation, *J. Cons. Stud.* **8**(9-10), 61-98 (2001).
171. K. Campbell, *Abstract Particulars* (Blackwell, Oxford,1990).
172. K. Gödel, Ůber formal unentscheidbare Satze der Principia Mathematica und verwandter System, *Monatshefte Mathem. Physik* **38**, 173 (1931).
173. O. Flanagan, *Consciousness Reconsidered* (MIT Press, Cambridge, MA, 1992).
174. D. Chalmers, First-person methods in the study of consciousness, *Consciousness Bulletin* (1999).
175. M. Lipkind, The morphogenetic origination of consciousness, in: Consciousness Research Abstracts of the 4th Conference *Toward a Science of Consciousness 2000*, Tucson, AZ, April 10-15, 2000, Abstract 323 (P2), pp. 135.
176. M. Lipkind, The hard problem of consciousness and the field concept: Attempt of formulation of an axiomatic protophenomenal fundamental, in: Consciousness Research Abstracts of the 5th Conference *Toward a Science of Consciousness 2002*, Tucson, AZ, April 8-12, 2002, Abstract 73 (P2), pp. 44.
177. R. Weingard, A philosopher looks at string theory, in: *Physics Meets Philosophy at the Planck Scale,* edited by C. Callender and N. Huggett (Cambridge University Press, Cambridge, 2001).
178. B. M. Mahan and R. J. Myers, *University Chemistry* (Benjamin/Cummings, 4th edn, Menlo Park, CA, 1987).
179. H. R. Pagels, *Perfect Symmetry* (Simon and Schuster, New York, 1985).
180. M. Kaku, *Hyperspace* (Oxford University Press, New York, 1994).
181. S.-P. Sirag, Consciousness: A hyperspace view. Appendix, in: *Roots of Consciousness*, 2nd edn., edited by J. Mishlove (Council Oak, Tulsa, OK,1993).
182. S.-P. Sirag, A mathematical strategy for a theory of consciousness, in: *Toward a Science of Cconsciousness*, edited by S. R. Hameroff, A.W. Kaczniak, and A.C. Scott (MIT Press, Cambridge, MA, 1996), pp. 579-588.

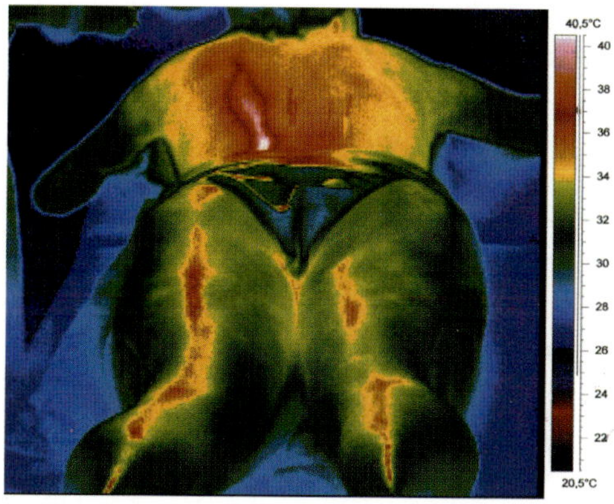

Figure 8a. Documentation of a part of the bladder meridians by moxibustion. They can be described in terms of bright solitons (see text).

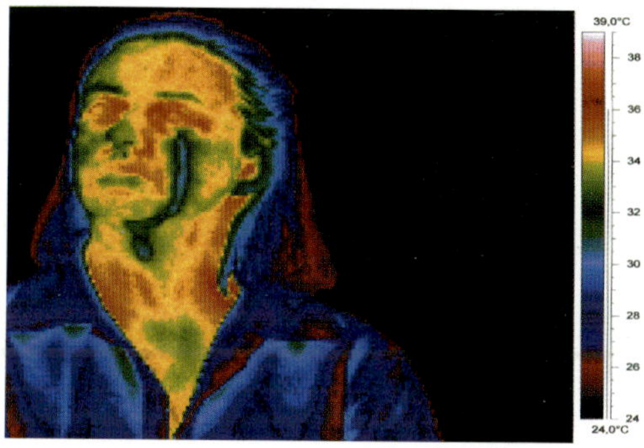

Figure 8b. Documentation of a part of the stomach meridian. There is no external excitation. This case corresponds to a black soliton where photon sucking from the surrounding tissue may take place.

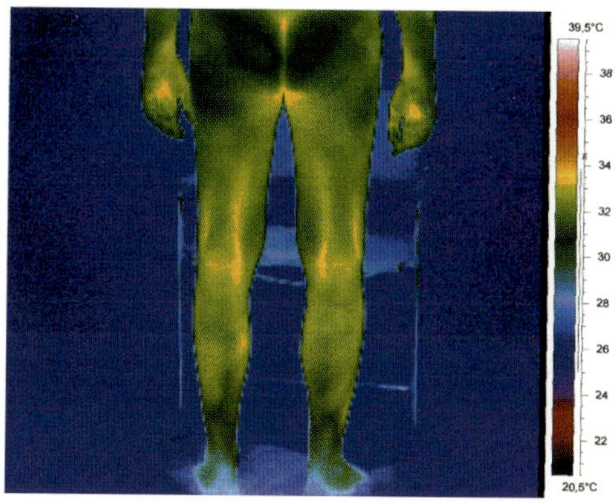

Figure 8c. Documentation of the superposition of bright and black solitons, giving rise to transparency. One can see three bones of the legs.

183. *Physics Meets Philosophy at the Planck Scale,* edited by C. Callender and N. Huggett (Cambridge University Press, Cambridge, 2001).
184. J. Kim, *Supervenience and Mind: Selected Philosophical essays* (Cambridge University Press, Cambridge, 1993).
185. W. Seager, Consciousness, information and panpsychism, *J. Cons. Stud.* **2**(3), 272-288 (1995).
186. J. R. Lukas, Minds, machines, and Gödel, *Philosophy* **36**, 120-124 (1961).
187. J. R. Lukas, *The Freedom of Will* (Oxford University Press, Oxford, 1970).
188. A. Turing, Computing machinery and intelligence, *Mind* **54**, No. 236, 433-424 (1950).
189. J. Searle, Minds, brains, and programs, *Behav. Brain Sci.* **3**, 417-424 (1980).
190. M. Beeson, *Foundations of Constructive Mathematics* (Springer-Verlag, Berlin, 1985).

20
VERNADSKY'S NOOSPHERE AND SLAVOPHILE SOBORNOST'

Some early concepts of field phenomena in social life

Marco Bischof*

1. INTRODUCTION

Vladimir I. Vernadsky (1863-1945) certainly is one of the exemplary pioneers of an interdisciplinary and holistic biophysics (Bischof, 2003). The concept of the "noosphere", one of the less conventional elements of his work, may also turn out to be valuable for the further development of this novel biophysical perspective. The noosphere concepts of Vernadsky and Pierre Teilhard de Chardin (1881-1955), although developed by both scientists from common discussions in Paris, where Teilhard and his friend Le Roy attended Vernadsky's lectures at the Sorbonne from 1920 to 1924, usually are considered to be fundamentally different (Levit, 2000, 2001). However, upon a closer look at the facts of Vernadsky's life and personality, the differences appear less fundamental, especially if we do not restrict ourselves to Vernadsky's published statements. Particularly, the notion of a field of collective consciousness, an important element of Teilhard's view of the noosphere, may also have been part of Vernadsky's concept of the noosphere. We therefore suggest that Vernadsky's published statements on the noosphere are a carefully controlled and greatly reduced expression of Vernadsky's real view of the noosphere, and that Vernadsky's ideas may also have been influenced by the Slavophile concept, of *sobornost'*. By discussing the history and relationship of these concepts I hope to further elucidate the concept of integral science and to show the possible significance of these philosophical concepts for a biophysical understanding of the field aspects of social life.

---

* Marco Bischof, International Institute of Biophysics, D-41472 Neuss, Germany, and Future Science & Medicine, D-12435 Berlin, Germany.

2. GENESIS OF THE NOOSPHERE CONCEPT

The concept of the noosphere was created in 1922 in Paris; it was suggested by the French mathematician and philosopher Edouard Le Roy (1870-1954) to characterize Vernadsky's ideas on the final phase of the evolution of the biosphere, when he attended Vernadsky's lectures on geochemistry at the Sorbonne (Paris University), together with his friend, the Jesuit, geologist, and paleontologist Pierre Teilhard de Chardin (Lapo, 1987; Grinevald, 1988; Samson & Pitt, 1999; Levit, 2000, 2002). Although Le Roy, who was the first to publish on the noosphere (1927), is cited by both Vernadsky and Teilhard for his work on the topic, he is generally underestimated as one of the creators of the concept of the noosphere (Samson & Pitt, 1999). Drawing on the philosophy of his teacher Henri Bergson, Le Roy advocated the view that the biological evolution was coming to an end and that with the appearance of man a new, spiritual phase of evolution had begun that led to the creation of a new sphere beyond the biosphere. In Paris, the two unorthodox Catholic thinkers, who were so closely collaborating in this period that it was impossible, according to Le Roy, to distinguish the contributions of each in their respective work (Le Roy, 1927), discussed the new scientific idea of the biosphere introduced by Vernadsky, and developed, together with Vernadsky, the concept of the noosphere (Grinevald, 1988). However, both Vernadsky and Teilhard only published their ideas on the noosphere much later in their lives. Teilhard wrote the works in which he described his ideas in the 1920's and 1930's, but as he was not allowed to publish them by the Catholic church, they were only printed posthumously (Teilhard de Chardin, 1959). Vernadsky did not use the term noosphere until the 1930's and only published some of his ideas toward the end of his life. Even then, in his last works he wrote, although the subject clearly occupied his thoughts in the years before his death in 1945, only a few suggestive passages on it (Vernadsky, 1945, 1988, 1997).

From the analysis of their published writings, it is usually concluded that the noosphere concepts of Vernadsky and Teilhard are very different in several essential points (Levit, 2000, 2002). Both considered life on Earth as a kind of superorganism whose development culminated in the formation of reflexive consciousness, and both were convinced that science had to play an essential role in the formation and development of the noosphere. However, the differences between their concepts began already in their views of the biosphere. While Vernadsky saw for instance an impermeable division between living an nonliving matter, for Teilhard such a clearly defined difference did not exist. Teilhard saw the noosphere as a specific, additional "thinking layer" generated by the processes of consciousness: a "new sheath" besides the biosphere which "since its formation in the Tertiary has spread over the world of plants and animals, outside and beyond the biosphere" (Teilhard, 1959). For Teilhard, the noosphere was a transitory state between nonreflexive life and what he-called the "Omega-Point", a psychic center deep in the core of our consciousness beyond space and time, toward which the noosphere would develop after its formation, in the last phase of evolution. Through the noosphere, the entire biosphere would be transformed into a new state; thought would completely permeate and transform it, the same way the prebiological spheres had been pervaded and transformed by life. According to Teilhard, consciousness has been present in matter from the beginning, although unconsciously, as the "inner side" of the material world, and has unfolded in the course of evolution. This unfoldment reaches its climax in man, and through human activity the entire planet and matter are "spiritualized". According to Teilhard, this process will also result, through global cooperation and communication, in the unification of the many individual consciousnesses to a single planetary consciousness, the "spirit of the earth". Teilhard called this process "the planetisation of man". The last phase of evolution is envisioned by Teilhard to consist in the disconnection of intelligence from its material matrix and in the development of a planetary field of pure consciousness.

Vernadsky's concept of the noosphere seems to differ fundamentally from the Teilhardian view. His point of depart was the statement that man is about to become a powerful geological force, by transforming the entire face of the planet and and nature (Lapo, 1987). He cultivates new plants and animals, takes millions of tons of raw materials from the earth and introduces them into the cycle of life. Vernadsky considers man as a part of the biosphere, in which he has a certain function, namely the creation and development of the noosphere as the culmination of the entire development of the earth; at the same time, the noosphere is the living space of man. In his publications, Vernadsky seems to see the main trend in the development of the noosphere in the optimization of human living conditions. The decisive role in this is assigned, as Vernadsky writes in his last book "Scientific Thought as a Planetary Phenomenon", written in 1937-1938, to scientific thought which he considered the main driving force for this transformation of the earth (Vernadsky, 1997). Through planned, systematic activity man would master nature, achieve a just distribution of wealth, and would finally develop a united humanity.

3. VERNADSKY RECONSIDERED

That much to Vernadsky's concept of the noosphere as it appears in his written statements and as it is represented by most of his scientific interpreters. However, there are indications that this interpretation possibly does not render Vernadsky's full intentions and that his concept of the noosphere may have been nearer to Teilhard's ideas than it may appear at the first glance. It is often assumed that only Teilhard has seen in the noosphere a kind of field that develops through the mental activity of humanity, while Vernadsky's noosphere only referred to the material realm. However, it has become obvious that Vernadsky's real personality and views do not wholly conform to the picture many Soviet interpreters have given of him. In spite of the tampering by censorship, it has been apparent since a long time that his work defies the appropriation by vulgar materialist interpretation that has consistently been attempted (Hagemeister, 1997). Although Vernadsky's outlook was characterized by a strong interest in the improvement of the human lot, in social justice and the belief in science as the principal instrument for realizing this goal, Vernadsky never has been a communist nor even a socialist in his life; he was a liberal. Also his characterization as a materialist and rationalist has to be revised. In his student days, it seems that Vernadsky even had "a leaning towards spiritualism", as some of his fellow students reported (G.Vernadsky, 1968; Grevs, 1921, quoted in Bailes, 1990). As Bailes writes, "The nature of his philosophical outlook is not altogether clear, but one fact is certain: Vernadsky, unlike most Russian radicals of the time, had great respect for the religious side of mankind's nature (even as a stimulus to science) and worked well with religious figures who shared his own ideals for a more progressive and democratic Russia. Vernadsky had spiritualist tendencies at times. There were both rationalist and mystical sides to his nature, although rationalism usually predominated" (Bailes, 1990). However, after the Revolution of 1917 such interests and tendencies had to remain more or less private. In the 1920s the Soviet Cheka (state security police) constantly observed all related activities; in August 1922, the Cheka forced many scholars, scientists, and writers to leave the country, among them N. Kondratiev, L.P. Karsavin, N.O. Lossky, N.A. Berdiaev, P.A. Sorokin, S. Bulgakov, S.L. Frank, and I.I. Lapshin. From 1926 onwards esotericists, mystics, occultists, Rosicrucians, theosophists, freemasons, etc., were subjected to very brutal reprisals as representatives of "passive counter-revolution". Later, especially in the time of Stalin, the situation was not much different. Therefore, many Russian intellectuals with such views that had not left the country had to hide and dissimulate their true opinions, not only to get their works through the censure, but to protect their lives (Bogomolov, 2000). Vernadsky's late works on natural philosophy that contained his ideas on the noosphere were generally mutilated by censorship up to 1988 (Hagemeister, 1997). He never fully expressed his ideas on the noosphere, at least in his published writing (Samson & Pitt, 1999).

4. THE PRIYUTINO BROTHERHOOD

Ultimately, Vernadsky's concept of the noosphere may have its origin in the membership of the scientist in the "Priyutino brotherhood", a spiritual community that existed almost for 60 years, from 1886 to 1941 (Bailes, 1990; Borisov, Perchenok & Roginsky, 1993). This community was inspired by the ideas of Lev Tolstoy, the Slavophile movement and the ideas of cosmism. It developed from a student circle (kruzhok) at St. Petersburg University which Vernadsky joined in 1882. It formed around the brothers Sergey and Fedor Oldenburg and consisted of between ten and twelve members, mostly sons of Tsarist generals and civil servants, and their fiancees. To the core group belonged, besides Vernadsky and his spouse, Natasha Egorovna Vernadskaya (nee Staritskaya), Sergei Fedorovich Oldenburg (1863-1934), orientalist and specialist in Buddhist studies, who was to become the founder of the *Bibliotheca Buddhica* (1897), a renowned international publication project to which many prominent scholars from all over the world would contribute, and permanent secretary of the USSR Academy of Sciences. His brother, Fedor Fedorovich Oldenburg, was a teacher. Prince Dimitry Ivanovich Shakhovskoi (1861-1939), historian and journalist, was to become a leading figure in the Zemstvo movement and the Cadet party, a member of the first People's Congress (Duma), and a minister in the provisional government of 1917. Ivan Michailovich Grevs

(1860-1941), son of a retired Tsarist officer and landowner, later became a noted literary historian and biographer of Turgenev. Alexander Alexandrovich Kornilov (1862-1925), son of the governor-general of Warsaw, became a historian and professor at Leningrad University.

This group set themselves the task of developing a common worldview and a common set of goals in life. The friendships Vernadsky formed in this circle not only lasted many decades, but were a strong support and stimulant for much of his later scientific career and work as a public figure. Although the members differed in many questions, they shared a common set of beliefs, such as a belief in modern knowledge and science and its transforming power for individuals and society, in the responsibility of the educated to spread their knowledge in order to improve the material and spiritual conditions of Russian life. These beliefs linked the group to the populist movement of the 1870's and later to the Russian liberal movement of the early 20th century. As Vernadsky explained in 1916, in the 1880s there existed, besides the socialist tendencies among the Russian students, other tendencies which in some respects were close to the latter, but in others quite different (Vernadsky, 1922). In their moral and democratic aims, the non-socialist tendencies did not differ from the socialists and agreed with them on the necessity to work for the impoverished mass of the people and in the rejection of the status quo. But they did not share with them the same attitude toward religion, art, philosophy, political life, and science. In the 1880's, when socialist groups began to form, circles also began to form of those other tendencies. The Oldenburg group was part of the democratic *zemstvo* movement which finally (after 1905) led to the Union of Liberation and the constitutional Democratic Party.

The brotherhood was dedicated to the ideal of "the self-development of the individual personality" as well as to serving people through cultural enlightenment (Vernadsky, 1922). The idea was that the individual could develop its talents only through service to others. The circle was influenced by the ideas of Lev Tolstoy and shared his interest in living a simple life and enriching their spiritual life by drawing closer to the Russian people. Their ideal was to work and produce as much as possible and use as little as possible, to treat the needs of others as if they were one's own. However, they considered that Tolstoy undervalued secular knowledge and was too much of a anarchist with regard to the state. They considered knowledge "one of the highest achievements of the human spirit", acknowledged the value of contemporary culture in general and believed in the necessity of law courts and the state. Unlike Tolstoy, they considered political reforms as essential. The members, while sometimes collaborating with the Tolstoyans, for example in providing relief during the famine of 1891, mainly were active in education, research, local self-government (the *zemstvos*), and eventually in the democratic liberalism of the Cadet party.

As students in the 1880s, the members of the Priyutino brotherhood, besides their studies, were especially active in two organizations, the Literary-Scientific Society of St.Petersburg University which existed from 1882 until it was closed down in 1887 by the government, and a study group associated with the St. Petersburg Committee on Literacy, an organization with populist affiliations whose activities were aimed at bringing adult education to the working class of the Russian capital. The Literary-Scientific Society, under the faculty sponsorship of Professor Orest Miller, was taken over by the Priyutino group in 1882 and turned into a lively center of intellectual life where a variety of ideas and viewpoints, from conservative and liberal to radical ones, could be discussed in a spirit of mutal respect and tolerance, something rare in these times. The society soon became a great success, comprising more than 300 members and scheduling dozens of lectures, discussions, readings of original works of criticism and poetry; it eventually also built a library of some five thousand volumes under the direction of Sergei Oldenburg that included many controversial books and often was more up-to-date than the university library. The society became a very important proving ground where the members of the Priyutino brotherhood (and other students as well) could display initiative in a relatively open atmosphere, could gain confidence and skills as public speakers, present their original work and discuss important ideas in European and Russian intellectual life. In his work for the Literary-Scientific Society, Vernadsky developed the organizational talents for which he (and other members of the Priyutino brotherhood) became known in later years. In the brotherhood and in the Literary-Scientific Society, Vernadsky also developed the basic ideas and convictions which should form the base of his later scientific and philosophical work. One of the models for his forming outlook was Orest Miller, the faculty sponsor nof the society (Bailes; 1990). A specialist in Russian literature, Miller was a Slavophile, a devoted Orthodox believer and an admirer of the novelist Dostoyevsky. An enthusiastic teacher and a warm man who always had an open ear and some help for his students, Miller provided a model of the good teacher for Vernadsky and his friends, in contrast to many other members of the faculty whose teaching was of a low quality and who were inaccessible to the students. As Bailes writes, Millers example probably also strengthened Vernadsky's interest in religion and his belief in cooperating with religious people who worked for progressive social change. Vernadsky retained a lifelong interest in religion, which associations with intellectuals like Orest Miller and later the mystic philosophers Vladimir Soloviev and Sergey Trubetskoi, one of his closest friends at Moscow University in the early 1890s, only strengthened.

Another influence on the Priyutino brotherhood were the experiences of the American and British communities of William Frey. Frey, a former Russian aristocrat whose original name was Vladimir K.Geins, had founded in 1871 the utopian community "Progressive Colony" in Cedar Vales in the US state of Kansas, and subsequently a number of other settlements in the United States and Great Britain. The meeting of the Oldenburg group with Frey, when he came to Russia in the winter of 1885/86 to meet Tolstoy and spent some time in St. Petersburg to speak to some private gatherings, was the last impulse for the foundation of the Priyutino brotherhood on January 7, 1886. Like Frey, the Priyutinians wanted to put into practice the idea of *sobornost'*– the unique concept of spiritual community developed by the Slavophiles of the early 19th century that has since remained the core idea of Russian religious philosophy. The Priyutino brotherhood is the only known non-ecclesiastical example of such an attempt. In its initial period (1880s-1890s), the brotherhood was mainly busy with transforming the group into a "collective personality" and to forge a common spiritual purpose. This was done by continuously sharing everything the members felt, experienced and thought, as well concerning the work as also private matters, in intense conversations and correspondences. In this way, the special nature of the forming community, as its was experienced by the individual members, was constantly reflected and manifested. Different proposals were discussed, how the *sobornost'* of the community could be secured by means of this or that lifestyle, including communal life.

Vernadsky also carried the principles of the brotherhood into the various scientific groupings he founded or to which he belonged, such as his mineralogical, geochemical and biogeochemical schools, the Radium Institute and the scientific Academies of the Ukraine and the USSR. In his "Philosophical Thoughts of a Naturalist" (1988) he writes that the transition to the noosphere presupposes "the community of all humanity, of humans as brethren". Throughout Vernadsky's work, one can clearly trace the author's interest in the "comradely, brotherly element in scientific organizations of the past and the present" which according to Vernadsky pave the way to the noosphere (Vernadsky, 1927). In Vernadsky's view, brotherhood should become the principle of relations between scholars asnd subsequently between all people on earth.

Although in the last phase of the brotherhood (1917-1941) the goal to make it into an influential national body was not achieved, the brotherhood still exerted a great influence within this scientific framework. The community of friends with its practical lived experiences formed a model and a first core for the social organization of the noosphere, as it also was one of the most important sources of Vernadsky's "scientific faith" that formed the foundation of this concept.

5. SLAVOPHILES AND *SOBORNOST'*

In order to understand the full significance of Vernadsky's membership in the Priyutino brotherhood for the noosphere concept, we now need to take a closer look at the Slavophile philosophy and its central concept of *sobornost'*. In the dispute on the specificity and historical role of the Russian culture that was conducted by Russian intellectuals in the early 19th century, Slavophiles and 'Westernizers' were the two main currents of thought (Lossky, 1951; Zenkovskii; 1953; Edie, Scanlan & Zeldin, 1965; Khomiakov & Kireevsky, 1998). While the Westernizers did not attribute any contribution to world culture to the Russian past, the Slavophiles were convinced that the Russian past contained important values, mainly Orthodox Christianity. Another characteristic of Slavophilism was the severe criticism of the social reality in the country.

One of the leading thinkers of this movement was Alexei Stepanovich Khomiakov (1804-1860) who has had and still has a strong influence on the Russian Orthodox church and religious philosophy with his concept of *sobornost'* (Khomiakov & Kireevsky, 1998). Central to Khomiakov's philosophy was the mystical experience of the "community in the Holy Spirit", for which he used the notion of *sobornost'* (from Russian *sobrat'*, collect, gather, unite). For him, love and the living experience of mystical unity took priority over the intellect. Love for him was a mode of knowing, and he considered the loving community the only practical possibility of finding truth.

He also saw the principle of *sobornost'* as the characteristic that distinguished Orthodox Christianity from Western "religious rationalism" which he criticized in its Catholic form because of the postulate of obedience to external authority and the resulting bondage of the believers, and in its protestant variant because of individualism. Khomiakov defined *sobornost'* as a spiritual community which is together voluntarily and out of love, and does neither subordinate to any external authority nor know individualistic separation. The concept also denotes the state of spiritual and psychic unity that in German is called "being one heart and one soul" and refers to the biblical "Where two or three have gathered together in my name, there I am in their midst" (Matthew 18:20). According to Khomiakov, in the state of *sobornost'* people and things are connected by a kind of energy field that communicates a mutual nonverbal exchange (communion) of being, experience and knowledge about each other. By this process, *sobornost'* makes "integral knowledge" possible.

Integral existence and integral knowledge were the central concerns of the philosophy of Ivan Kireevsky (1806-1856), the second main protagonist of Slavophile thought. He argued that humanity must strive to develop all its potential abilities in a harmonic and balanced manner, in order to achieve an adequate mutual relation and hierarchical order of forces and abilities. According to Kireevsky, a harmonical existence resulted from "the concentration, the gathering, and the wholeness of inner forces". He believed it was necessary to subordinate rationalistic understanding, which alone would lead only to blind and distorted understanding, to faith, an "understanding of the heart" that made accessible, through the perception of God, a "higher light" which only integrates the lower faculties into an integral whole. Kireevsky's notion of faith was quite different from the ordinary understanding. Faith for him was man's highest cognitive faculty, and even more than that, because it "encompasses the wholeness of human existence" and "appears only in the moments of this wholeness, of this integrity". Thus, the main task of the religious intellect is to "collect all the separate parts of the soul into one force, to find that inner center of being where the mind and the senses, the beautiful and the true, the willed and the unexpected, the just and the charitable, conscience and reason, all merge in a living unity, where the total range of the spirit becomes a living unity". In this way, the essential human personality will be "re-established in its original indivisibility". Only faith can give such an unitary view, not reason and analysis whose dominance must be overcome. True knowledge could only flow from an integral being, as reason was subordinated not to logical understanding, but to the heart. Kireevsky therefore discriminates between "reason" which for him is an integral faculty, and "intellect" which for him denotes the analytical, rationalistic understanding, the "fallen reason". Kireevsky's "living, integral knowledge" is a knowledge that overcomes the subject-object separation; it transcends mere rational knowledge and is based on the unity of the cognizing subject with the object of cognition which is reached in the act of knowing and makes it possible. Because integral reason can only be achieved in an organic, integral society, Kireevsky's notion of knowledge necessarily has a social or interpersonal dimension, which is covered by Khomiakov's concept of *sobornost'*.

The concepts of integral knowledge and of *sobornost'* later became the corner-stones in various social and philosophical movements in the late 19th and early 20th centuries in Russia (Kornblatt & Gustafsson, 1996). Around 1880, the revival of the controversy about Russia's relationship to the West led to the emergence of neo-Slavophilism. This debate mainly took place in St.Petersburg where Vernadsky studied at this time. After a first attempt in 1879-1880, the Philosophical Society was founded in 1897 by Vladimir S. Soloviev and Ernst L.Radlov, among others, and in the Moscow Psychological Society, founded in 1884, and for the first time in the country a professional, academic philosophy came into existence (Nethercott, 1995). The debate took a similar form as in the early 1800's, and the differences between "neo-Slavophils" and "neo-Westernizers" culminated around 1910 in a veritable polemic. The main objective of the new generation of Slavophiles – besides Soloviev, Nikolai Berdiaev, Sergei Bulgakov, Nikolai O. Lossky, Vladimir Ern, Pavel A. Florensky, Sergei and Evgeny N. Trubetskoi, and Ernst L. Radlov – soon became to criticize the deficiencies of the Slavophile theory of knowledge and to improve on it.

6. *SOBORNOST'* IN SOLOVIEV AND LATER PHILOSOPHERS

Kireevsky's and Khomiakov's program of evaluating human knowledge and culture from the viewpoint of "integral reason" and *sobornost'* which these authors had only sketched in its barest principles, perhaps found its most fruitful elaboration in the work of Vladimir Sergeievich Soloviev (1853-1900) (Radloff, 1925; Edie, Scanlan & Zeldin, 1965; Katasonov, 2002). Writing in the last quarter of the 19th century, he developed the fragmentary insights of the early Slavophile thinkers into a veritable philosophical system. Based on a concept of integrality that was far more comprehensive than that of his predecessors, he attempted to create a "Great Synthesis", also called "universal religion" by him, that claimed to reevaluate the entire human knowledge and to redefine metaphysics on the base of a modern interpretation of secret models of antiquity. His starting point was also the critique of the abstract knowledge of Western rationality and the development of an alternative theory of knowledge based on the possibility of the immediate knowledge of reality by mystical intuition. He agreed with Kireevsky that in the "fallen" nature of contemporary man the natural, organic unity of being had disintegrated into disparate faculties such as reason, faith, the aesthetic sense, etc., which sometimes were at war with one another. He therefore critized as well "dogmatic theology" as also narrow rationalistic and empirical conceptions of science and philosophy. He called for a return to a genuine "religious worldview".

However, for him, like for his Slavophile predecessors, religion and science, faith and reason were not necessarily contradictory and mutually exclusive. He saw himself in the tradition of the church fathers of Eastern Christianity in which it always has been taken for granted that reason – particularly scientific reason – is one of the faculties that grow out of the general root of faith, and therefore could never be in conflict with this root (Katasonov, 2002). Along with other faculties of the soul – such as the aesthetic and moral senses – reason would always remain rooted in and connected with its spiritual roots. The present conflict of faith and reason was seen as evidence for the fallen nature of humanity, its descent from grace and wholeness that must be overcome by a conscious and constant striving for the unity of all aspects of creation: material, moral, rational and spiritual. Rather than opposed to knowledge, faith was vital to it, because Soloviev believed that knowledge can only reach its goal if all elements of faith are included within it. Faith is a witnessing of the connection of the whole and its parts, man and God, earth and heaven. On the other hand, faith also must be justified in the light of reason.

In order to achieve the "living, integral thinking" (here Soloviev uses Slavophile terminology), the disparate faculties which alone only can give relative and insufficient knowledge, must be united. Empirism and rationalism by themselves are incapable of knowing real being or the "integral whole", which can only be known by mystical, intuitive perception. Reason and rationalistic thought are only adequate for the control of knowledge, not for its acquisition. Full truth is only attained by the combination of empirical experience, thinking and mystical knowledge. Also in Soloviev's philosophy, love takes a pivotal place. Love is the re-establishment of the presence of God in the material world (*sobornost'*); only by love, the surrendering of egoism, wholeness can be restored and the living knowledge of reality can be attained. Love is also the means to overcome the fallenness of the present human condition.

According to Soloviev, our fallen nature is a consequence of human freedom. He explains it by the double nature of the Godhead which has a "dark ground" with a certain autonomy although it remains enfolded in the Godhead. The apostasy of this dark principle – the "World-Soul" – from divine unity is an act of freedom and constitutes the fundamental evil in Soloviev's worldview. God in his love, however, strives to re-unite this fallen ground with himself. This effort is the cause of the cosmic and historical world process which will finally culminate in the victory of divine grace over the fundamental evil and death. An important role in this process is played by humanity, whose collective consciousness or soul is identical to the world-soul. Humanity must voluntarily realize the divine principle of unity and wholeness and thereby complete the historical process.

Soloviev's ideas found a deep resonance in Russian culture. Both Tolstoy and Dostoyevsky were profoundly influenced by his work, which also played a major role in the development of Russian philosophy. With the Slavophile legacy the idea of brotherhood and of a "new living community" based on *sobornost'* was also the main inspiration for Fedor M. Dostoyevsky's (1821-1881) principal work "The Brothers Karamazov". While creating the book, Dostoevsky not only studied the ideas of the Slavophiles (mainly in the form of Vladimir Soloviev's philosophy), but also immersed himself in the ideas of Nikolai F.Fedorov (1828-1903), the founder of the cosmism movement who at that time worked on a manuscript with the title of *"The Problem of Brotherhood or Relatedness, and of the Reasons for the Unbrotherly, Dis-Related, or Unpeaceful State of the World, and the Means for the Restoration of Relatedness"* (Fedorov, 1965). Later Fedorov developed his *"Philosophy of the Common Task"* (Fedorov, 1906-1913) in which he argued that the union of "brothers" would make the overcoming of all injustice and all suffering and the salvation and transformation of the natural world, including man himself, possible. Then the universe, by mobilization of all scientific and technical knowledge and skills, could be transformed into a perfect artificial paradise. In the 1980s, the "cosmism" inspired by Fedorov's ideas became a broad and significant intellectual movement in post-communist Russia in which the idea that the universe and humanity undergo a meaningful development toward a collective planetary consciousness, a noosphere, plays a central role (Hagemeister, 1997).

Another follower of Soloviev who continued the Slavophile tradition in Russian philosophy, was Prince Sergei Nikolaevich Trubetskoi (1862-1905), an intimate friend of Vernadsky, member of the Priyutino brotherhood in the 1890s, and Vernadsky's close comrade-in-arms in the fight for academic freedom at Moscow University. The religious mystic and critical idealist philosopher founded the Moscow Philosophical Society and in 1905 was rector of Moscow University for a few months till his death. In the center of his philosophy is the idea of the *sobornost`* nature of human consciousness, derived from the ideas of Khomiakov, Kireevsky, and Soloviev. According to Berdiaev, "his work *'On the nature of consciousness'* is permeated with Slavophile spirit and develops the Slavophile idea of *sobornost'* in its epistemological aspect" (Berdiaev, 1998). Trubetskoi summarized the essence of his outlook in the famous formula "In all acts of theoretical and moral character we are joined with all others". According to him, human consciousness is neither personal nor impersonal; each individual consciousness is based on a "universal consciousness" on which it depends. There exists a "universal, all-embracing perception whose subject may be neither a limited individual being nor the Absolute Being, but merely that psycho-physical being which is as universal as time and space and furthermore does not dispose of any attributes of Absolute Being. It is the cosmic being or the world in its psychic foundation, i.e., what Plato called the *'World Soul'* " (Zenkovsky, 1953).

From the later Russian religious philosophers who have further elaborated the concepts of the Slavophiles and of Soloviev, I would like to mention three more. N.O. Lossky, N.A. Berdiaev, and S.L. Frank were at the same time important continuators of the Slavophile tradition of thinking and also played a considerable role in the diffusion of Bergson's philosophy in Russia. Bergson was also the main influence in Edouard Le Roy's and Teilhard de Chardin's worldview, and even Vernadsky has aknowledged his influence. Nikolai O. Lossky (1870-1965), the founder of Russian "intuitivism" (Nethercott, 1995), studied mathematics, physics, physiology and philosophy at St.Petersburg University from 1891-1900. His early philosophical development was influenced by the panpsychism of Alexei A. Kozlov (1831-1901), through his friend Sergei Askoldov-Alekseiev, natural son of the Leibnizian philosopher. Lossky already started to develop a theory of knowledge centered in in the notion of intuition as normal mode of perception and cognition before he first encountered Bergson's philosophy. In his first works he tried to reconcile empirism with the mystical traditions. His principal thesis was that reality, including inner world and external reality, is knowable in an immediate manner from within; this knowledge is not limied to sensual data, but includes the supra-sensible. Because this knowledge is much vaster than mere empirical knowledge, he first called his theory "universalist, or mystic, empiricism"; however, when Lossky published his doctoral dissertation in 1906, he changed the name into "intuitivism" (Lossky, 1906). For Lossky, intuition designated the normal modes of perception which he believed possessed the characteristics of immediate contemplation of the real, and not a special faculty that was a privilege of the talented. With his theory of intuitivism he intended to revalidate the religious experience while at the same time keeping up the validity of scientific observation; intuition was the faculty on which both scientific observation of the external world and and the experience of the inner worlds was based; he did not consider the latter to be illusory and purely subjective. Thus, Lossky gave intuition a meaning quite different from that of Bergson. It was in 1908 that Lossky discovered Bergson who was pointed out to him by Askoldov. In 1909, Bergson's *"Creative Evolution"* (Bergson, 1919) was published as the first Russian translation of this author. In 1914, several of Bergson's works were published in Russian, at the same time as Freud's books and Soloviev's Collected Works. In the same year Lossky wrote the book *"The Intuitive Philosophy of Bergson",* in which he wanted to make clear the differences between Bergson's and his own philosophy.

Bergson's philosophy appealed to those Russian intellectuals who opposed positivism and the reduction of the individual and humanity to the mere material conditions; it resonated with the intense spiritual movement of the period. In the eyes of many Russian thinkers, Bergson had, with his notion of "integral experience", designating his belief in man and a novel vision of his relation to God, situated himself very near to the Slavophile and Solovievian quest for "integral knowledge". Likewise with his emphasis on intuition as the principal way to knowledge, as according to Radlov (1916) intuitivism is a general chraracteristic of Russian philosophical thought, and Bergson now provided the Slavophile tradition with a better expression for the special cognitive faculty they previously had called faith. Bergson's ideas also were discussed in Russia in connection with the interest of certain intellectuals in the Catholic modernist movement, because those who wanted to revive and reform the Orthodox faith were confronted with very similar problems. In this movement, Bergson's ideas took an important place and Bergson's friend and pupil Edouard Le Roy, who has also coined the term "noosphere", was one of its principal protagonists and spokesmen. In the discussion, Le Roy's book *"Dogme et Critique"* which was published in Russian in 1915 with an introduction by N.Berdiaev, played a central role (Le Roy, 1907). Lossky's work announced a new direction in Russian philosophical thought. His theory of knowledge inspired the most diverse philosophical and religious, as well as scientific, orientations of his contempraries. Among those influenced by his work were N.A. Berdiaev and S.L. Frank.

One of the Russian thinkers who received Bergson's ideas well was Nikolai A. Berdiaev (1874-1948) who likewise believed in intuition as the essential method of obtaining knowledge. Although he criticized the "biologism" of Bergson, Berdiaev integrated the French philosopher's concepts in his own vision of the "organic and dynamic nature" of religious experience and the creative nature of reality, and tried to combine Bergson's "*élan vital*" with the ideas of the Slavophile Khomiakov about whom he published a monograph in 1910. In his introduction to Le Roy's *"Dogma and Critique"* Berdiaev states that, in the same way that the Catholic modernist movement had been inspired by Bergson's vision of life as a creative and inexhaustible force, Khomiakov had, a hundred years ago, defended the concept of *sobornost'* whose characteristic principles also constituted an active and dynamic principle (see Le Roy, 1907). He connected the modernist perception of Bergson's philosophy to Slavophile thought and saw the convergence of the two philosophies in the idea of the dynamic participation of man in his pact with God.

Particularly relevant for our topic is the work of Semyon L. Frank (1877-1950) who attempted, from the mid-1910s to the 1940s, to develop a system of social philosophy based on this tradition of Russian metaphysical philosophy and the concept of *sobornost'* (Goerdt, 1995). Frank saw in Lossky's intuitivist approach the keystone of a theory of knowledge and being. Quoting the work of Edouard Le Roy, *"Dogma and Critique"*, he agreed to the Frenchman's view that the meaning and the value of religious faith escapes any theoretical determination or evaluation. According to Frank, there was, besides the discursive, logical thinking which was dissociated from psychic life, another type of knowledge taking the form of experience, of "emotional intuition". This was a knowledge that "acquires its power of persuasion exactly in the energy which completely surrounds the personality; it is a knowledge that is not the impersonal subordination of thought to the norms of logic and gnoseology, it is rather the personal revelatory creation of truth (*pravda*) from the depths of feeling" (Frank, 1910).

Concerning his social philosophy, Frank was convinced that any theory of society must be founded on the understanding of the nature of man, and that this understanding was not complete without an inclusion of man's relationship to the roots of his existence in the metaphysical ground of existence. In Frank's view, the human self had two very different dimensions. The empirical self, the experiential unity of our psychic life that forms an intimate unity with bodily experience, was rooted in the deeper, intuitively experienced inner self, in the "I Am" of pure and immediate self-awareness, where man was connected to Being. This intuitive self was not static, but rather "a creative source, a potency that constantly strives to actualization and reveals its true nature only in this striving". The relationship of empirical and intuitive self was like the relation of surface and depth. The bodily self is merely "the imperceptible opening of a large subterranean pit", of the intuitive self, that was "a cosmos by itself" which, in its last depth, "immediately borders on the core of Being itself, or rather flows into it" (Frank, 1929-1930). "In this deep dimension of our existence that is open to intuitive observation, all humans are rooted in a common soil, are connected to each other by deep subterranean passages or roots, such that one single life permeates all of them. Only through this deeper, subterranean unity of life and essence the outer encounter and interaction (between people) is possible that is assumed in the ordinary view and that remains unexplainable from its viewpoint". The fact that human bodies are spatially separate and human souls are bound to the body "cannot refute the spiritual unity in which and by which human beings live inwardly and which alone makes possible their outer communalization".

Therefore, according to Frank social reality is also characterized by the two aspects of its "mechanical-outer" and its "organic-inner" structure, *"obshchestvennost"* (sociality) and "*sobornost'*" (communality or conciliarity). From the present discrepancy between the empirical-objective, bodily self and the intuitive self results a corresponding disunion between sociality and *sobornost'* in contemporary society. While in the life of external sociality, isolated selves opposed to each other and seeking the satisfaction of their needs and desires treat each other as objects, the selves in the invisible inner life of the deep *sobornost'* dimension are always inseparably connected with the original unity of the 'we'. The state of *sobornost'* is, in the words of Frank, "the organic-indestructible unity of 'me' and 'thou'growing from the original unity of the 'we'". Here the self is the "inner-organic member of a whole, in such a way that the whole not only enfolds it from without, but permeates it from from within and constitutes its inner life". It is the "spiritual nurture through which the personality lives inside, its inner wealth, its personal dignity". Like Teilhard de Chardin, Frank is convinced that "on the highest levels of human spiritual and cultural existence, *sobornost'* may even, as a purely spiritual entity, separate from its natural-cosmic primeval ground (the "cosmic sobornost'")" (Frank, 1987). "Thus, the visible society in every given epoch lives through its invisible inner, time-transcending *sobornost'*".

Frank proposes a "philosophy of the 'we'" that is based on the awareness of the primary, original character of community, as an alternative to the "philosophy of the self" prevailing since the days of Descartes; the latter is "barring the possibility of a philosophical penetration of the nature of social life" (Frank, 1929-1930). "'Me' and 'thou' do not arise from an external encounter, but are only conceivable in the unity of a 'we' – the 'we' is not a subsequent synthesis, but the original unity. The self is unthinkable without the 'thou' relationship, because it is only constituted in it, born within it". The 'we' is also "not a sum, not a plural of different members, but a true organic unity", and it is "in principle or potentially infinite, it can encompass all being", not only the entire humanity can become a 'we', but also "all creatures can be included in it, and even the unity with God can be thought as a we".

7. FIELD ASPECTS OF SOCIAL LIFE

Let me summarize: These Russian philosophers propose that we are only at the surface isolated individuals separated from each other, and that our personality and our consciousness have a deeper dimension where we are connected to each other by a collective consciousness or common psyche that is the ground and root of our individual selves. According to Frank, this connection has to do with "the energy which completely surrounds the personality", i.e., some type of field around our bodies. At the first glance this concept seems to contradict Teilhard's and Le Roy's understanding of the noosphere – and maybe also Vernadsky's unpublished notion of it, if we assume that he privately accepted the noosphere to entail such a field of collective consciousness – where a field of collective consciousness does not exist before individual consciousness, but is generated by a joining of minds as the final result of evolution. However, these two views can easily be reconciled if we assume that this collective consciousness, while it may have always existed as a potential or unconscious connection, is only now emerging as a consciously perceived and utilized field.

Even if it appears rather strange and in contradiction to our most cherished notions of personality, the concept of a collective consciousness is not unique to Russian philosophy and has a long and well established place not only in Oriental, but also in Western thinking. As Merlan has shown, the tradition of monopsychism, i.e., the doctrine that all souls are ultimately one, and the variants of mono-noism or solmentalism – teaching that there is only one mind or consciousness, and what is called thinking in the individuals is a participation in the thinking of this universal consciousness – can be traced back to the Neoplatonist Plotinos and were also held in one or the other form by Averroes, Kant, Schopenhauer and Wittgenstein, among others (Merlan, 1963). Also, Husserl and the quantum physicist Schroedinger were inclined to accept this view. Such a notion would also fit well into the new "participative worldview" of total interconnectedness that is considered by many a necessary consequence of the ecological worldview and can also be derived from quantum mechanics (Braud, 1992; Skolimowski, 1994; Laszlo, 2003). Similarly, it would answer to a proposal by Alex Comfort. The well known author, biologist and gerontologist has suggested that the traditional Western Cartesian positional, focused, homuncular perspective that leads to viewing reality as objective, should be replaced by an non-positional, observerless perspective, as it derives from the "oceanic" field perception of mystical experience and is proposed in the models of Buddhist and Hindu ontology, and leads to a field-theoretic, oceanic, unfocussed and thingless worldview (Comfort, 1979a, b).

As we like to discuss the usefulness of the noosphere and sobornost' concepts to science, particularly to the development of an integrative biophysics, we must naturally ask if there is any scientific evidence for such a view. Here, I will only summarize the evidence already described more completely in other publications (Bischof, 1995, 1998, 2003a) which is of several different types. Firstly, the existence of different types of physical fields emitted by the human organism is now well established (Bigu, 1976). They include electromagnetic fields from the optical range (biophotons, 200-800 nm) to the radio, microwave and ELF ranges. These weak but coherent fields probably form a "field body" enveloping the solid body within a range of decimeters (Zhang, 2003), however, being highly coherent they could well reach much further and serve to connect the individual with other individuals. Recent research has brought much new evidence to light about the strongest electromagnetic field produced by the body, the field generated by the heart (McCraty, Atkinson & Rein, 1993; Tiller, McCraty & Atkinson, 1996; McCraty, Atkinson, Tomasino & Tiller, 1997). We also should not forget the well known electromagnetic fields produced by the brain, of which the electric fields measured in electroencephalography (EEG) are an expression. There is evidence that all these endogenous electromagnetic fields become more coherent when a person experiences meditative states, deep feelings of love, care, appreciation or other positive emotions, or in situations of entrainment between two persons through healing, intense attention and listening, or being in love (McCraty, Atkinson and Rein, 1993; McCraty,

Atkinson, Tomasino and Tiller, 1997; Wackermann, 2004). Coherence in the electrocardiogram (ECG) power spectrum of heart rate variability indicates sympathico-vagal balance in the autonomic nervous system (Tiller, McCraty & Atkinson, 1996). An exchange of some type of energy seems to occur between individuals when people touch or are in proximity, indicated by the registration of the ECG signal of one person in another person's EEG and elsewhere on the other person's body (McCraty, Atkinson, Tomasino & Tiller, 1997). The collective field state of *sobornost'* actually may be a phenomenon of field coherence between people analogous to the coherence of electromagnetic fields.

Finally, we may not be able to explain all the interpersonal field phenomena by electromagnetic, or other known physical fields. There are many reports on subjective experiences of a field-like interpersonal connection, even in the scientific literature, pointing to the possible existence of new types of fields of non-electromagnetic nature. Although such observations usually have been dismissed or explained by other mechanisms, there are now also a number of scientific experiments giving evidence, to some extent, for the physical reality of some of these phenomena. An important example of such observations are the studies of nonverbal behaviour that have shown a synchrony of the body motion of speakers and listeners with the speech pattern (Condon & Sander, 1974; Hatfield et al., 1994), which probably serves to establish empathic resonance. Such a synchronization between mother and child may be the basis and origin of human bonding and communication (Condon & Sander, 1974). A related phenomenon is the well established phenomenon of *"emotional contagion"* (Hatfield et al., 1994), or *"transmission of affect"* (Brennan, 2004). Psychiatrists and psychotherapists have always been familiar with the *"praecox feeling"*, the field-like aura displayed by their patients announcing impending psychosis or schizophrenic episodes, and have been well aware of the contagious nature of these states (Deane, 1961; Ihle, 1962). The phenomenon of *"transference"* between therapist and patient is equally well-known and has led a number of authors to the hypothesis of an *"interpersonal field"* (Schwartz-Salant, 1988). In mutual hypnosis two persons create a common psychic field which in the more advanced stages can turn into a shared hallucinatory or dreamlike reality (Tart, 1969). Families may, according to some psychotherapists, possess a common unconscious and shared emotional field (Taub-Bynum, 1984). All these observations suggest the possible existence of interpersonal fields which may be an essential element of human communication.

A number of experimental studies give evidence that such interpersonal fields may have some physical basis. Investigations of distant influence between persons and distant correlations between brain states of spatially separated persons indicate the possibility of long-range field connections between these persons, of distant effects of states of consciousness and of the transfer of information between two distant persons (Braud, 2003; Schmidt, 2002; Wackermann, 2004). Experiments on the influence of fields of unknown nature generated by group events "with a high degree of subjective resonance between participants" by the Princeton Engineering Anomalies (PEAR) Lab of Princeton University, NJ, USA, The Global Consciousness Project of Roger Nelson, and others show a significant degree of influence on the random output of portable random event generators (REG) and indicate the presence of a field within such groups (Nelson et al., 1996, 1998). These experiments suggest that the interpersonal fields on which the observed effects are based may be fields of physical, but non-electromagnetic nature.

As physical particles and fields can be considered to emerge from the quantum vacuum, the concepts of the vacuum or the zero-point field are used by some to denote a kind of universal background field or matrix in which all material objects and fields are rooted and which connects them. Because of its consciousness-like properties, the vacuum is also identified by some as the universal noetic field underlying collective consciousness (as a popular example of this view see McTaggart, 2001). However, some of those who do not see the necessity to unduly physicalize consciousness, do postulate the existence of a universal field of consciousness as the foundation of the physical universe (Hagelin, 1987; Goswami, 1993).

These are only examples; we can conclude that there is indeed some evidence for the view that human individuals are not isolated, completely self-contained existences, but are in many ways and on multiple levels connected with each other and with the environment. This obviously is the case on the material level, as ecological science has shown in the past decades. Recent research has produced some evidence that it is also the case on the level of electromagnetic fields. But here we are more interested in the material demonstrating human interconnectedness on the level of psychic, emotional, and cognitive-intellectual activity. The possible existence of such an invisible dimension of human connectedness is now supported by enough evidence from a number of different disciplines, but this evidence has never been summarized in a coherent picture that would also make obvious its fundamental consequences for the social sciences and for psychology.

In the 19th and early 20th centuries, a number of field theories have been proposed in social psychology and sociology, among them the psychological field theories of the gestalt theorists (Lewin, 1951; Ash, 1995), and the phenomenological psychologists (Gurwitsch, 1964). Some of the founders of sociology in the 19th century, mainly Gabriel de Tarde, Emile Durkheim and Gustave Le Bon, actually had conceptions of social mechanisms based on fields and collective forces. Durkheim wrote that society is a "system of active forces" or "collective tendencies" that are "intangible and unconscious" and have a "mind of their own" independent of the individual (Durkheim, 1951). According to Tarde, much of social interaction is based on a process of social contagion, in that tendencies, beliefs, desires and behaviours spread through society by imitation (Clark, 1969). However, now that ecology and quantum theory have sensitized us for such a perspective of interconnectedness and that we know much more about the field nature of the human organism and of its interaction with others, a field picture of social processes has become much more plausible and should be elaborated more fully.

I believe all this material calls for a field theory of social interaction that brings all the evidence into a coherent picture. An ongoing research project of this author is devoted to this task. I consider the investigation of interpersonal fields and the construction of a theory of social physics a legitimate subject of integrative biophysics.

REFERENCES

Ash, M.G., 1995, *Gestalt Psychology in German Culture 1890-1967*. Cambridge University Press, Cambridge.

Bailes, K.E., 1990, *Science and Russian Culture in an Age of Revolutions. V.I. Vernadsky and His Scientific School, 1863-1945*. Indiana University Press, Bloomington.

Berdiaev, N., 1998, Extract from *Aleksei Stepanovich Khomiakov*, in: Khomiakov & Kireevsky, *On Spiritual Unity – A Slavophile Reader*. Lindisfarne Books, Hudson, N.Y., pp. 326-350.

Bergson, H., 1919, *Creative Evolution*. Macmillan, London.

Bigu, J., 1976, On the biophysical basis of the human "aura", *J. Res. Psi Phenomena* **1**: 8-43.

Bischof, M., 1995, *Biophotons – The Light in Our Cells* (in German). Zweiztausendeins, Frankfurt am Main.

Bischof, M., 1998, Holism and field theories in biology – Non-molecular approaches and their relevance to biophysics, in: *Biophotons*, J.J. Chang, J. Fisch, and F.A. Popp, eds., Kluwer Academic Publishers, Dordrecht, pp. 375-394.

Bischof, M., 2003a, Introduction to integrative biophysics, in: *Integrative Biophysics*, F.A. Popp and L.V. Belousov, eds., Kluwer Academic Publisher, Dordrecht, pp. 1-115.

Bischof, M., 2003b, Vom integralen Weltbild zu einer neuen Wissenschaft, in: *Im Dialog über die Seele*, M.Utsch, and J. Fischer, eds., LIT-Verlag, Münster, pp. 3-51.

Bogomolov, N.A., 2000, *Russian Literature at the Beginning of the 20th Century and Occultism* (in Russian). Moscow, pp. 412-443.

Borisov, V.M., Perchenok, F.F., and Roginsky, A.B., 1993, Community as the source of Vernadsky's concept of noosphere, *Configurations* **1**: 415-43.

Braud, W., 1992, Human interconnectedness: Research indications, *ReVision* **14**: 140-148.

Braud, W., 2003, *Distant Mental Influence*. Hampton Roads Publishiing, Charlottesville, VA.

Brennan, T., 2004, *The Transmission of Affect*. Cornell University Press, Ithaca.

Clark, T.N., 1969, *Gabriel Tarde on Communication and Social Influence*. University of Chicago Press, Chicago.

Comfort, A., 1979a, *I and That - Notes on the Biology of Religion*. Crown Publishers, New York.

Comfort, A., 1979b, The cartesian observer revisited: ontological implications of the homuncular illusion, *Journal of Social and Biological Structures* **2**: 211-223.

Condon, W.S., Sander, L.A., 1974, Neonate movement is synchronized with adult speech. *Science* **183**, 99-101.

Deane, W.N., 1961, The reactions of a nonpatient to a stay on a mental hospital ward. *Psychiatry* **24**: 61-68.

Durkheim, E., 1951, *Suicide – A Study in Sociology*. Free Press, Glencoe, Illinois.

Edie, J.M., Scanlan, J.P., and Zeldin, M.B., eds., 1965, *Russian Philosophy*. Quadrangle Books, Chicago. 3 vols.

Fedorov, N.F., 1906-1913, *Philosophy of the Common Task*. 2 vols. Verny. Reprint Farnborough, Hants., England 1970.

Fedorov, N.F., 1965, The question of brotherhood or relatedness, and of the reasons for the unbrotherly, dis-related, or unpeaceful state of the world, and of the means for the restauaration of relatedness, in: Edie, J.M., Scanlan, J.P., and Zeldin, M.B., eds., 1965, *Russian Philosophy*. Quadrangle Books, Chicago, vol.III, pp.16-54.

Frank, S.L., 1910, The philosophy of religion of William James, *RM*, **2**: 155-164.

Frank, S.L., 1929-1930, Ich und Wir. Zur Analyse der Gemeinschaft, in: *Der russische Gedanke*, **1**, pp. 49-62.

Frank, S.L., 1987, *The Spiritual Foundations of Society*. Ohio University Press, Athens.

Goerdt, W., 1995, *Russian Philosophy* (in German). Verlag Karl Alber, Freiburg.

Goswami, A., 1993, *The Self-Aware Universe*. J.P.Tarcher/Putnam.

Grandpierre, A., 2001, Measurement of collective and social fields of consciousness, *World Futures* **57**: 85-94.

Grevs, I.M., 1921, *Byloe* **16**, pp. 137-166.

Grinevald, J., 1988, A history of the idea of biosphere, in: *Gaia, the Thesis and the Implications*, P.Bunyard and E.Goldsmith, eds., Wadebridge Ecological Centre, Camelford, Cornwall, pp. 1-34.

Gurwitsch, A., 1964, *The Field of Consciousness*. Duquesne University Press, Pittsburgh.

Hagelin, J.S., 1987, Is consciousness the unified field ? A field theorist's perspective, *Modern Science and Vedic Science* **1**: 29-87.

Hagemeister, M., 1997, Russian cosmism in the 1920's and today, in: Rosenthal, B.G., ed., *The Occult in Russian and Soviet Culture*, Cornell University Press, Ithaca, pp. 185-202.

Hatfield, E., Cacioppo, J.T., and Rapson, R.L., 1994, *Emotional Contagion*. Cambridge University Press, Cambridge.

Ihle, G., 1962, Das "Praecoxgefühl" in der Diagnostik der Schizophrenie. *Archiv für Psychiatrie* **203**: 385-406.

Katasonov, V., 1999, Integral reason: Science and religion in Russian culture. *Science & Spirit*, **9** (3).

Khomiakov, A., and Kireevsky, I., 1998, *On Spiritual Unity – A Slavophile Reader*. Lindisfarne Books, Hudson N.Y.

Kornblatt, J.D., and Gustafsson, R.F., 1996, *Russian Religious Thought*. University of Wisconsin Press, Madison.

Lapo, A.V., 1987, *Traces of Bygone Biospheres*. Synergetic Press, Oracle, Arizona, & London/Mir Publishers, Moscow.

Laszlo, E., 2003, *The Connectivity Hypothesis. Foundations of an Integral Science of Quantum, Cosmos, Life, and Consciousness*. State University of New York Press, Albany.

Le Roy, E., 1907, *Dogme et Critique*. Bloud et Cie., Paris. Russian Translation: *Dogmat i Kritika*. Moscow 1915.

Le Roy, E., 1927, *L'Exigence Idéaliste et le Fait de l'Évolution*. Boivin, Paris.

Le Roy, E., 1928, *Les Origines Humaines et l'Évolution de l'Intelligence*. Boivin, Paris. Levit, G.S., 2000, The biosphere and the noosphere theories of V.I.Vernadsky and P.Teilhard de Chardin: A methodological essay, *Archives Internationales d'Histoire des Sciences* **50**: 160-176.

Levit, G.S., 2001, *Biogeochemistry – Biosphere – Noosphere. The Growth of the Theoretical System of Vladimir Ivanovich Vernadsky*. (Studien zur Theorie der Biologie, Vol.4). VWB – Verlag für Wissenschaft und Bildung, Berlin.

Lewin, K., 1951, *Field Theory in Social Science*. Harper & Row, New York.

Lossky, N.O., 1906, The Foundations of Intuitivism (in Russian). English translation: *The Intuitive Basis of Knowledge*. Macmillan, London 1919.

Lossky, N.O., 1951, *History of Russian Philosophy*. International Universities Press, New York.

McCraty, R., Atkinson, M., and Rein, G., 1993, ECG spectra: The measurement of coherent and incoherent frequencies and their relatiuonship to mental and emotional states, in: *Proceedings of the Third Annual Conference of the International Society for the Study of Subtle Energy and Energy Medicine*, Monterey, CA, 44-48.

McCraty, R., Atkinson, M., Tomasino, D., and Tiller, W.A., 1997, The electricity of touch: Detection and measurement of cardiac energy exchange between people, in: *Proceedings of the Fifth Appalachian Conference on Neurobehavioral Dynamics: Brain and Values*, Lawrence Erlbaum, Mahwah, NJ, pp.1-14.

McMenamin, M., 1998, *The Garden of Ediacara*. Columbia University Press, New York.

McTaggart, L., 2001, *The Field*. HarperCollins Publishers, London.

Merlan, P., 1963, *Monopsychism – Mysticism – Metaconsciousness*. Martinus Nijhoff, The Hague.

Nelson, R.D., Bradish, G.J., Dobyns, Y.H., Dunne. B.J., and Jahn, R.G., 1996, FieldREG anomalies in group situations. *Journal of Scientific Exploration* **10**: 111-141.

Nelson, R.D., Jahn, R.G., Dunne. B.J., Dobyns, Y.H., and Bradish, G.J., 1998, FieldREG II: Consciousness field effects: Replications and explorations. *Journal of Scientific Exploration* **12**: 425-454.

Nethercott, F., 1995, *Une Rencontre Philosophique - Bergson en Russie (1907-1917)*. Éditions L'Harmattan, Paris.

Radlov, E.L., 1916, The Slavophile theory of knowledge (in Russian), *ŽMNP*, February, pp.153-165.

Radloff (Radlov), E.L., 1925, *Russian Philosophy* (in German). Ferdinand Hirt, Breslau.

Samson, P.R., and Pitt, D., eds., 1999, *The Biosphere and Noosphere Reader – Global Environment, Society, and Change*. Routledge, London.

Schmidt, S., 2002, *Aussergewöhnliche Kommunikation ? Eine kritische Evaluation des parapsychologischen Standardexperiments zur direkten mentalen Interaktion* (Transpersonale Studien, 6). Bibliotheks- und Informationssystem der Universität Oldenburg, Oldenburg.

Schwartz-Salant, N., 1988, *The Borderline Personality – Vision and Healing*. Chiron Publications, Wilmette, Illinois.

Skolimowski, H., 1994, *The Participatory Mind*. Arkana – Penguin Books, London.

Tart, C., 1969, Psychedelic experiences associated with a novel hypnotic procedure, mutual hypnosis, in: *Altered States of Consciousness*, C. Tart, ed.. Wiley, New York, pp. 291-308.

Taub-Bynum, E.B., 1984, *The Family Unconscious – An Invisible Bond*. Quest Books, Wheaton, Illinois.

Teilhard de Chardin, P., 1959, *The Phenomenon of Man*. Harper & Brothers, New York (original French edition 1955).

Tiller, W.A., McCraty, R., and Atkinson, M., 1996, Cardiac coherence: A new non-invasive measure of autonomic system order, *Alternative Therapies* **2**: 52-65.

Vernadsky, G., 1968, The Priyutino brotherhood (in Russian), *Novyi Zhurnal* (New York) **93**: 148-170.

Vernadsky, V.I., 1922, Iz proshlogo, in: *Ocherki i Rechi Akad.V.I. Vernadskogo*, Vol. 2. Petrograd, p.106.

Vernadsky, V.I., 1927, Works on the history of knowledge, in: *The USSR Academy of Sciences for Ten Years*, Leningrad, p.156.

Vernadsky, V.I., 1945, The biosphere and the noosphere, *American Scientist* **33**: 2-12.

Vernadsky, V.I., 1984, *Life and Activities in the Ukraine*. Kiev, p.184.

Vernadsky, V.I., 1988, *Philosophical Thoughts of a Naturalist*. Moscow.

Vernadsky, V.I., 1997, *Scientific Thought as a Planetary Phenomenon*. Nongovernmental Ecological V.I. Vernadsky Foundation, Moscow.

Vernadsky, V.I., 1998, *The Biosphere – Complete Annotated Edition*. Copernicus/Springer Verlag, New York.

Wackermann, J., 2004, Dyadic correlations between brain functional states: Present facts and future perspectives, *Mind and Matter* **2**: 105-122.

Zenkovsky, V.V., 1953, *A History of Russian Philosophy*. Columbia University Press, New York.

Zhang, C.L., 2003, Electromagnetic body versus chemical body. *Network* **81**: 7-10.

21
A LIFE THAT LINKED MITOGENETIC RAYS AND BIOPHOTONS

A brief tribute to Professor Anna Alexandrovna Gurvich

L.V. Beloussov, M. Lipkind, F.-A. Popp, and V.L. Voeikov*

As a kind of epilogue to the 3rd Alexander Gurwitsch Conference, having as one of its aims to draw a direct historical line from the first Crimean experiments of Alexander Gurwitsch to the modern biophotonics, we, four members of the International Institute of Biophysics (IIB), decided to remember a person whom we had a privilege to know and who made more than anybody else for this line to be really conducted. This was Alexander Gurwitsch's daughter, Professor Anna Alexandrovna Gurvich (A.A.G.).

A.A.G. was born on April 24th, 1909, in St Petersburg. In 1927-1931, she graduated from Moscow State University and after that started to work in her father's labs, firstly in the Institute of Experimental Medicine, Leningrad (up to 1941), and then (1943-1948) in the Academy of Medical Sciences, Moscow. A group photo given below relates to this latter period. After her father's retirement in 1948 (hastened by Lysenkoist attacks), she became the Head of the largely diminished Group of Mitogenesis in the Institute of Pathophysiology, Acad. Med Sci. She held this position until she passed in September 26th, 1993. She was Professor, DSc.

The main A.A.G. investigations were devoted to a refined analysis of a functional state of neural and muscular systems (including brain) with the use of mitogenetic radiation (biophoton emission in UV range detected mostly by the yeast budding technique). She was a master of an ingenious procedure of so-called mitogenetic spectral analysis performed by counting buds in the samples of yeast cultures arranged along a spectrograph slit at the positions corresponding to different wavelengths in UV range. The data registered by means of such approach were related to the functional state of the neuromuscular system *in situ* with a minimal operational intrusion. In such a way, the results of

* L.V. Beloussov. Department of Embryology, Faculty of Biology, Moscow State University, Moscow 119899 Russia. M. Lipkind, Unit of Molecular Virology, Kimron Veterinary Institute, P.O. Box 12, Beit Dagan, Israel 50250. F.-A. Popp, International Institute of Biophysics, Neuss, Germany. V.L. Voeikov, Department of Bioorganic Chemistry, Faculty of Biology, Moscow State University, Moscow 119899 Russia.

A Mitogenetic group from the Institute of Experimental Medicine, Moscow, in 1948. Alexander Gurwitsch is in the first row second from right, Anna Gurvich third from right.

unique significance were obtained. Namely, A.A.G. has demonstrated essential diversity of mitogenetic spectra, which were specific for particular neuro-muscular complexes in resting state, while changing (also specifically) after excitation. She elegantly demonstrated that the radiation spectra of the brain tissue depended upon the visual images displayed to the experimental animal. Another finding concerned the rhythmic mode of the radiation. The remarkable results were related to a regular evolution of mitogenetic spectra of radiation from muscle during early aging (demonstrated on the rabbit model). It was found that this evolution was associated with the establishment of neuromuscular connections, which was expressed by a fusion of several narrow spectral lines into a small number of more wide ones. In the spirit of Anna Gurwitsch's father's way of scientific exploration, all the obtained results were interpreted in the view of general comprehension of biological regularities. This was a basis for the conclusion that the nerve functioning is not reduced to a mere impulse conduction according to "all-or-nothing" principle. By Anna Gurwitsch's view, the "working substrate" of the nerve impulse (and the muscle contraction as well) consists of "non-equilibrial" (unbalanced) molecular constellations. Consequently, the mitogenetic radiation from the neuro-muscular system is degradational by its internal mechanism, i.e., caused by degradation of the "non-equilibrial" molecular constellations, which provided what she defined as "common energetic levels". Another example of the influence of general biological conception on creating working hypotheses and the respective experimental approach was the idea of induction of a new spectral pattern after cross transplantation of two nerves into respective muscles. In full expectation with theoretical prerequisites, under influence of the "new" muscles, the radiation spectra of the transplanted nerves were changed appropriately acquiring a new specific "imprint".

Thus, intensive experimental work was performed "under guidance" of general theoretical interpretation based, essentially, on Alexander Gurwitsch's theory of cellular field, whose influence on the molecular level was expressed in vectorization of the molecular processes in living systems contrary to their chaotic movements.

Along with these investigations, A.A.G. made much effort for elaborating the methods of physical detection of mitogenetic radiation. Due to imperfection of then (1960-1980) techniques in Russia, the results were restricted, but they paved a way for further generations of investigators.

Under difficult and contradictory political conditions in the post-war Soviet Union, A.A.G. tried her best for restoring and enlarging international scientific relations which were so effective in the first decades after her father's discovery of mitogenetic rays. As soon as she got first information about the works by Fritz Popp's biophotonic group in Worms, she started and continued regular correspondence and exchange of news with her colleagues (with the active aid of Dr Lipkind who himself established at that time direct research contacts with Dr Popp's group). In the last years of her life, Anna Gurwitsch was in a close contact with the group in Kaiserslautern, sustained by mutual respect and attachment. Two papers of Anna Gurwitsch (see a list below) were expressions of this first rapprochement. Then, during participation in the colloquium at Moscow University, Fritz-Albert Popp had the occasion to meet Anna Gurwitsch personally at her home. He was deeply impressed by her decent, wise, and warm personality, and he confessed that this meeting was one of the highlights of his life. Anna Gurwitsch was the connector between Gurwitsch's original approach to field of mitogenetic radiation and the more modern and fashionable development of biophotons and biophotonics. In spite of her age, she was very open to new approaches and new information, and actively participated in the starting cooperation between just established biophotonic group of the Moscow State University and the IIB, which soon became irreversible and highly fruitful. A.A.G. passed a year before the First International Alexander Gurwitsch Conference took in Moscow. Until her last days she participated in a full scale in the Conference organization and prepared a report that was read on the Conference session.

This little fragile woman was a really heroic person keeping on her shoulders after the retirement of Alexander Gurwitsch all the burden of managing his line in very difficult conditions. She also made a lot of effort promoting the publication of still unpublished works of her father. A.A.G. was highly estimated and respected by several most outstanding personalities in Russian science.

At the same time, she was a simple and very kind person, having many friends of different ages (including a lot of young), and with a refined feeling in arts. She was a gifted amateur painter, making nice etudes of the Crimean land. Classical music meant a lot in her life. In her last years she was a close friend of a zoologist, philosopher, and a writer, Professor Kuzin, who himself was the best friend of the famous Russian poet Ossip Mandelstam. In this way, A.A.G. linked also a "Silver Age" of Russian poetry with the latest generations.

A.A.G. has written several dozens of papers and a book (in Russian): *A Problem of Mitogenetic Radiation as an Aspect of Molecular Biology.*

(Leningrad, Medizina). 1968. Some of her findings she summarized in the following English papers:

A.A. Gurwitsch (1988) A historical review of the problem of mitogenetic radiation. *Experientia* 44: 545-549.

A.A. Gurvich (1992) Mitogenetic radiation as an evidence of nonequilibrium properties of living matter. In: *Recent Advances in Biophoton Research and its Applications* (F.-A. Popp, K.H. Li, and Q. Gu, eds.) World Scientific Singapore-New Jersey-London-Hong Kong pp. 457-468.

INDEX

A
Acupuncture 77, 79, 80
Aging
  blood 51, 52
  leaf 60, 68
  parameter 134
Antioxidant 98, 178, 222
ATP 6, 92-93, 142-143
Amplification
  of a signal 6, 136, 140, 245
  resonance 74
Arrhenius
  plot 70
  law 72
Autocorrelation 143-154

B
Bioenergetics 89-102
Biological
  effects 5, 105, 174
  functions 94
  processes 43, 102, 121
  rhythms 193
  structures 48-61, 106
Biophotons 1, 14, 31, 69, 130, 188, 191, 293, 301
Biopolymers 110
Biosphere 105, 167, 200, 280
Bonds
  chemical 11, 95, 142
  hydrogen 91
  macroergic 142

C
Cancer 132, 135, 142,
Cell
  bacterial 32, 159-161
  culture 15, 17, 98, 145-156
  cycle 169-174
  division 2-15, 82, 99, 141, 145, 174
  mitosis 2, 6
  suspension 19, 101, 168, 169, 171
  tumor 98, 129-135
  yeast 5, 10, 142, 160, 168-174
Coherence 14, 17, 22-31, 47-68, 68, 80, 123, 139, 155, 250, 294
Coherent
  modes 26, 76
  radiation 14, 27, 31, 47
  states 23, 26, 67-72, 76
Combustion 89, 99
Communication 281, 294
  chemical 159
  intercellular 17, 166
  interspecies 162, 165
  long-range 27
  optical 153
  process 62
Consciousness 235, 279
Cooperativity 14
Coupling 14, 24, 27, 67-69, 147, 155

D
Decay 132, 135, 206
  hyperbolic 19, 24, 25, 65
  parameters 19, 32, 39, 136
  radioactive 225, 232
Development
  embryological 2, 10, 80, 83, 145, 155, 265
  noosphere 280
  oscillations 100
  spores 162
Distribution
  count 34, 42
  curves 152
  function 54
  intensity 59, 187
  spatial 178, 261, 258
  spectral 14, 133, 124
  statistical 2, 4, 226
DL
  See Luminescence, delayed
  dynamics 46, 60, 85, 91, 108, 139, 146, 203, 207-212, 251
  relaxation 19, 131

E
Electrodynamics 14, 77, 80
Electron
  excited state 98
  oxygen reduction 93, 96, 100, 135, 229, 259

transport chain 65, 94-96, 129
electrons 7, 93, 96, 99, 229, 270
Emission
(bio)photon 5, 14, 18, 24, 31, 34, 38, 65, 73, 99, 130, 139-156, 177-188, 204-214, 299
degradative 9, 10-12, 139, 147, 154
spontaneous 8, 28, 142, 143, 177, 179
Energy
conservation law 11, 246,
density 93, 101, 142
expenditure 7, 97
gap 26, 69, 72, 99
levels 11, 15, 76, 78, 82, 156
of activation 7, 70
of electronic excitation 4, 8, 10, 89, 99, 142
potential 11, 77
supply 11, 90, 92
transfer 12, 31, 94, 9
Excitation
collective 72, 142
water 102
wave 140

F
Field 292-295
morphogenetic 2, 35, 256, 265,
electromagnetic 24, 140, 170, 191, 250
magnetic 97, 105-121, 171, 175, 191, 225, 232
electrical 12, 76, 135
biological 12, 129, 236-239, 257, 301
coherent 26, 56, 60-62, 76, 79, 82-84, 293
photon 27, 38, 67-69
autonomous 252-254
Fluctuations 12, 13, 31, 34, 38-42, 13, 179, 187, 188, 193, 266
Fluorescence 9, 65
Free radicals 8, 13, 94-99
Free will 235-260
Function
Bessel 81
correlation 52, 59, 60
hyperbolic 81

I
Imaging
System 178
Interaction
field-matter 26, 55, 69
hydrophobic 105-121
long-range 13, 35, 47, 50, 62, 159-165
optical 49, 61, 153, 203-212
Interference 16, 37-43, 47, 238
constructive and destructive 20-24
Irradiation 6, 8, 31, 47-68, 85, 130-132, 144, 168-171, 203-210

L
Laser 14, 31, 47, 48, 52-54, 57, 59, 60, 62, 63, 66, 67, 71, 73, 76, 129-133, 135, 137, 213, 214, 223
Light
coherent 51, 53, 144
Lipid
bilayer 105
peroxidation 13, 16, 47, 94, 130
Liquid
Crystal 132
Water 90, 91, 110
Luminescence 55
delayed 17, 24, 27, 65, 129, 144, 172, 203

M
Magnetic fields 97, 105, 106, 117, 118, 175, 191, 194, 200
Mammalian cells 73, 129, 131, 133, 135, 137, 174
Mechanical
action 3, 10
energy 93, 95
forces 142, 147, 154, 203, 238, 254, 292
Melanoma 131-135, 137
Membrane 104, 115, 117, 118, 133, 152-154, 163, 164, 166, 171, 176, 177, 182, 187, 227, 255-257, 286
Meosis 83
Microwaves 22, 157, 191
Mitogenetic rays, MGR 1, 3-13, 15, 16, 99, 103, 105, 137, 157, 299-302
Mitosis 2, 6, 83, 171, 172, 256
Monochromatic fields 24
irradiation 55
light 73
Morphogenetic field 2, 33, 238
Morphogenesis 62, 83, 157, 191, 223, 236, 256
Muscle 204, 212, 300

N
Non-equilibrium
molecular constellations 2, 10
phase transitions 76, 87

state 12, 132
thermodynamics 32
Non-radiative transfer of energy 203
Nuclease 9
Nuclei, nucleus (in cells) 2, 6, 137, 163, 175, 254-256
Nucleic 105
Nucleotides 169, 170

O
Optical interactions 61, 153, 156, 203, 205, 206, 209, 212
Oscillation(s) 17, 24, 27, 65-70, 72, 73, 76, 93, 100, 101, 104, 122, 127, 140, 141, 147-149, 155, 157, 213, 216-223, 251, 252
Oxygen 7, 8, 13, 89, 93-104, 130

P
Pathological
case 30
physiology 98
processes 199
states 51
Phase
correlation 48, 196
shift 196-198
Photomultiplier 13, 66, 67, 130, 132, 136, 153, 177-179, 181-183, 187, 188, 204, 205, 210
Photon
storage 16, 23
sucking 17-19, 21-27, 29, 31, 69
trapping 23, 68
Photosynthesis 73, 94
Plant
shoots 213, 215, 217, 219, 221, 223
roots 4, 34, 84, 213-217, 219-221, 223
Poisson
statistics 225, 226, 233
Proliferation 130, 166

Q
Quantum 7, 54, 96, 99, 34, 38, 45, 67, 83, 85, 250
coherence 274, 276
correlations 137
efficiency 132, 136, 178
electrodynamics 157
mechanics 76, 86, 250, 275, 293
medicine 75, 76, 79, 84, 85, 87
noise 251
optics 73, 129
physics 14, 96, 129, 237, 251
state 33, 35, 43
theory 21, 32, 251, 295
transitions 233
uncertainty 243-246
vacuum 295

R
Radiation (see also Mitogenetic radiation, UV radiation) 14, 16, 31, 32, 47-54, 56, 58-63, 76, 79, 80, 85, 87, 99, 103, 104, 128-132, 136, 137, 144, 156, 157, 166-171, 175, 191, 200, 203, 204, 206, 210, 211, 300, 301
coherent 47, 54
of degradation (degradational, degradative) 9, 10, 12, 33, 142, 154
field 28, 46
infrared 27
pressure 22
Radiation-less
propagation 4
transfer 211
Radicals
free 7, 8, 15, 94-100, 103, 104, 111, 141, 142
Relaxation 17, 19, 24, 25, 27, 32, 63, 65-70, 131, 132, 142, 143, 157, 212, 232
Resonator 22, 23
Resonance 79, 84, 85, 87, 141, 250, 252, 253, 276

S
Saccharomyces cerevisiae 137, 167-169, 172-175
Scattering, scattered irradiation 52-56, 60-62, 69
Seeds 14, 49, 50, 160, 203-206, 208-212, 302
Self-organization 75, 84, 86, 100, 200, 225
Spectra 5, 6, 8, 10, 11, 13, 72, 75, 93, 110, 112, 113, 115, 117, 122, 124, 126, 132, 137, 178, 195, 203, 204, 207-211, 229, 297, 300
frequency (Fourier) 139, 143-147, 149-151, 144-146
Spectral analysis 127, 131, 132-135, 137, 188, 299
decomposition 33, 34, 38-42, 43-45
density 22, 79

intensity 28, 133
profiles 43
range 105, 136, 143
shifts 111, 116
Squeezed state 31, 33-37, 39, 41, 43-46
Subradiance 22, 143, 149, 150, 153-155
Superradiance 22
Synchrony, synchronization 2, 82, 120, 139, 142, 154, 175, 192, 194, 196, 197, 198-200, 221, 222, 227, 228, 251, 252, 275, 294, 296

T
Temperature 10, 11, 15, 38, 39, 50, 56, 59, 66, 69, 70, 72, 73, 94, 95, 98, 106, 107, 131, 132, 136, 137, 178, 179, 214
Triplet state 96, 99, 236

U
Ultraweak photon emission 31, 73, 130, 131, 134, 135, 137, 139, 156
Uncertainty principle 14, 20, 21, 244-246
UV (ultra-violet) light, spectral range, radiation 4, 5, 7, 12-15, 93-95, 99, 103, 130, 131, 134, 137, 142, 159, 164, 165, 203, 204, 206-208, 210, 211, 299

V
Vectorial
factor 12
field 236, 244
flow 141
Vibrational
energy 93, 103, 127
excitation 96
modes 156

W
Water 6-10, 15, 38, 44, 75, 76, 89-103, 106-113, 115, 117, 119-122, 137, 141, 152, 155, 160, 161, 168, 175, 203, 204, 208, 209, 212-223

Y
Yeasts 5, 10, 14, 33, 137, 142, 134, 160, 166-175, 299